Stetson University
3 4369 00454955 4

W9-AGM-864

Encyclopedic Dictionary of International Finance and Banking

Jae K. Shim
Michael Constas

St. Lucie Press
Boca Raton London New York Washington, D.C.

Library of Congress Cataloging-in-Publication Data

Shim, Jae K.
Encyclopedic dictionary of international finance and banking / Jae K. Shim and Michael Constas.
p. cm.
ISBN 1-57444-291-0 (alk. paper)
1. International finance—Encyclopedias. 2. Banks and banking, International—Encylopedias. 3. International economic relations—Encyclopedias. I. Constas, Michael, 1952- . II. Title.

HG3880 .S55 2001
332'.042'068—dc21 2001001297

International Standard Book Number 1-57444-291-0
Library of Congress Card Number 2001001297
Printed in the United States of America 2 3 4 5 6 7 8 9 0
Printed on acid-free paper

Preface

WHAT THIS BOOK WILL DO FOR YOU

The *Encyclopedic Dictionary of International Finance and Banking* is written and compiled for working professionals engaged in the fields of international finance, global trade, foreign investments, and banking. It may be used for day-to-day practice and for technical research. The *Encyclopedic Dictionary* is a practical reference of proven techniques, strategies, and approaches that are successfully used by professionals to diagnose multinational finance and banking problems. The book covers virtually all important topics dealing with multinational business finance, investments, financial planning, financial economics, and banking. It also covers such topics as computers, quantitative techniques and models, and economics as applied to international finance and banking. The *Encyclopedic Dictionary* will benefit practicing financial analysts, CFOs, controllers, financial managers, treasurers, money managers, fund managers, investment analysts, and professional bankers, among others.

The subjects are explained with

- Clear definitions and explanations, including step-by-step instructions
- Exhibits and statistical data, as needed
- Charts, exhibits, and diagrams, where appropriate
- Checklists
- Practical applications

The *Encyclopedic Dictionary* will enlighten the practitioner by presenting the most current information, offering important directives, and explaining the technical procedures involved in the aforementioned dynamic business disciplines. This reference book will help you diagnose and evaluate financial situations faced daily. This library of international finance and banking will answer nearly every question you may have. Real-life examples are provided, along with suggestions for handling everyday problems. The *Encyclopedic Dictionary* applies to large, medium, or small multinational companies. It will help you to make smart decisions in all areas of international finance and banking. It should be used as an advanced guide for working professionals, rather than as a reference guide for laymen or a glossary of international finance and banking terms.

The *Encyclopedic Dictionary* is a handy reference for today's busy financial executive. It is a working guide to help you quickly pinpoint

- What to look for
- How to do it
- What to watch out for
- How to apply it in the complex world of business
- What to do

You will find ratios, formulas, examples, applications, exhibits, charts, and rules of thumb to help you analyze and evaluate any business-related situation. New, up-to-date methods and techniques are included. Throughout, you will find this *Encyclopedic Dictionary* practical, comprehensive, quick, and useful. In short, this is a veritable cookbook of guidelines, illustrations, and how-to's for you, the modern decision maker. The uses of this handbook are as varied as the topics presented. Keep it handy for easy reference throughout your busy day.

There are approximately 570 major topics in international finance, banking, and investments covered in the *Encyclopedic Dictionary*, as well as numerous related entries. Where appropriate, there is a cross-reference to another entry to explain the topic in greater detail. The entries are listed in alphabetical order for easy reference. There are approximately 120 examples and 110 exhibits to help explain the material. The *Encyclopedic Dictionary* is so comprehensive that almost any subject area of interest to financial executives, as well as other interested parties, can be found.

About The Authors

JAE K. SHIM, Ph.D., is Professor of Business at California State University, Long Beach. He received his M.B.A. and Ph.D. degrees from the University of California at Berkeley (Haas School of Business). He is also Chief Financial Officer (CFO) of a Los Angeles–based multinational firm.

Dr. Shim is a coauthor of *Encyclopedic Dictionary of Accounting and Finance; Handbook of Financial Analysis, Forecasting, and Modeling; Managerial Accounting; Financial Management; Strategic Business Forecasting; Barron's Accounting Handbook; Financial Accounting; The Vest-Pocket CPA; The Vest-Pocket CFO*, and the best selling *Vest-Pocket MBA*. Dr. Shim has 45 other professional and college books to his credit.

Dr. Shim has also published numerous refereed articles in such journals as *Financial Management, Advances in Accounting, Corporate Controller, The CPA Journal, CMA Magazine, Management Accounting, Econometrica, Decision Sciences, Management Science, Long Range Planning, OMEGA, Journal of Operational Research Society, Journal of Business Forecasting*, and *Journal of Systems Management*. He was a recipient of the *1982 Credit Research Foundation Outstanding Paper Award* for his article on financial management.

Michael Constas, Ph.D., J.D., is a Professor of Business at California State University, Long Beach. Before teaching, he was a partner in a major California law firm. Dr. Constas received his Ph.D., J.D., and M.B.A. from U.C.L.A. He has published numerous articles in the area of investments in academic and professional journals. He is a coauthor of *The International Investment Source Book* (with Dr. Shim). Dr. Constas is an author of *Private Real Estate Syndications*, which is part of the collections at the libraries of our nation's leading universities.

Notes and Abbreviations

KEY NOTES

1. This book has the following features:
   - Plenty of examples and illustrations
   - Useful strategies and checklists
   - Ample number of exhibits (tables, figures, and graphs)
2. Foreign exchange rate quotations—*direct* or *indirect*—may confuse some readers. Indirect quotes are used more widely in examples throughout the book.
3. *Selling forward* means "buy a forward contract to *sell* a given currency," and *buying forward* means "buy a forward contract to *buy* a given currency." As a matter of terminology, *selling forward* or *buying forward* could mean the same transaction. For example, a contract to deliver dollars for British pounds in 180 days might be referred to as *selling dollars forward for pounds* or *buying pounds forward for dollars*.

ABBREVIATIONS USED IN THIS TEXT

A$	Australian Dollar
£	British Pound
C$	Canadian Dollar
DM	Deutsche Mark
€	Euro
FFr	French Franc
IRS	Internal Revenue Service
¥	Japanese Yen
LC	Local Currency
MNC	Multinational Corporation
SFr	Swiss Franc
¢	U.S. cent
$	U.S. Dollar
U.S.$	U.S. Dollar

A

ABSOLUTE PURCHASING POWER PARITY

See PURCHASING POWER PARITY.

ABSOLUTE RATE

An interest rate that is determined without reference to an index or funding base such as LIBOR or U.S. treasury rates. For example, rather than LIBOR + 0.75%, the bid is expressed as 10.375%.

ACCOUNTING EXPOSURE

See TRANSLATION EXPOSURE.

ACCOUNTING FOR MULTINATIONAL OPERATIONS

At the beginning of the 21st century, the world economy has become truly internationalized and globalized. Advances in information technology, communications, and transportation have enabled businesses to service a world market. Many U.S. companies, both large and small, are now heavily engaged in international trade. The foreign operations of many large U.S. multinational corporations now account for a major percentage (10 to 50%) of their sales and/or net income.

The basic business functions (i.e., finance/accounting, production, management, marketing) take on a new perspective when conducted in a foreign environment. There are different laws, economic policies, political framework, and social/cultural factors that all have an effect on how business is to be conducted in that foreign country. From an accounting standpoint, global business activities are faced with three realities:

1. Accounting standards and practices differ from country to country. Accounting is a product of its own economic, legal, political, and sociocultural environment. Because this environment changes from country to country, the accounting system of each country is unique and different from all others.
2. Each country has a strong "accounting nationalism." It requires business companies operating within its borders to follow its own accounting standards and practices. Consequently, a foreign company operating within its borders must maintain its books and records and prepare its financial statements in the local language, use the local currency as a unit of measure, and be in accordance with local accounting standards and procedures. In addition, the foreign company must comply with the local tax laws and government regulations.
3. Cross-border business transactions often involve receivables and payables denominated in foreign currencies. During the year, these foreign currencies must be translated (converted) into the local currencies for recording in the books and records. At year-end, the foreign currency financial statements must be translated (restated) into the parent's reporting currency for purposes of consolidation. Both the recording of foreign currency transactions and the translation of financial statements require the knowledge of the exchange rates to be used and the accounting treatment of the resulting translation gains and losses.

The biggest mistake a company can make in international accounting is to not be aware of, or even worse, to ignore these realities. It should know that differences in accounting standards, tax laws, and government regulations do exist and that these differences need to be an integral part of formulating its international business plan.

A. Accounting for Foreign Currency Transactions

International business transactions are cross-border transactions; therefore, two national currencies are usually involved. For example, when a United States corporation sells to a corporation in Germany, the transaction can be settled in U.S. dollars (the seller's currency) or in German marks (the buyer's currency).

A.1. Transactions Denominated in U.S. Currency

When the foreign transaction is settled in U.S. dollars, no measurement problems occur for the U.S. corporation. As long as the U.S. corporation receives U.S. dollars, the transaction can be recorded in the same way as a domestic transaction.

EXAMPLE 1

A U.S. firm sells on account equipment worth $100,000 to a German company. If the German company will pay the U.S. firm in U.S. dollars, no foreign currency is involved and the transaction is recorded as usual:

Accounts Receivable	100,000	
Sales		100,000
(*To record sales to German company*)		

A.2. Transactions Denominated in Foreign Currency

If the transaction above is settled in German marks, however, the U.S. corporation will receive foreign currency (German marks) that must be translated into U.S. dollars for purposes of recording on the U.S. company's books. Thus, a foreign currency transaction exists when the transaction is settled in a currency other than the company's home currency.

A foreign currency transaction must be recorded in the books of accounts when it is begun (date of transaction), then perhaps at interim reporting dates (reporting date), and finally when it is settled (settlement date). On each of these three dates, the foreign currency transaction must be recorded in U.S. dollars, using the spot rate on that date for translation.

A.3. Accounting at Transaction Date

Before any foreign currency transaction can be recorded, it must first be translated into the domestic currency, using the spot rate on that day. For the U.S. company, this means that any receivable and payable denominated in a foreign currency must be recorded in U.S. dollars.

EXAMPLE 2

Assume a U.S. firm purchases merchandise on account from a French company on December 1, 20X1. The cost is 50,000 French francs, to be paid in 60 days. The exchange rate for French francs on December 1 is $.20. Using the exchange rate on December 1, the U.S. firm translates the FFr 50,000 into $10,000 and records the following entry:

Dec. 1	Purchases	10,000	
	Accounts Payable		10,000

[*To record purchase of merchandise on account (FFr 50,000 × $.20 = $10,000).*]

A.4. Accounting at Interim Reporting Date

Foreign currency receivables and payables that are not settled at the balance sheet date are adjusted to reflect the exchange rate at that date. Such adjustments will give rise to foreign exchange gains and losses that are to be recognized in the period when exchange rates change.

EXAMPLE 3

Assume the same facts as in Example 2 and that the U.S. corporation prepares financial statements as of December 31, 20X1 when the exchange rate for the French franc is $0.22. The U.S. firm will make the following adjusting entry:

Dec. 31	Foreign Exchange Loss	1,000	
	Accounts Payable		1,000
	[To adjust accounts payable to current exchange rate (FrF 50,000 × $0.22 = $11,000; $11,000 – $10,000 = $1,000).]		

A.5. Accounting at Settlement Date

When the transaction is settled, if the exchange rate changes again, the domestic value of the foreign currency paid on the settlement date will be different from that recorded on the books. This difference gives rise to translation gains and losses that must be recognized in the financial statements.

EXAMPLE 4

To continue our example, assume that the payable is paid on February 1, 20X2 when the exchange rate for the French franc is $0.21. The settlement will be recorded as follows:

Feb. 1	Accounts Payable	11,000	
	Cash		10,500
	Foreign Exchange Gain		500
	[To record payment of accounts payable (FrF 50,000 × $0.21 = $10,500) and foreign exchange gain.]		

To summarize: In recording foreign currency transactions, SFAS 52 adopted the two-transaction approach. Under this approach, the foreign currency transaction has two components: the purchase/sale of the asset and the financing of this purchase/sale. Each component will be treated separately and not netted with the other. The purchase/sale is recorded at the exchange rate on the day of the transaction and is not adjusted for subsequent changes in that rate. Subsequent fluctuations in exchange rates will give rise to foreign exchange gains and losses. They are considered as financing income or expense and are recognized separately in the income statement in the period the foreign exchange fluctuations happen. Thus, exchange gains and losses arising from foreign currency transactions have a direct effect on net income.

B. Translation of Foreign Currency Financial Statements

When the U.S. firm owns a controlling interest (more than 50%) in another firm in a foreign country, special consolidation problems arise. The subsidiary's financial statements are usually prepared in the language and currency of the country in which it is located and in accordance with the local accounting principles. Before these foreign currency financial statements can be consolidated with the U.S. parent's financial statements, they must first be adjusted to conform with U.S. GAAP (Generally Accepted Accounting Principles) and then translated into U.S. dollars.

Two different procedures may be used to translate foreign financial statements into U.S. dollars: (1) translation procedures and (2) remeasurement procedures. Which one of these two procedures is to be used depends on the determination of the functional currency for the subsidiary.

B.1. The Functional Currency

SFAS 52 defines the functional currency of the subsidiary as the currency of the primary economic environment in which the subsidiary operates. It is the currency in which the subsidiary realizes its cash flows and conducts its operations. To help management determine the functional currency of its subsidiary, SFAS 52 provides a list of six salient economic indicators regarding cash flows, sales price, sales market, expenses, financing, and intercompany transactions. Depending on the circumstances:

- The functional currency can be the local currency. For example, a Japanese subsidiary manufactures and sells its own products in the local market. Its cash flows, revenues, and expenses are primarily in Japanese yen. Thus, its functional currency is the local currency (Japanese yen).
- The functional currency can be the U.S. dollar. For foreign subsidiaries that are operated as an extension of the parent and integrated with it, the functional currency is that of the parent. For example, if the Japanese subsidiary is set up as a sales outlet for its U.S. parent, i.e. it takes orders, bills and collects the invoice price, and remits its cash flows primarily to the parent, then its functional currency would be the U.S. dollar.

The functional currency is also the U.S. dollar for foreign subsidiaries operating in highly inflationary economies (defined as having a cumulative inflation rate of more than 100% over a three-year period). The U.S. dollar is deemed the functional currency for translation purposes because it is more stable than the local currency.

Once the functional currency is determined, the specific conversion procedures are selected as follows:

- If foreign currency is the functional currency, use translation procedures.
- If U.S. dollar is the functional currency, use remeasurement procedures.

B.2. Translation Procedures

If the local currency is the functional currency, the subsidiary's financial statements are translated using the current rate method. Under this method:

- All assets and liabilities accounts are translated at the current rate (the rate in effect at the financial statement date);
- Capital stock accounts are translated using the historical rate (the rate in effect at the time the stock was issued);
- The income statement is translated using the average rate for the year; and
- All translation gains and losses are reported on the balance sheet, in an account called "Cumulative Translation Adjustments" in the stockholders' equity section.

The purpose of these translation procedures is to retain, in the translated financial statements, the financial results and relationships among assets and liabilities that were created by the subsidiary's operations in its foreign environment.

EXAMPLE 5

Assume that the following trial balance, expressed in the local currency (LC) is received from a foreign subsidiary, XYZ Company. The year-end exchange rate is 1 LC = $.1.50, and the average exchange rate for the year is 1 LC = $1.25. Under the current rate method, XYZ Company's trial balance would be translated as in Exhibit 1 which shows the translation procedures applied to XYZ Company's trial balance. Note that the translation adjustment is reflected as an adjustment of stockholders' equity in U.S. dollars.

EXHIBIT 1
Translation Procedures
XYZ COMPANY
Trial Balance
12/31/01

	Local Currency			U.S. Dollars	
	Debit	**Credit**	**Exchange Rate**	**Debit**	**Credit**
Cash	LC 5,000		(1 LC = $ 1.50)	$7,500	
Inventory	15,000		"	22,500	
Fixed Assets	30,000		"	45,000	
Payables		LC 40,000	"		$60,000
Capital Stock		4,000	Historical rate		5,000
Retained Earnings		6,000	to balance		10,000
Sales		300,000	(1 LC = $1.25)		375,000
Cost of Goods Sold	210,000		"	262,500	
Depreciation Expense	5,000		"	6,250	
Other Expenses	85,000		"	106,250	
	LC 350,000	LC 350,000		$450,000	$450,000

B.3. Remeasurement Procedures

If the U.S. dollar is considered to be the functional currency, the subsidiary's financial statements are then remeasured into the U.S. dollar by using the temporal method. Under this method:

- Monetary accounts, such as cash, receivables, and liabilities, are remeasured at the current rate on the date of the balance sheet;
- Nonmonetary accounts, such as inventory, fixed assets, and capital stock, are remeasured using the historical rates;
- Revenues and expenses are remeasured using the average rate, except for cost of sales and depreciation expenses that are remeasured using the historical exchange rates for the related assets; and
- All remeasurement gains and losses are recognized immediately in the income statement.

The objective of these remeasurement procedures is to produce the same U.S. dollar financial statements as if the foreign entity's accounting records had been initially maintained in the U.S. dollar. Exhibit 2 shows these remeasurement procedures applied to XYZ Company's trial balance. Note that the translation gain/loss is included in the income statement.

EXHIBIT 2
Remeasurement Procedures
XYZ COMPANY
Trial Balance
12/31/01

	Local Currency			U.S. Dollars	
	Debit	**Credit**	**Exchange Rate**	**Debit**	**Credit**
Cash	LC 5,000		(1 LC = $1.50)	$7,500	
Inventory	15,000		(1 LC = $1.30)	19,500	
Fixed Assets	30,000		(1 LC = $0.95)	28,500	
Payables		LC 40,000	(1 LC = $1.50)		$60,000
Capital Stock		4,000	—		5,000
Retained Earnings		6,000			7,000
Sales		300,000	(1 LC = $1.25)		375,000
Cost of Goods Sold	210,000		(1 LC = $1.30)	273,000	
Depreciation Expense	5,000		(1 LC = $0.95)	4,750	
Other Expenses	85,000		(1 LC = $1.25)	106,250	
				439,500	447,000
Translation Gain/Loss				7,500	
	LC 350,000	LC 350,000		$447,000	$447,000

C. Interpretation of Foreign Financial Statements

To evaluate a foreign corporation, we usually analyze its financial statements. However, the analysis of foreign financial statements needs special considerations:

1. We often have the tendency of looking at the foreign financial data from a home country perspective. For example, a U.S. businessman has the tendency of using U.S. GAAP to evaluate the foreign financial statements. However, U.S. GAAP are not universally recognized and many differences exist between U.S. GAAP and the accounting principles of other countries (industrialized or nonindustrialized).
2. Because of the diversity of accounting principles worldwide, we have to overcome the tendency of using our home country GAAP to evaluate foreign financial statements. Instead, we should try to become familiar with the foreign GAAP used in the preparation of these financial statements and apply them in our financial analysis.
3. Business practices are culturally based. Often they are different from country to country and have a significant impact on accounting measurement and disclosure practices. Therefore, local economic conditions and business practices should be taken into consideration to correctly analyze foreign financial statements.

D. Harmonization of Accounting Standards

The diversity of accounting systems is an obstacle in the development of international trade and business and in the efficiency of the global capital markets. Many concerted efforts have been made to reduce this diversity through the harmonization of accounting standards. Also, as international business expands, there is a great need for international accounting standards that can help investors make decisions on an international scale. The agencies working toward the harmonization of accounting standards are:

D.1. The International Accounting Standards Committee (IASC)

The IASC was founded in 1973. At that time, its members consisted of the accountancy bodies of Australia, Canada, France, Ireland, Japan, Mexico, the Netherlands, the United Kingdom, the United States, and West Germany. Since its founding, membership has grown to around 116 accountancy bodies from approximately 85 countries.

IASC's fundamental goal is the development of international accounting standards. It is also working toward the improvement and harmonization of accounting standards and procedures relating to the presentation and comparability of financial statements (or at least through enhanced disclosure, if differences are present). To date, it has developed a conceptual framework and issued a total of 32 International Accounting Standards (IAS) covering a wide range of accounting issues. It is currently working on a project concerned with the core standards in consultation with other international groups, especially the International Organization of Securities Commissions (IOSCO), to develop worldwide standards for all corporations to facilitate multilisting of foreign corporations on various stock exchanges.

D.2. The International Federation of Accountants (IFAC)

IFAC was founded in 1977 by 63 accountancy bodies representing 49 countries. By 1990, IFAC membership had grown to 105 accountancy bodies from 78 different countries. Its purpose is to develop "a coordinated worldwide accountancy profession with harmonized standards." It concentrated on establishing auditing guidelines to help promote uniform auditing practices throughout the world. It also promoted general standards for ethics, education, and accounting management.

In addition to the IASC and IFAC, there are a growing number of regional organizations involved in accounting harmonization at the regional level. These organizations include, among others, the Inter-American Accounting Association established in 1949, the ASEAN Federation of Accountants (AFA) established in 1977, and the Federation des Experts Comptables Europeens (FEE), created by the merger in 1986 of the former Union Europeenne des Experts Comptables Economiques et Financiers (UEC) and the Groupe d'Etude (GE).

D.3. The European Economic Community (EEC)

The EEC, although not an accounting body, has made great strides in harmonizing the accounting standards of its member countries. During the 1970s, it began the slow process of issuing EEC directives to harmonize the national accounting legislation of its member countries. The directives must go through a three-step process before they are finalized. First, they are proposed by the EEC Commission and presented to the national representatives of the EEC members. Second, if the proposal is satisfactory to the nations, it is adopted by the commission. Finally, it must be issued by the Council of Ministers of the EEC, before it can be enforced on the members.

The most important directives in the harmonization of accounting standards among EEC members are:

- The Fourth Directive (1978), regarding the layout and content of annual accounts, valuation methods, annual report, publicity, and audit of public and private company accounts;
- The Seventh Directive (1983), regarding the consolidation of accounts for certain groups of enterprises; and
- The Eighth Directive (1984), regarding the training, qualification, and independence of statutory auditors.

ACCOUNTS RECEIVABLE MANAGEMENT

Accounts receivable management is the strategy used by some MNCs to adjust their accounts receivable (A/R) to reduce *currency risk* and to optimally time fund transfers. Various hedging alternatives are available, including forward and money-market hedges. Operating and financial strategies can also be used to minimize currency risk exposure. For example, in countries

where currency values are likely to drop, financial managers of the subsidiaries should avoid giving excessive trade credit. If accounts receivable balances are outstanding for an extended time period, interest should be charged to absorb the loss in purchasing power. *Note:* a net asset position (i.e., assets minus liabilities) is not desirable in a weak or potentially depreciating currency. In this case, you should expedite the disposal of the asset. Likewise, you should lag or delay the collection against a net asset position in a strong currency.
See also TRANSACTION EXPOSURE.

ACU

See ASIAN CURRENCY UNIT.

ADB

See ASIAN DEVELOPMENT BANK.

ADJUSTED PRESENT VALUE

Adjusted present value (APV) is a type of *net present value (NPV)* analysis by multinational companies in capital budgeting. A foreign investment project that is financed differently from that of the parent firm could be evaluated using this approach. In APV, operating cash flows are *discounted separately* from (1) the various tax shields provided by the deductibility of interest and other financial charges and (2) the benefits of project-specific concessional financing. Each component cash flow is discounted at a rate appropriate for the risk involved. Typically, the operating cash flows from the project are discounted at the *all-equity rate* plus any financing side effects discounted at *all-debt rate.*
Or

$$\text{APV} = -I + \sum_{t=1}^{T} \frac{CF_t}{(1+k)^t} + \sum_{t=1}^{T} \frac{FIN_t}{(1+kf)^t}$$

where $-I$ = the initial investment or cash outlay, CF_t = estimated cash flows in t ($t = 1,\ldots T$), k = the discount rate on those cash flows, FIN_t = any additional financial effect on cash flows, and kf = the discount rate applied to the financial effects. Possible financial effects include the tax shield arising from depreciation charges, subsidies, credit terms, interest savings, or penalties associated with project-specific financing. A project that is financed differently from that of the parent company should be evaluated with APV.

For example, consider a parent firm whose capital structure is 60% equity and 40% debt that is evaluating the financial feasibility of a potential foreign subsidiary whose capital structure would be only 40% equity and 60% debt. Discounting the potential subsidiary's estimated net operating cash flows by the parent's *weighted average cost of capital (WACC)* could be inappropriate.

APV divides the present value analysis into two components: (1) the operating cash flows which are customarily considered the only relevant cash flows and (2) the financial effects such as interest expense tax shields resulting from the use of debt in the financing of the project. Each component cash flow is discounted by its appropriate cost of all-equity and all-debt discount rates, respectively.

EXAMPLE 6

Suppose MYK Gold Miners has an opportunity to enter a small, developing country and apply its new gold recovery technique to some old mines that no longer yield profitable amounts of ore under conventional mining. MYK estimates that the cost of establishing the foreign operation will be \$12 million. The project is expected to last for two years, during which period the operating cash flows from the new gold extracted will be \$7.5 million per year. In addition, the new operating unit will allow the company to repatriate an additional \$1 million per year in funds that have been tied up in the developing country by capital controls. If MYK applies a discount rate of 6% to operating cash flows and 10% to the funds that will be freed from controls, then the APV is:

$$\begin{aligned}
\text{APV} &= -\$12 + \$7.5/(1+0.06) + \$7.5/(1+0.06)^2 + \$1/(1+0.10) + \$1/(1+0.10)^2 \\
&= -12 + 7.5\ \text{T4}(6\%, 2\ \text{years}) + 1\text{T4}(10\%, 2\ \text{years}) \\
&= -12 + 7.5(1.8334) + 1(1.7355) \\
&= -12 + 15.49 = 3.49
\end{aligned}$$

where T4 = present value of an annuity of \$1. (See Table 4 in the Appendix.)

Thus, the APV of the gold recovery project equals \$3.49 million. The firm can compare this value to the APV of other projects it is considering in order to budget its capital expenditures in the optimum manner.

EXAMPLE 7

Am-tel Corporation is an MNC which owns a foreign subsidiary named Ko-tel. It has the following operating cash flows:

	Year					
	0	**1**	**2**	**3**	**4**	**5**
Operating cash flows (in thousands)	−11,000.0	1,274.4	1,881.4	2,578.3	3,378.8	11,343.4

Assume Ko-tel was capitalized with 40% equity capital from Am-tel and the remaining 60% debt, the Am-tel's capital structure was 40% debt and 60% equity, the cost of debt was 12.12%, the cost of equity was 18%, and U.S. taxes were 34%. The weighted average cost of capital would be:

$$\text{WACC} = (0.40)\ (12.12\%)\ (1 - 0.34) + (0.60)\ (18\%) = 14\%.$$

The regular NPV approach yields:

$$\begin{aligned}
\text{NPV} &= -\$11{,}000.0 + \$1{,}274.4/(1+0.14) + \$1{,}881.4/(1+0.14)^2 \\
&\quad + \$2{,}578.3/(1+0.14)^3 + \$2{,}378/(1+0.14)^4 + \$11{,}343.4/(1+0.14)^5 \\
&= -\$11{,}000.0 + \$1{,}274.4\ \text{T3}(14\%, 1) + \$1{,}881.4\ \text{T3}(14\%, 2) \\
&\quad + \$2{,}578.3\ \text{T3}(14\%, 3) + \$2{,}378\ \text{T3}(14\%, 4) + \$11{,}343.4\ \text{T3}(14\%, 5) \\
&= -\$11{,}000.0 + \$1{,}274.4(0.8772) + \$1{,}881.4(0.7695) \\
&\quad + \$2{,}578.3(0.6750) + \$2{,}378(0.5921) + \$11{,}343.4(0.5194) \\
&= -\$11{,}000.0 + \$12{,}197.83 \\
&= \$1{,}197.83
\end{aligned}$$

where T3 = present value of \$1. (See Table 3 in the Appendix.)

In contrast, APV would decompose the valuation into the above operating cash flows and the tax shields arising from the use of debt in the foreign subsidiary. The operating cash flows from above would then be discounted by the cost of equity for a similar project undertaken with 100% equity. For illustration purposes here, we use the firm's current cost of equity:

(Continued)

$$\text{NPV (operating cash flows)} = -\$11{,}000.0 + \$1{,}274.4/(1+0.18) + \$1{,}881.4/(1+0.18)^2 + \$2{,}578.3/(1+0.18)^3 + \$2{,}378/(1+0.18)^4 + \$11{,}343.4/(1+0.18)^5 = -\$298.48$$

The tax shields resulting from the use of debt in Ko-tel are found by estimating the annual interest expense on $727,200 ($6,000,000 in debt at 12.12% per year), and the Ko-tel local tax savings resulting from interest expense deductions of $218,160 (30% local income tax on $727,200) over the life of the project.

$$\text{NPV (tax shield)} = \$218.16/(1+0.1212) + \$218.16/(1+0.1212)^2 + \$218.16/(1+0.1212)^3 + \$218.16/(1+0.1212)^4 + \$218.16/(1+0.1212)^5 = \$703.79$$

The total APV would be

$$\text{APV} = \$1{,}197.83 + \$703.79 = \$1{,}901.62$$

The resulting APV could be a proper approach to the valuation of the cash flows when the project is financed differently from that of the parent firm. Although the operating cash flows are valued lower (higher discount rate of straight equity applied to them), the tax shields resulting from the increased used of debt in the subsidiary (discounted at the cost of debt) offset the loss in equity-financed cash flows. *Note:* Although APV is a viable method of analysis it is not as widely used in practice as the traditional method using a weighted average cost of capital (WACC).

See also NET PRESENT VALUE.

ADR

See AMERICAN DEPOSITORY RECEIPTS.

AD VALOREM TARIFF

An ad valorem tariff is a tariff assessed as a percentage of the value of the goods cleared through customs. *Ad valorem* means "according to value." A 5% *ad valorem* tariff means the tariff is 5% of the value of the merchandise.

ADVISING BANK

An advising bank is a *corresponding bank* in the beneficiary's country to which the *issuing bank* sends the *letter of credit*.

See also CORRESPONDENT BANK; ISSUING BANK; LETTERS OF CREDIT.

AGENCY FOR INTERNATIONAL DEVELOPMENT

The Agency for International Development (AID) is a U.S. government agency founded by President Kennedy in 1961 whose mission is to promote social and economic development in the Third World. It has been responsible for assisting transition to market-based economies in East Europe; establishment of a regulatory framework for securities markets in Indonesia, Jordan, and Sri Lanka; road construction and maintenance in Latin America and Southern Asia; and agricultural research and farm credits worldwide. AID fields workers worldwide and administrative officers in Washington, D.C. identify worthy projects and then ask U.S. industries to submit proposals. The winners receive government support.

AID

See AGENCY FOR INTERNATIONAL DEVELOPMENT.

ALL-EQUITY BETA

All-equity beta is the beta associated with the unleveraged cash flows of a capital project or company. It is determined as follows:

$$b^* = \frac{b}{1 + (1 - t)(DE)}$$

where b = a firm's beta, t = tax rate, and DE = debt-equity ratio.

EXAMPLE 8

If the beta of a firm's stock is 1.2, and it has a debt-equity ratio of 60% and a tax rate of 34%, then its all-equity beta, b^*, is 0.93;

$$0.93 = 1.3/(1 + 0.66 \times 0.6)$$

ALL-EQUITY (DISCOUNT) RATE

This is the discount rate that reflects only the business risks of a capital project and separates them from the effects of financing. This rate applies directly to a project that is financed entirely with owners' equity.

ALL-IN-RATE

Rate used in charging clientele for accepting *banker's acceptances* that consists of the interest rate for the discount and the commission.
See also BANKER'S ACCEPTANCE.

AMERICAN DEPOSITORY RECEIPTS

An American depository receipt (ADR) is a certificate of ownership, issued by a U.S. bank, representing a claim on underlying foreign stocks. ADRs may be traded in lieu of trading in the actual underlying shares. The bank issues all ADRs, not the corporation's stock certificate, to an American investor who buys shares of that corporation. The stock certificate is kept at the bank. The process of ADRs works as follows: a foreign company places shares in trust with a U.S. bank, which in turn issues depository receipts to U.S. investors. The ADRs are, therefore, claims to shares of stock and are essentially the same as shares. The depository bank performs all clerical functions—issuing annual reports, keeping a shareholder ledger, paying and maintaining dividend records, etc.—allowing the ADRs to trade in markets just as domestic securities trade. ADRs are traded on the NYSE, AMEX, and OTC markets as a share in stock, minus the voting rights. Examples of ADRs are Hanson, Cannon, and Smithkline Beecham. ADRs have become an increasingly convenient and popular vehicle for investing internationally. Investors do not have to go through foreign brokers, and information on company operations is usually available in English. Therefore, ADRs are good substitutes for *direct foreign investment*. They are bought and sold with U.S. dollars, and they pay their dividends in dollars. Further, the trading and settlement costs that apply in some foreign markets are waived. The certificates are issued by depository banks (for example, the Bank of New York). ADRs, however, are not for everyone. Disadvantages are the following:

1. ADRs carry an element of currency risk. For example, an ADR based on the stock of a British company would tend to lose in value when the dollar strengthens against the British pound, if other factors were held constant. This is because as the pound weakens, fewer U.S. dollars are required to buy the same shares of a U.K. company.
2. Some thinly traded ADRs can be harder to buy and sell. This could make them more expensive to purchase than the quoted price.
3. You may face problems obtaining reliable information on the foreign companies. It may be difficult to do your own research in selecting foreign stocks. For one thing, there is a shortage of data: the annual report may be all that is available, and its reliability is questionable. Furthermore, in many instances, foreign financial reporting and accounting standards are substantially different from those accepted in the U.S.
4. ADRs can be either sponsored or unsponsored. Many ADRs are not sponsored by the underlying companies. Nonsponsored ADRs oblige you to pay certain fees to the depository bank. The return is reduced accordingly.
5. There are a limited number of issues available for only a small fraction of the foreign stocks traded internationally. Many interesting and rewarding investment opportunities exist in shares with no ADRs. For quotations on ADRs, log on to **www.adr.com** by J.P. Morgan.

See also AMERICAN SHARES; GLOBAL REGISTERED SHARES.

AMERICAN SHARES

Instead of buying foreign stocks overseas, investors can purchase foreign equities traded in the United States typically in two ways: (1) *American Depository Receipts (ADRs)* and American shares. American shares are securities certificates issued in the U.S. by a transfer agent acting on behalf of the foreign issuer. The foreign issuer absorbs part or all of the handling expenses involved.

See also AMERICAN DEPOSITORY RECEIPTS; GLOBAL REGISTERED SHARES.

AMERICAN TERMS

American terms are foreign exchange quotations for the U.S. dollar, expressed as the U.S. dollar price per unit of foreign currency. For example, U.S. $0.00909/yen is an American term. It is also called *American basis* or *American quote*. American terms are normally used in the interbank market of the U.K. pound sterling, Australian dollar, New Zealand dollar, and Irish punt. Sterling is quoted as the foreign currency price of one pound. The relationship between American terms and European terms and between direct and indirect can be summarized as follows:

American Terms	European Terms
U.S. dollar price of one unit of foreign currency (e.g., U.S. $0.00909/¥)	Foreign currency price of one U.S. dollar (e.g., ¥110/$)
A direct quote in the U.S.	A direct quote in Europe
An indirect quote in Europe	An indirect quote in the U.S.

American terms are used in many retail markets (e.g., airports for tourists), on the foreign currency futures market in Chicago, and on the foreign exchange options market in Philadelphia.

ANALYSIS OF FOREIGN INVESTMENTS

Also called *international capital budgeting*, foreign investment decisions are basically capital budgeting decisions at the international level. Capital budgeting analysis for foreign as compared with domestic projects introduces the following complications:

1. Cash flows to a project and to the parent must be differentiated.
2. National differences in tax systems, financial institutions, financial norms, and constraints on financial flows must be recognized.
3. Different inflation rates can affect profitability and the competitive position of an affiliate.
4. Foreign exchange-rate changes can alter the competitive position of a foreign affiliate and the value of cash flows between the affiliate and the parent.
5. Segmented capital markets create opportunities for financial gains and they may cause additional costs.
6. Political risk can significantly change the value of a foreign investment.

The foreign investment decision requires two major components:

1. *The estimation of the relevant future cash flows.* Cash flows are the dividends and possible future sales price of the investment. The estimation depends on the sales forecast, the effects on exchange rate changes, the risk in cash flows, and the actions of foreign governments.
2. *The choice of the proper discount rate (cost of capital).* The cost of capital in foreign investment projects is higher due to the increased risks of:
   (a) Currency risk (or foreign exchange risk)—changes in exchange rates. This risk may adversely affect sales by making competing imported goods cheaper.
   (b) Political risk (or sovereignty risk)—possibility of nationalization or other restrictions with net losses to the parent company.

The methods of evaluating multinational capital budgeting decisions include *net present value (NPV), adjusted present value (APV),* and *internal rate of return (IRR).*

EXAMPLE 9

In what follows, we will illustrate a case of multinational capital budgeting. We will analyze a hypothetical foreign investment project by a U.S. manufacturing firm in Korea. The analysis is based on the following data gathered by a project team.

Product. The company (to be called *Ko-tel* hereafter) is expected to be a wholly owned Korean manufacturer of customized integrated circuits (ICs) for use in computers, automobiles, and robots. Ko-tel's products would be sold primarily in Korea, and all sales would be denominated in Korean won.

Sales. Sales in the first year are forecasted to be Won 26,000 million. Sales are expected to grow at 10% per annum for the foreseeable future.

Working capital. Ko-tel needs gross working capital (that is, cash, receivables, and inventory) equal to 25% of sales. Half of gross working capital can be financed by local payables, but the other half must be financed by Ko-tel or by Am-tel, the parent company.

Parent-supplied components. Components sold to Ko-tel by Am-tel have a direct cost to Am-tel equal to 95% of their sales price. The margin is therefore 5%.

(*Continued*)

Depreciation. Plant and equipment will be depreciated on a straight-line basis for both accounting and tax purposes over an expected life of 10 years. No salvage value is anticipated.

License fees. Ko-tel will pay a license fee of 2.5% of sales revenue to Am-tel. This fee is tax-deductible in Korea but provides taxable income to Am-tel.

Taxes. The Korean corporate income tax rate is 35%; the U.S. rate is 38%. Korea has no withholding tax on dividends, interest, or fees paid to foreign residents.

Cost of capital. The cost of capital (or minimum required return) used in Korea by companies of comparable risk is 22%. Am-tel also uses 22% as a discount rate for its investments.

Inflation. Prices are expected to increase as follows.

Korean general price level:	+9% per annum
Ko-tel average sales price:	+9% per annum
Korean raw material costs:	+3% per annum
Korean labor costs:	+12% per annum
U.S. general price level:	+5% per annum

Exchange rates. In the year in which the initial investment takes place, the exchange rate is Won 1050 to the dollar. Am-tel forecasts the won to depreciate relative to the dollar at 2% per annum.

Dividend policy. Ko-tel will pay 70% of accounting net income to Am-tel as an annual cash dividend. Ko-tel and Am-tel estimate that over a five-year period the other 30% of net income must be reinvested to finance working capital growth.

Financing. Ko-tel will be financed by Am-tel with a $11,000,000 purchase of Won 10,503,000,000 common stock, all to be owned by Am-tel.

In order to develop the normal cash flow projections, Am-tel has made the following assumptions.

1. Sales revenue in the first year of operations is expected to be Won 26,000 million. Won sales revenue will increase annually at 10% because of physical growth and at an additional 9% because of price increases. Consequently, sales revenue will grow at (1.1) (1.09) = 1.20, or 20% per annum.
2. Korean raw material costs in the first year are budgeted at Won 4,000 million. Korean raw material costs are expected to increase at 10% per annum because of physical growth and at an additional 3% because of price increases. Consequently, raw material cost will grow at (1.1) (1.03) = 1.13, or 13% per annum.
3. Parent-supplied component costs in the first year are budgeted at Won 9,000 million. Parent-supplied component costs are expected to increase annually at 10% because of physical growth, plus an additional 5% because of U.S. inflation, plus another 4% in won terms because of the expected deterioration of the won relative to the dollar. Consequently, the won cost of parent-supplied imports will increase at (1.1) (1.05) (1.04) = 1.20 or 20% per annum.
4. Direct labor costs and overhead in the first year are budgeted at Won 5,000 million. Korean direct labor costs and overhead are expected to increase at 10% per annum because of physical growth and at an additional 12% because of an increase in Korean wage rates. Consequently, Korean direct labor and overhead will increase at (1.1) (1.12) = 1.232 or 12.32% per annum.
5. Marketing and general and administrative expenses are budgeted at Won 4,000 million, fixed plus 4% of sales.
6. Liquidation value. At the end of five years, the project (including working capital) is expected to be sold on a going-concern basis to Korean investors for Won 9,000 million, equal to $7045.1 million at the expected exchange rate of Won 1,277.49/$. This sales price is free of all Korean and U.S. taxes and will be used as a terminal value.

Given the facts and stated assumptions, the beginning balance sheet is presented in Exhibit 3, while Exhibit 4 shows revenue and cost projections for Ko-tel over the expected five-year life of the project.

EXHIBIT 3
Beginning Balance Sheet

		Millions of Won	Thousands of Dollars
Assets			
1	Cash balance	650	619
2	Accounts receivable	0	0
3	Inventory	1050	1000
4	Net plant and equipment	7000	6667
5	Total	8700	8286
Liabilities and Net Worth			
6	Accounts payable	700	667
7	Common stock equity	8000	7619
8	Total	8700	8286

EXHIBIT 4
Sales and Cost Data

Item		Year 1	2	3	4	5
1	Total sales revenue	26000	31174	37378	44816	53734
2	Korean raw material	4000	4532	5135	5818	6591
3	Components purchased from Am-tel	9000	10811	12986	15599	18737
4	Korean labor and overhead	5000	6160	7589	9350	11519
5	Depreciation	700	700	700	700	700
6	Cost of sales [(2) + (3) + (4) + (5)]	18700	22203	26410	31466	37548
7	Gross margin [(1) – (6)]	7300	8971	10968	13350	16187
8	License fee [2.5% of (1)]	650	779	934	1120	1343
9	Marketing, general, and administrative	5040	5247	5495	5793	6149
10	EBIT* [(7) – (8) – (9)]	1610	2945	4538	6437	8694
11	Korean income taxes (35%)	564	1031	1588	2253	3043
12	Net income after Korean taxes [(10) – (11)]	1046	1914	2950	4184	5651
13	Cash dividend [70% of (12)]	733	1340	2065	2929	3956

* EBIT = earnings before interest and taxes

Exhibit 5 shows how the annual increase in working capital investment is calculated. According to the facts, half of gross working capital must be financed by Ko-tel or Am-tel. Therefore, half of any annual increase in working capital would represent an additional required capital investment.

Exhibit 6 forecasts project cash flows from the viewpoint of Ko-tel. Thanks to healthy liquidation value, the project has a positive NPV and an IRR greater than the 22% local (Korean) cost of capital for projects of similar risk. Therefore, Ko-tel passes the first of the two tests of required rate of return.

EXHIBIT 5
Working Capital Calculation

Item		Year				
		1	2	3	4	5
1	Total revenue	26000	31174	37378	44816	53734
2	Net working capital needs at year-end [25% of (1)]	6500	7794	9344	11204	13434
3	Less year-beginning working capital	1700	6500	7794	9344	11204
4	Required addition to working capital	4800	1294	1551	1860	2230
5	Less working capital financed in Korean by payables	2400	647	775	930	1115
6	Net new investment in working capital	2400	647	775	930	1115

EXHIBIT 6
Cash Flow Projection—NPV and IRR for Ko-tel

Item		Year					
		0	1	2	3	4	5
1	EBIT [Exhibit 4, (10)]		1610	2945	4538	6437	8694
2	Korean income taxes (35%)		564	1031	1588	2253	3043
3	Net income, all equity basis		1046	1914	2950	4184	5651
4	Depreciation		700	700	700	700	700
5	Liquidation value						9000
6	Half of addition to working capital		2400	647	775	930	1115
7	Cost of project	–8000					
8	Net cash flow	–8000	–654	1967	2874	3954	14236
9	IRR	0.26765 = 26.77%					
10	NPV = PV (at 22%) – I	$1,421.37					

Does Ko-tel also pass the second test? That is, does it show at least a 22% required rate of return from the viewpoint of Am-tel? Exhibit 7 shows the calculation for expected after-tax dividends from Ko-tel to be received by Am-tel. For purposes of this example, note that Am-tel must pay regular U.S. corporate income taxes (38% rate) on dividends received from Ko-tel. However, the U.S. tax law allows Am-tel to claim a tax credit for income taxes paid to Korea on the Korean income that generated the dividend. The process of calculating the regional income in Korea is called "grossing up" and is illustrated in Exhibit 7, lines (1), (2), and (3).

This imputed Korean won income is converted from won to dollars in line (5). Then the U.S. income tax is calculated at 38% in line (6). A tax credit is given for the Korean income taxes paid, as calculated in line (7). Line (8) then shows the net additional U.S. tax due, and line (10)

EXHIBIT 7
After-tax Dividend Received by Am-tel

Item	Year 0	1	2	3	4	5
In Millions of Won						
1 Cash dividend paid [Exhibit 4, (13)]		733	1340	2065	2929	3956
2 A 70% of Korean income tax [Exhibit 2, (11)]		394	721	1112	1577	2130
3 Grossed-up dividend [(1) + (2)]		1127	2061	3177	4506	6086
4 Exchange-rate (won/$)	1050.00	1075.20	1101.00	1127.43	1154.49	1182.19
In Thousands of Dollars						
5 Grossed-up dividend [(3)/(4) × 1000]		1048.2	1872.3	2817.7	3902.7	5147.8
6 U.S. tax (38%)		398.3	711.5	1070.7	1483.0	1956.2
7 Credit for Korean taxes [(2)/(4) × 1000]		366.9	655.3	986.2	1365.9	1801.7
8 Additional U.S. tax due [(6) − (7), if (6) is larger]		31.4	56.2	84.5	117.1	154.4
9 Excess U.S. tax credit [(7) − (6), if (7) is larger		0.0	0.0	0.0	0.0	0.0
10 Dividend received by Am-tel after all taxes [(1)/(4) × 1000 − (8)]		649.9	1160.8	1747.0	2419.7	3191.6

shows the net dividend received by Am-tel after the additional U.S. tax is paid. Finally, Exhibit 8 calculates the rate of return on cash flows from Ko-tel from the viewpoint of Am-tel. However, Ko-tel fails to pass the test because it has a negative NPV and an IRR, below the 22% rate of return required by Am-tel.

EXHIBIT 8
NPV and IRR for Am-tel

Item	Year 0	1	2	3	4	5
In Millions of Won						
1 License fee from Ko-tel (2.5%) [Exhibit 4, (8)]		650	779	934	1120	1343
2 Margin on exports to Ko-tel [5% of (3) in Exhibit 4]		450	541	649	780	937
3 Total receipts		1100	1320	1583	1900	2280

(*Continued*)

4	Exchange rate (won/$)	1050.00	1092.00	1135.68	1181.11	1228.35	1277.49
In Thousands of Dollars							
5	Pre-tax receipts [(3)/(4) × 1000]		1007.3	1162.2	1340.9	1547.1	1784.7
6	U.S. taxes (38%)		382.8	441.6	509.5	587.9	678.3
7	License fees and export profits, after tax		624.5	720.6	831.4	959.2	1106.7
8	After-tax dividend [Exhibit 7, (10)]		649.9	1160.8	1747.0	2419.7	3191.6
9	Project cost	−11000.0					
10	Liquidation value						7045.1
11	Net cash flow	−11000.0	1274.4	1881.4	2578.3	3378.8	11343.4
12	IRR	0.1714 = 17.14%					
13	NPV = PV (at 22%) − I	($1,549.20)					

A. What-if Analysis

So far the project investigation team has used a set of "most likely" assumptions to forecast rates of return. It is now time to subject the most likely outcome to sensitivity analyses. As many probabilistic techniques are available to test the sensitivity of results to political and foreign exchange risks as are used to test sensitivity to business and financial risks. But it is more common to test sensitivity to political and foreign exchange risk by simulating what would happen to net present value and earnings under a variety of "what if" scenarios. Spreadsheet programs such as Excel can be used to test various scenarios (see Exhibit 9).

EXHIBIT 9
NPV Profiles for Ko-tel and Am-tel—Sensitivity Analysis

Discount rate (%)	0	4	8	12	16
Project pt of view (Ko-tel)	$14,378.57	10,827.01	7,958.71	5,621.75	3,702.09
Parent pt of view (Am-tel)	$9,456.37	6,468.67	4,043.44	2,056.77	415.48

20	22	24	28	32	36	40
$2,113.17	1,421.37	788.65	(322.83)	(1,261.35)	(2,058.46)	(2,739.20)
$ (951.26)	(1,549.20)	(2,097.87)	(3,066.53)	(3,890.23)	(4,594.99)	(5,201.51)

Exhibit 10 depicts an NPV graph of various scenarios.

EXHIBIT 10
NPV Profiles for Ko-tel and Am-tel Sensitivity Analysis

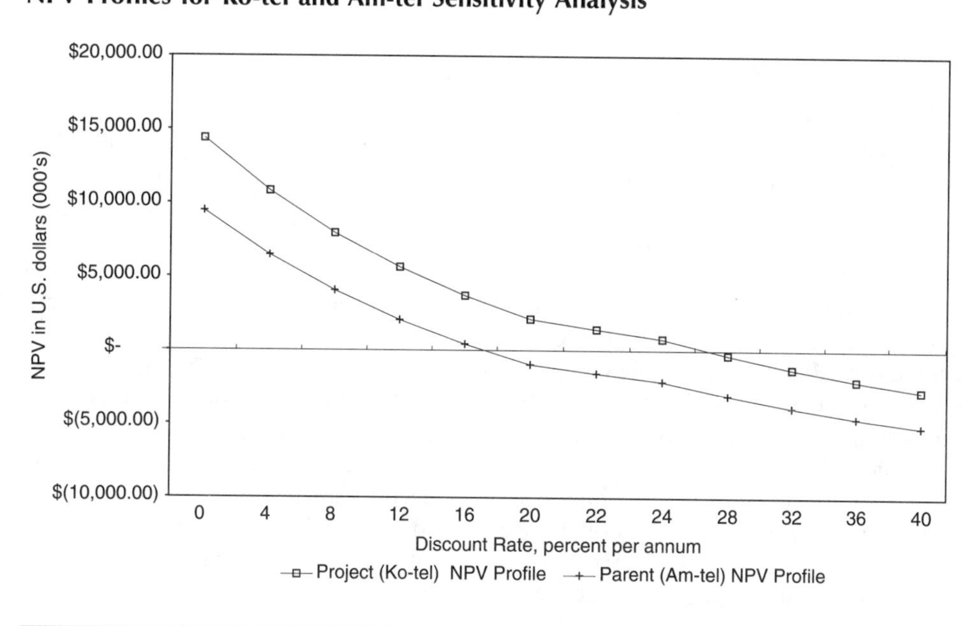

ANNUAL PERCENTAGE RATE

Different types of investments use different compounding periods. For example, most bonds pay interest semiannually. Some banks pay interest quarterly. If an investor wishes to compare investments with different compounding periods, he or she needs to put them on a common basis.

The annual percentage rate (APR), or effective annual rate, is used for this purpose and is computed as follows:

$$\text{APR} = (1 + r/m)^m - 1.0$$

where r = the stated, nominal, or quoted rate, and m = the number of compounding periods per year.

EXAMPLE 10

Assume that a bank offers 6% interest, compounded quarterly, then the APR is:

$$\text{APR} = (1 + .06/4)^4 - 1.0 = (1.015)^4 - 1.0 = 1.0614 - 1.0 = .0614 = .0614 = 6.14\%$$

This means that if one bank offered 6% with quarterly compounding, while another offered 6.14% with annual compounding, they would both be paying the same effective rate of interest.

Annual percentage rate (APR) also is a measure of the cost of credit, expressed as a yearly rate. It includes interest as well as other financial charges such as loan and closing costs and fees. A lender is required to tell a borrower the APR. The APR provides a good basis for comparing the cost of loans, including mortgage plans.

A/P

1. In accounting, abbreviation for "accounts payable."
2. In international trade and finance documentation, abbreviation for "authority to purchase" or "authority to pay."

APPRECIATION OF THE DOLLAR

Also called *strong dollar, strengthening dollar*, or *revaluation of a dollar*, appreciation of the dollar refers to a rise in the foreign exchange value of the dollar relative to other currencies. The opposite of appreciation is weakening, deteriorating, or depreciation of the dollar. Strictly speaking, revaluation refers to a rise in the value of a currency that is pegged to gold or to another currency. A strong dollar makes Americans' cash go further overseas and reduces import prices—generally good for U.S. consumers and for foreign manufacturers. If the dollar is overvalued, U.S. products are harder to sell abroad and at home, where they compete with low-cost imports. This helps give the U.S. its huge trade deficit. A weak dollar can restore competitiveness to American products by making foreign goods comparatively more expensive. But too weak a dollar can spawn inflation, first through higher import prices and then through spiraling prices for all goods. Even worse, a falling dollar can drive foreign investors away from U.S. securities, which lose value along with the dollar. A strong dollar can be induced by interest rates. Relatively higher interest rates abroad will attract dollar-denominated investments which will raise the value of the dollar. Exhibit 11 summarizes the impacts of changes in foreign exchange rates on the multinational company's products and services.

EXHIBIT 11
The Impacts of Changes in Foreign Exchange Rates

	Weak Currency (Depreciation/devaluation)	Strong Currency (Appreciation/revaluation)
Imports	More expensive	Less expensive
Exports	Less expensive	More expensive
Payables	More expensive	Less expensive
Receivables	Less expensive	More expensive
Inflation	Fuel inflation by making imports more costly	Low inflation
Foreign investment	Discourage foreign investment. Lower return on investments by international investors.	High interest rates could attract foreign investors.
The effect	Raising interests could slow down the economy	Reduced exports could trigger a trade deficit

The amount of appreciation or depreciation is computed as the fractional increase or decrease in the home currency value of the foreign currency or in the foreign currency value of the home currency:

With *Direct Quotes* (exchange rate expressed in home currency):

$$\text{Percent change} = \frac{\text{Ending rate} - \text{beginning rate}}{\text{Beginning rate}} \times 100$$

With *Indirect Quotes* (exchange rate expressed in foreign currency):

$$\text{Percent change} = \frac{\text{Beginning rate} - \text{ending rate}}{\text{Ending rate}} \times 100$$

EXAMPLE 11

An increase in the exchange rate from $0.64 (or DM1.5625/$) to $0.68 (or DM1.4705) is equivalent to a DM appreciation of 6.25% [($0.68 – $0.64)/$0.64 = 0.0625] (direct quote) or [(DM1.5625 – DM1.4705)/DM1.4705 = 0.0625] (indirect quote). This also means a dollar depreciation of 5.88% [($0.64 – $0.68)/$0.68 = –0.0588].

See also DEPRECIATION OF THE DOLLAR.

ARBITRAGE

Arbitrage is the simultaneous purchase or sale of a commodity in different markets to profit from unwarranted differences in prices. That is, it involves the purchase of a commodity, including foreign exchange, in one market at one price while simultaneously selling that same currency in another market at a more advantageous price, in order to obtain a risk-free profit on the price differential. Profit is the price differential minus the cost. If exchange rates are not equal worldwide, there would be profit opportunity for simultaneously buying a currency in one market while selling it in another. This activity would raise the exchange rate in the market where it is too low, because this is the market in which you would buy, and the increased demand for the currency would result in a higher price. The market where the exchange rate is too high is one in which you sell, and this increased selling activity would result in a lower price. Arbitrage would continue until the exchange rates in different markets are so close that it is not worth the costs incurred to do any further buying and selling. When this situation occurs, we say that the rates are "transaction costs close." Any remaining deviation between exchange rates will not cover the costs of additional arbitrage transactions, so the arbitrage activity ceases.

EXAMPLE 12

Suppose ABC Bank in New York is quoting the German mark/U.S. dollar exchange rate as 1.4445—55 and XYZ Bank in Frankfurt is quoting 1.4425—35. This means that ABC will buy dollars for 1.4445 marks and will sell dollars for 1.4455 marks. XYZ will buy dollars for 1.4425 marks and will sell dollars for 1.4435 marks. This presents an arbitrage opportunity. An arbitrager could buy $1 million at XYZ's ask price of 1.4435 and simultaneously sell $1 million to ABC at their bid price of 1.4445 marks. This would earn a profit of DM0.0010 marks per dollar traded, or DM10,000 would be the total arbitrage profit. If such a profit opportunity existed, the demand to buy dollars from XYZ would cause them to raise their ask price above 1.4435, while the increased interest in selling dollars to ABC at their bid price of 1.4445 marks would cause them to lower their bid. In this way, arbitrage activity pushes the prices of different traders to levels where no arbitrage profits are earned.

Exhibit 12 illustrates bounds imposed on spot rates by arbitrage transactions. As can be seen, there is strong arbitrage opportunity between banks A and B: you can buy cheap from A at its ask price, and resell at a high bid rate to B. In contrast, if the A's quote is A′, you cannot profitably buy from either A′ or B and sell to the other.

EXHIBIT 12
Bounds Imposed on Spot Rates by Arbitrage Transactions

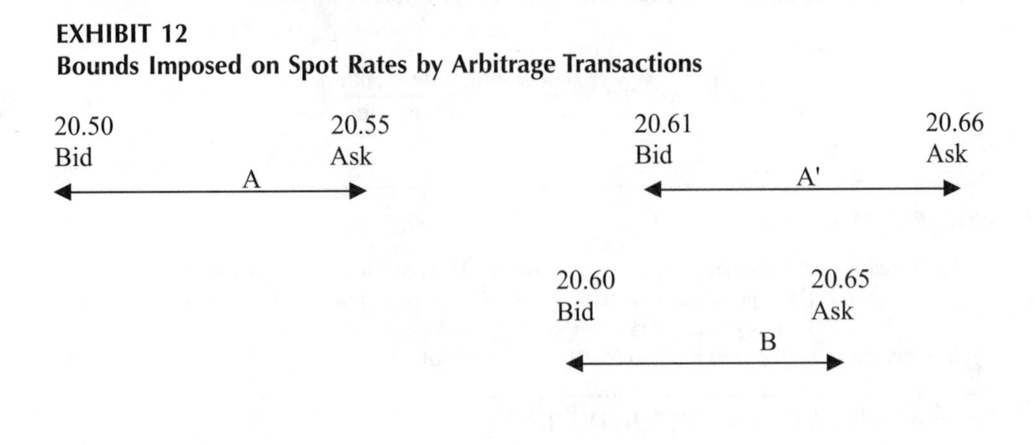

See also SIMPLE ARBITRAGE; FOREIGN EXCHANGE ARBITRAGE; TRIANGULAR ARBITRAGE; COVERED INTEREST ARBITRAGE.

ARBITRAGE PRICING MODEL (APM)

The *Capital Asset Pricing Model (CAPM)* assumes that required rates of return depend only on one risk factor, the stock's *beta*. The Arbitrage Pricing Model (APM) disputes this and includes any number of risk factors:

$$r = r_f + b_1 RP_1 + b_2 RP_2 + \cdots + b_n RP_n$$

where

r = the expected return for a given stock or portfolio
r_f = the risk-free rate
b_i = the sensitivity (or reaction) of the returns of the stock to unexpected changes in economic forces i ($i = 1,...n$)
RP_i = the market risk premium associated with an unexpected change in the ith economic force
n = the number of relevant economic forces

Roll and Ross suggest the following five economic forces:

1. Changes in expected inflation
2. Unanticipated changes in inflation
3. Unanticipated changes in industrial production
4. Unanticipated changes in the yield differential between low- and high-grade bonds (the default-risk premium)
5. Unanticipated changes in the yield differential between long-term and short-term bonds (the term structure of interest rates)

EXAMPLE 13

Suppose returns required in the market by investors are a function of two economic factors according to the following equation, where the risk-free rate is 7 percent:

$$r = 0.07 + b_1(0.04) + b_2(0.01)$$

ABC stock has the reaction coefficients to the factors such that $b_1 = 1.3$ and $b_2 = 0.90$. Then the required rate of return for the ABC stock is

$$r = 0.07 + (1.3)(0.04) + (0.90)(0.01) = 0.113 = 11.3\%$$

ARBITRAGE PROFITS

Profits obtained by an *arbitrageur* through an *arbitrage* process are called *arbitrage points.*

EXAMPLE 14

You own $10,000. The dollar rate on the Japanese yen is ¥106/$. The Japanese yen rate is given in Exhibit 13 below.

EXHIBIT 13
Selling Quotes for the Japanese Yen in New York

Country	Contract	$/Foreign Currency
Japan (yen)	Spot	0.009465
	30-day	0.009508
	90-day	0.009585

Note that the Japanese yen rate is ¥106/$, while the (indirect) New York rate is 1/0.009465 = ¥105.65/$. Assuming no transaction costs, the rates between Japan and New York are out of line. Thus, arbitrage profits are possible: (1) Because the yen is cheaper in Japan, buy $10,000 worth of yens in Japan. The number of yens would be $10,000 × ¥106/$ = ¥1,060,000. (2) Simultaneously sell the yens in New York at the prevailing rate. The amount received upon the sale of the yens would be: ¥1,060,000 × $0.009465 = $10,032.90. The net gain is $10,032.90 – $10,000 = $32.90.

EXAMPLE 15

You own $10,000. The dollar rate on the DM is 1.380 marks.

Country	Contract	U.S. Dollar Equivalent (Direct)	Currency per U.S.$ (Indirect)
Germany	Spot	.7282	1.3733
(mark)	30-day future	.7290	1.3716
	90-day future	.7311	1.3677

Based on the table above, are arbitrage profits possible? What is the gain (loss) in dollars? The dollar rate on the DM is 1.380 marks, while the table (indirect New York rate) shows 1.3733 (1/.7282) marks. Note that the rates between Germany and New York are out of line. Thus, arbitrage profits are possible. Since the DM is cheaper in Germany, buy $10,000 worth of marks in Germany. The number of marks purchased would be 13,800 ($10,000 × 1.380). Simultaneously sell the marks in New York at the prevailing rate. The amount received upon sale of the marks would be $10,049.16 (13,800 marks × $.7282/DM) = $10,049.16. The net gain is $49.16, barring transactions costs.

ARBITRAGEUR

An arbitrageur is an individual or business that exercises *arbitrage* seeking to earn risk-free profits by taking advantage of simultaneous price differences in different markets.

ARITHEMETIC AVERAGE RETURN VS. COMPOUND (GEOMETRIC) AVERAGE RETURN

It is one thing to calculate the return for a single holding period but another to explain a series of returns over time. If you keep an investment for more than one period, you need to understand how to derive the average of the successive rates of return. Two approaches to multiperiod average (mean) returns are the *arithmetic average return* and the *compound (geometric) average return*. The arithmetic average return is the simple mean of successive one-period rates of return, defined as:

$$\text{Arithmetic average return} = 1/n \sum r_t$$

where n = the number of time periods and r = the single holding period return in time t. *Caution:* The arithmetic average return can be misleading in multiperiod return computations.

A better accurate measure of the actual return obtained from an investment over multiple periods is the *compound (geometric) average return*. The compound return over n periods is derived as follows:

$$\text{Compound average return} = \neq \sqrt[n]{(1+r_1)(1+r_2)\cdots(1+r_n)} - 1$$

EXAMPLE 16

Assume the price of a stock doubles in one period and depreciates back to the original price. Dividend income (current income) is nonexistent.

	Time periods		
	t = 0	t = 1	t = 2
Price (end of period)	\$40	\$80	\$40
HPR	—	100%	–50%

The arithmetic average return is the average of 100% and –50%, or 25%, as indicated below:

$$\frac{100\% + (-50\%)}{2} = 25\%$$

However, the stock bought for \$40 and sold for the same price two periods later did not earn 25%; it earned zero. This can be illustrated by determining the compound average return. Note that $n = 2$, $r_1 = 100\% = 1$, and $r_2 = -50\% = -0.5$.

Then,

$$\begin{aligned}\text{Compound return} &= \sqrt{(1+1)(1-0.5)} - 1 \\ &= \sqrt{(2)(0.5)} - 1 \\ &= \sqrt{1} - 1 = 1 - 1 = 0\end{aligned}$$

EXAMPLE 17

Applying the formula to the data below indicates a compound average of 11.63 percent, somewhat less than the arithmetic average of 26.1 percent.

(1) Time	(2) Price	(3) Dividend	(4) Total return	(5) Holding period return (HPR)
0	$100	$–		
1	60	10	–30(a)	–0.300(b)
2	120	10	70	1.167
3	100	10	–10	–0.083

(a) $10 + ($60 – $100) = $–30
(b) HPR = $–30/$100 = –0.300

The arithmetic average return is

$$(-0.300 + 1.167 - 0.083)/3 = .261 = 26.1\%$$

but the compound return is

$$\sqrt[3]{[(1-0.300)(1+1.167)+(1+0.083)]} - 1 = 0.1163, \text{ or } 11.63\%.$$

See also TOTAL RETURN; RETURN RELATIVE.

ARM'S-LENGTH PRICING

Arm's-length pricing involves charging prices to which an unrelated buyer and seller would willingly agree. In effect, an arm's-length price is a *free market price*. Although a transaction between two subsidiaries of an MNC would not be an arm's-length transaction, the U.S. Internal Revenue Code requires arm's-length pricing for internal goods transfers between subsidiaries of MNCs.

See also INTERNATIONAL TRANSFER PRICING.

ARM'S-LENGTH TRANSACTION

An arm's-length transaction is a transaction between two or more unrelated parties. A transaction between two subsidiaries of an MNC would not be an arm's-length transaction.

See also ARM'S-LENGTH PRICING.

ASIAN CURRENCY UNIT

Asian Currency Unit (ACU) is a division of a Singaporean bank that deals in foreign currency deposits and loans.

ASIAN DEVELOPMENT BANK

Created in the late 1960s, the Asian Development Bank is a financial institution for supporting economic development in Asia. It operates on similar lines as the *World Bank*. Member countries range from Iran to the United States of America.

See also INTERNATIONAL MONETARY FUND; WORLD BANK.

ASIAN DOLLAR MARKET

Asian dollar market is the market in Asia in which banks collect deposits and make loans denominated in U.S. dollars.

ASIAN DOLLARS

Similar to Eurodollars, Asian dollars are U.S. dollar-denominated deposits kept in Asian-based banks.

ASKED PRICE

See ASKED RATE.

ASKED RATE

Also called *ask rate, selling rate, or offer rate*. The price at which a dealer is willing to sell foreign exchange, securities, or commodities.

See also BID RATE.

ASSET MANAGEMENT OF BANKS

A commercial bank earns profits for stockholders by having a positive spread in lending and through leverage. A positive spread results when the average yield on earning assets exceeds the average cost of deposit liabilities. A high-risk asset portfolio can increase profits, because the greater the risk position of the borrower, the larger the risk premium charged. On the other hand, a high-risk portfolio can reduce profits because of the increased chance that parts of it could become "nonperforming" assets. Favorable use of leverage (the bank's capital-asset ratio is falling) can increase the return on owners' equity. A mix of a high-risk portfolio and high leverage could result, however, in insolvency and bank failure. It is extremely important for banks to find an optimal mix.

A bank is also threatened with insolvency if it has to liquidate its asset portfolio at a loss to meet large withdrawals (a "run on the bank"). This can happen, because, historically, a large proportion of banks' liabilities come from demand deposits and, therefore, are easily withdrawn. For this reason, commercial bank asset management theory focuses on the need for liquidity. There are three theories:

1. *The commercial loan theory.* This theory contends that commercial banks should make only short-term self-liquidating loans (e.g., short-term seasonal inventory loans). In this way, loans would be repaid and cash would be readily available to meet deposit outflows. This theory has lost much of its credibility as a certain source of liquidity, because there is no guarantee that even seasonal working capital loans can be repaid.
2. *The shiftability theory.* This is an extension of the commercial loan theory stating that, by holding money-market instruments, a bank can sell such assets without capital loss in the event of a deposit outflow.
3. *The anticipated-income theory.* This theory holds that intermediate-term installment loans are liquid because they generate continuous cash inflows. The focus is not on short-term asset financing but on cash flow lending.

It is important to note that contemporary asset management hinges primarily on the shiftability theory, the anticipated-income theory, and liability management.

See also LIABILITY MANAGEMENT OF BANKS.

ASSET MARKET MODEL

The asset market model is a model that attempts to explain how a foreign exchange rate is determined. It states that the exchange rate between two currencies stands for the price that exactly balances the relative supplies of, and demands for, assets denominated in those currencies. Within the family of asset market models, there are two basic approaches: (1) In the *monetary approach*, the exchange rate for any two currencies is determined by relative money demand and money supply between the two countries. Relative supplies of domestic and foreign bonds are unimportant. (2) The *portfolio-balance approach* allows relative bond supplies and demands, as well as relative money-market conditions, to determine the exchange rate.

AUTOMATIC ADJUSTMENT MECHANISM

Automatic adjustment mechanism is the automatic response of an economy that is triggered by a *balance of payment* imbalance. When a trade deficit exists under *flexible exchange rates*, a currency devaluation generally occurs to revitalize exports and reduce imports. Under *fixed exchange rates*, domestic inflation is expected to be below a foreign counterpart, which leads to relatively cheaper domestic products, thereby escalating exports and plummeting imports.

B

BACK-TO-BACK FINANCING

An intercompany loan arranged through a bank.

See also BACK-TO-BACK LOANS.

BACK-TO-BACK LETTER OF CREDIT

Back-to-back letter of credit is one type of *letter of credit* (L/C). It is a form of *pretrade financing* in which the exporter employs the importer's L/C as a means for securing credit from a bank, which in turn supports its L/C to the exporter with the good chance of ability to repay that the importer's L/C represents.

BACK-TO-BACK LOANS

Also called *link financing, parallel loan*, or *fronting loan*, a back-to-back loan is a type of *swaps* used to raise or transfer capital. It may take several forms:

1. A loan made by two parent companies, each to the subsidiary of the other. As is shown in Exhibit 14, each loan is made and repaid in one currency, thus avoiding foreign exchange risk. Each loan should have the right to offset, which means that if either subsidiary defaults on its payment, the other subsidiary can withhold its repayment. This eliminates the need for parent company guarantees.
2. A loan in which two multinational companies in separate countries borrow each other's currency for a specific period of time and repay the other's currency at an agreed maturity. The loan is conducted outside the foreign exchange market and often channeled through a bank as an intermediary.

EXHIBIT 14
Back-to-Back Loan by Two Parent Companies

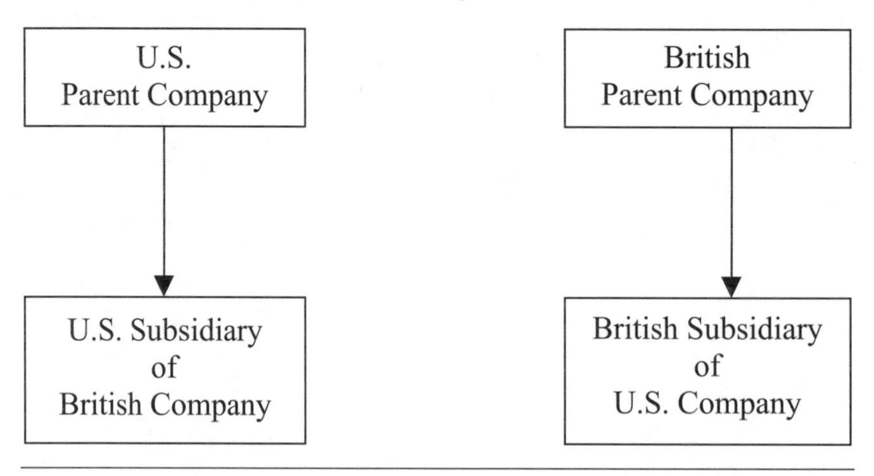

3. An intercompany loan in which two affiliates located in separate countries borrow each other's currency for a specific period of time and repay the other's currency at an agreed maturity. These loans are frequently channeled through a bank. Back-to-back loans are often used to finance affiliates located in countries with high interest rates or restricted capital markets or with a danger of *currency controls* and different tax rates applied to loans from a bank. They contrast with a *direct intercompany loan* which does not involve an intermediate bank. The loan process is depicted in Exhibit 15.

EXHIBIT 15
Back-to-Back Loan by Two Affiliates

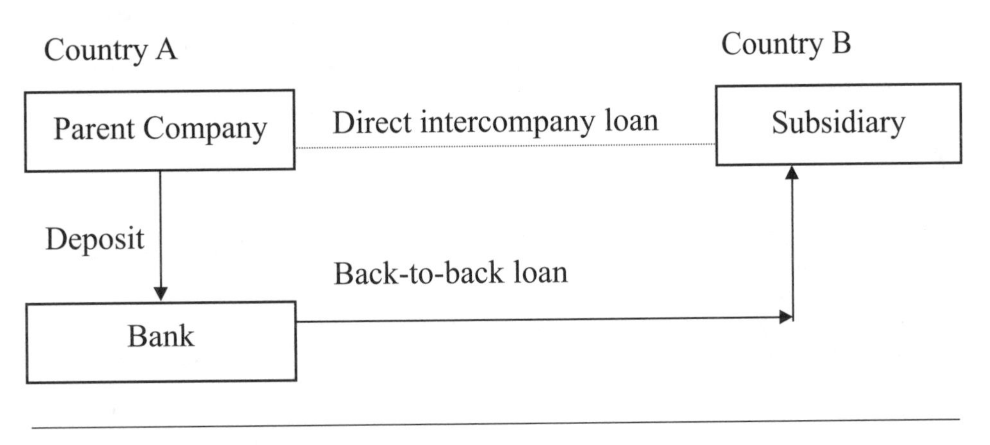

BAHT

Thailand's currency.

BALANCE OF PAYMENTS (BOP)

The balance of payments (BOP) is a systematic record of a country's receipts from, or payments to, other countries. In a way, it is like the balance sheets for businesses, only on a national level. The reference you see in the media to the *balance of trade* usually refer to goods within the goods and services category of the current account. It is also known as *merchandise* or *"visible" trade* because it consists of tangibles such as foodstuffs, manufactured goods, and raw materials. "Services," the other part of the category, is known as *"invisible" trade* and consists of intangibles such as interest or dividends, technology transfers, and others (e.g., insurance, transportation, financial). When the net result of both the current account and the capital account yields more credits than debits, the country is said to have a surplus in its balance of payments. When there are more debits than credits, the country has a deficit in the balance of payments. Exhibit 16 presents the components of each and their interrelationships. Data is collected by the U.S. Customs Service. Figures are reported in seasonally adjusted volumes and dollar amounts. It is the only nonsurvey, nonjudgmental report produced by the *Department of Commerce*. The balance of payments appears in *Survey of Current Business*.

EXHIBIT 16
Balance of Payments Accounts

Sources	Uses	Balance Account (Sources minus Uses)
1. Current Transactions		
• Exports of goods	• Imports of goods	• Trade balance
• Exports of services	• Imports of services	• Invisible balance
Inward unilateral transfers	Outward unilateral transfers	• Net inward transfers
• Private	• Private	
• Public	• Public	
		CA = Current account balance = Net inflow from current transactions
2. Capital Transactions		
• Classified as private versus government		
• Classified by type of transaction:		
• Inward portfolio investment	• Outward portfolio investment	• Net inward investment
• short term	• short term	
• long term	• long term	
• Inward direct investment	• Outward direct investment	• Net inward investment
		KA = Capital account balance = Net inflow from capital transactions
3. Settling Items		
3A. Central bank transactions		
• Decreases in foreign reserves	• Increases in foreign reserves	• Net decreases in foreign reserves (−)ΔRFX
3B. Errors and omissions		
• Unrecorded inflows	• Unrecorded outflows	• Errors, omissions (E&O)
	Grand total of BOP	**0 = CA + KA − ΔRFX + E&O**

BALANCE OF PAYMENTS ACCOUNTING

The *balance of payments* (BOP) statement is based on a double-entry bookkeeping system that is used to record transactions. Every transaction is recorded as if it consisted of an exchange of something for something else—that is, as a debit and a credit. As a general rule, currency inflows are recorded as *credits*, and outflows are recorded as *debits*. Exports of goods and services are recorded as *credits*. In the case of imports, goods and services are normally acquired for money or debt. Hence, they are recorded as *debits*. Where items are given rather than exchanged, special types of counterpart entries are made in order to furnish the required offsets. Just as in accounting, the words *debits* and *credits* have no value-laden meaning—either good or bad. They are merely rules or conventions; they are not economic truths. Under the conventions of double-entry bookkeeping, an increase in the assets of an entity is always recorded as a debit and an increase in liabilities as a credit. Thus a debit records (1) the import of goods and services, (2) increase in assets, or (3) reductions in liabilities. A credit records (1) the export of goods and services, (2) a decrease in assets, or (3) increases in liabilities.

The balance of payments statement is traditionally divided into three major groups of accounts: (1) current accounts, (2) capital accounts, and (3) official reserves accounts. We will define these accounts and illustrate them with some transactions. The double-entry system used in the preparation of the balance of payments allows us to see how each transaction is financed and how international transactions usually affect more than one type of account in the balance of payments. The illustrative transactions presented here are for the U.S. in the year 2001.

A. Current Accounts

The current accounts record the trade in goods and services and the exchange of gifts among countries. The trade in goods is composed of exports and imports. A country increases its exports when it sells merchandise to foreigners. This is a source of funds and a decrease in real assets. A country increases its imports when it buys merchandise from foreigners. This is a use of funds and an acquisition of real assets.

EXAMPLE 18

A U.S. manufacturer exports $5,000 in goods to a customer in Greece. According to the sales terms, this account will be paid in 90 days. In this case two things happen. The merchandise export, a reduction in real assets, provides an increase in external purchasing power—a credit entry. But the exporter is financing the transaction for 90 days; that is, the exporter's accounts receivable have increased by $5,000. The company has made a short-term investment abroad. This acquisition of a short-term asset or claim represents a use of the country's external purchasing power—a debit entry. In the U.S. balance of payments accounts, this transaction will appear as shown below:

	Debit	Credit
Increase in short-term claims on foreigners (the accounts receivable)	$5,000	
Exports		$5,000

The *trade in services* includes interest and dividends, travel expenses, and financial and shipping charges. Interest and dividends received measure the services that the country's capital has rendered abroad. Payments received from tourists measure the services that the country's hotels and shops provided to visitors from other countries. Financial and shipping charges to foreigners measure the fees that the financial community and ship owners charged to foreigners for the special services they rendered. In these cases the nation gave the service of assets it possessed (for example, a hotel) to foreigners. Thus, these transactions are a source of external purchasing power, in contrast to the preceding cases, when the country's residents are the recipients of the services from foreign-owned assets and the given country loses purchasing power to the rest of the world.

EXAMPLE 19

A Japanese resident visits the U.S. Upon his arrival, he converts his yen into $2,500 worth of dollars at the airport bank. When the visitor departs, he has no dollars left. In this case, the U.S. provided services (such as hotel room and meals) to foreigners amounting to $2,500. In exchange for these services, U.S. banks now have $2,500 worth of yen. The willingness of U.S. banks to hold the yen balances—a liability of the Japanese government—provided the required financing for the Japanese tourist. The services that the U.S. provided to the Japanese are clearly a source of purchasing power for the U.S.—a credit entry. However, the accumulation of yen in U.S. banks is an increase in U.S. holdings of foreign financial obligations—a use of purchasing power, and thus a debit entry. In the U.S. balance of payments this transaction will appear as shown below:

	Debit	Credit
Increase in short-term claims on foreigners (the yen holdings)	$2,500	
Receipts for travel services to foreigners		$2,500

The exchange of gifts among countries is recorded in the *unilateral transfers account.* This account is also labeled *remittances* or *unrequited transfers*. A typical entry in this account is the money that emigrants send home. Another example is a gift that one country makes to another. When a country makes a gift, it can be said that it is acquiring an asset which we may call *goodwill.* As with any other asset acquisition, the gift represents a use of external purchasing power.

EXAMPLE 20

A U.S. resident who left his family in Hungary sends a $1,000 check to his wife in Hungary. The gift that the U.S. resident sent is a unilateral or unrequited transfer. For accounting purposes it can be treated as a purchase of goodwill, that reduces U.S. purchasing power (a debit entry). However, this gift was made possible by the credit or financing that the Hungarians extended to the U.S. when they accepted a financial obligation (a check) in U.S. dollars from a U.S. resident. This latter part of the transaction, an increase in liabilities to foreigners, is a source of external purchasing power (a credit entry). The entry for this transaction in the U.S. balance of payments is shown below:

	Debit	Credit
Gifts to foreigners	$1,000	
Increase in short-term liabilities to foreigners (the check)		$1,000

B. Capital Accounts

The *capital accounts* record the changes in the levels of international financial assets and liabilities. The various classifications within the capital account are based on the original term to maturity of the financial instrument and on the extent of the involvement of the owner of the financial asset in the activities of the security's issuer. Accordingly, the capital accounts are subdivided into *direct investment, portfolio investment*, and *private short-term capital flows. Direct investment* and *portfolio investment* involve financial instruments that had a maturity of more than 1 year when issued initially. The distinction between direct investment and portfolio investment is made on the basis of the degree of management involvement. Considerable management involvement is presumed to exist in the case of direct investment (usually a minimum of 10% ownership in a firm), but not of portfolio investment.

EXAMPLE 21

A U.S. resident buys a $3,000 bond newly issued by a German company. The payment is made with a check drawn on a U.S. bank account. As a result the U.S. resident now owns a German bond, and the German company owns U.S. dollar deposits. U.S. acquisition of the German bond (a financial asset) implies a decrease in U.S. external purchasing power; the long-term investments or claims on foreigners must be debited. However, the dollar balances that the German company now owns represent an increase in U.S. liabilities to foreigners, which increases U.S. foreign purchasing power; the account short-term liabilities to foreigners must be credited. Two interpretations are possible here. We can say that the purchase of the German bond was financed with short-term liabilities issued by the U.S., or we can say that the purchase of short-term dollar instruments by the Germans was financed by their issuing a long-term bond. In the U.S. balance of payments this transaction will appear as shown as follows:

	Debit	**Credit**
Increase in long-term claims on foreigners (the German bond)	$3,000	
Increase in short-term liabilities to foreigners (the dollar deposits)		$3,000

Short-term capital movements involve financial paper with an original maturity of less than 1 year. In the previous examples, payment or financing of various transactions was made with either currency or a short-term financial note (except for the alternative interpretation of the financing of Example 21). Payments in U.S. dollars were called changes in U.S. short-term liabilities to foreigners. Payments in foreign currency were called changes in U.S. short-term claims on foreigners. These accounts are part of the short-term capital accounts. The given examples produced a net increase in short-term claims on foreigners (a debit of $7,500), and a net increase in short-term liabilities to foreigners (a credit of $4,000). A different type of entry in these accounts is presented in the next example.

EXAMPLE 22

A Swiss bank buys $6,000 worth of U.S. Treasury bills. It pays by drawing on its dollar account with a U.S. bank. The sale of Treasury bills to a foreigner is equivalent to U.S. borrowing external purchasing power from foreigners, an increase in liabilities to foreigners (a credit entry). However, the purchase is paid by reducing another debt that the U.S. had to foreigners (U.S. dollars in the hands of foreigners). This reduction in U.S. liabilities is a use of funds (a debit entry). In the U.S. balance of payments the transactions will be entered as shown in the following:

	Debit	**Credit**
Decrease in short-term liabilities to foreigners (the dollar account)	$6,000	
Increase in short-term liabilities to foreigners (the Treasury bill)		$6,000

C. Official Reserve Accounts

Official reserve accounts measure the changes in international reserves owned by the country's monetary authorities, usually the central bank, during the given period. International reserves are composed mainly of gold and convertible foreign exchange. Foreign exchange reserves are financial assets denominated in such currencies as the U.S. dollar, which are freely and easily convertible into other currencies, but not in such currencies as the Indian rupee, because the Indian government does not guarantee the free conversion of its currency into others and not much of an exchange market exists. An increase in any of these financial assets constitutes a use of funds, while a decrease in reserve assets implies a source of funds. In some situations, this fact seems to run against intuitive interpretations, as when we say that an increase in gold holdings is a use of funds (signified by a minus sign or debit in the U.S. balance of payments). However, an increase in gold holdings is a use of funds in the sense that the U.S. might have chosen to purchase an alternative asset such as a bond issued by a foreign government. In order to be considered part of official reserves, the financial asset must be owned by the monetary authorities. The same asset in private hands is not considered part of official reserves. In addition, the country's own currency cannot be considered part of its reserve assets; a country's currency is a liability of its monetary authorities. Changes in these liabilities are reported in the short-term capital account, as illustrated previously.

EXAMPLE 23

An exchange trader is worried about a recent economic forecast anticipating an increased rate of inflation in the U.S. As a result, she sells $4,700 of U.S. dollars against marks (she buys marks). The transaction is done with the U.S. central bank (the Federal Reserve System). One reason the central bank may have wanted to be a party to this transaction is to support the exchange rate of the U.S. dollar (in order to prevent the possible decline in the value of the U.S. dollar that could result from the sale of the dollars by the trader). When the central bank purchases the dollars, there is a decrease in the U.S. liabilities to foreigners (a debit entry). The central bank pays for these dollars with marks it maintained as part of the country's foreign exchange reserves. The central bank is financing the support of the exchange rate with its reserves. The decrease in the level of reserves (a financial asset) represents a credit entry. In U.S. balance of payments this transaction will appear as indicated below:

	Debit	Credit
Decrease in short-term liabilities to foreigners (the dollars)	$4,700	
Decrease in official exchange reserves (the marks)		$4,700

D. The Balance of Payments Statement

Exhibit 17 summarizes the transactions discussed in the examples of this section, together with some additional transactions, in a balance of payments statement for the U.S. The additional transactions are the following:

1. A foreign car, priced at $4,000 equivalent, is purchased. Payment is made with foreign currency held by the importer in the U.S.
2. A foreigner's fully owned subsidiary in the U.S. earns $2,000 in profits after taxes. These profits are kept as part of retained earnings in the subsidiary.
3. A U.S. resident receives a $500 check in guilders as a gift from a cousin who lives abroad.
4. A U.S. company purchases 30% of a foreign candy store for $4,500. Payment is made in U.S. dollars.
5. A U.S. resident sells a $5,000 bond issued by a U.S. company to a French investor. Payment is made in U.S. dollars.
6. The U.S. central bank purchases $5,000 worth of gold to be kept as part of foreign reserves. Payment is made in U.S. dollars.

EXHIBIT 17
Balance of Payments for U.S. for the Year 2000*

(+: Sources of funds; –: Uses of funds)

Current Accounts			
Merchandise account			
Exports	$5,000		
Imports	–4,000		
Balance on merchandise trade		$1,000	
Service account			
Receipts for interest and dividends, travel, and financial charges	2,500		
Payments for interest and dividends, travel, and financial charges	2,000		
Balance in invisibles (services)		500	
Balance of trade in goods and services			$1,500
Unilateral transfers			
Gifts received from foreigners	500		
Gifts to foreigners		–1,000	
Balance in unilateral transfers			–500
Current accounts balance			1,000
Capital Accounts			
Long-term capital flows			
Direct investment			
U.S. investment abroad (+: decrease; –: increase)	–4,500		
Foreigners' investment in U.S. (+: increase; –: decrease)	2,000	–2,500	
Portfolio investment			
U.S. claims on foreigners (–: decrease; +: increase)	–3,000		
U.S. liabilities to foreigners (+: increase; –: decrease)		5,000	2,000
Balance on long-term capital			–500
Basic balance			500
Private short-term capital flows			
U.S. claims on foreigners (+: decrease; –: increase)	–4,000		
U.S. liabilities to foreigners (+: increase; –: decrease)	3,800		
Balance on short-term private capital			–200
Overall balance			$300
Official Reserves Accounts			
Gold exports less imports (–)			$ –5,000
Decrease or increase (–) in foreign exchange			4,700
Balance on official reserves			$–300

* This is also the format followed by the International Monetary Fund in its "analytic presentation" of balance of payments tables which appears in *The Balance of Payments Yearbook*.

Each of the tables shown in a balance of payments represents the total of the transactions affecting the given account during the reporting period. However, these totals are not calculated from entries such as the ones we have discussed. In our examples, we recorded a debit and a credit for each international transaction. In practice, the data reported in the balance of payments are gathered from sources that often are concerned with only a portion of the transactions discussed above. For example, the data presented in the import account are often collected from customs declarations, while the financing of these transactions appears largely among the data for changes in foreign assets and liabilities reported by financial institutions. That is why we often find an additional account in the balance of payments statement called *errors and omissions*.

The accounts in the balance of payments are often presented in a format similar to the one shown in Exhibit 17. Entries appear under the three major groupings of accounts discussed in the preceding section: current accounts, capital accounts, and official reserve accounts. The statement often supplies totals for these major groups of accounts, as well as for some of their components. In addition, as one reads from top to bottom, the typical presentation of the balance of payments provides cumulative running subtotals, usually called *balances*.

In Exhibit 17 the *trade balance in goods and services* shows a positive balance of $1,500. The sources of external purchasing power exceeded the uses on the trade accounts by $1,500. This balance is composed of a positive balance in trade in merchandise of $1,000 and a positive balance in trade in services of $500. When we add the negative balance of $500 in unilateral transfers to the balance of trade in goods and services, we obtain the *balance on the current accounts*. In the U.S., the current accounts balance is a surplus of $1,000.

In the long-term capital account, the U.S. had a deficit in direct investments. While foreigners invested $2,000 in the U.S. (the U.S. increased its liabilities to foreigners—a source of funds for the U.S.), the U.S. made direct investments in foreign countries in the amount of $4,500 (the U.S. acquired financial assets—a use of funds for the U.S.). Many of these investments involved acquiring whole ventures in other countries. Although in some cases the ownership had to be shared with others, the direct investor retained a substantial share (at least 10%) of the total ownership and, presumably, management. The deficit in the direct investment accounts of the U.S. was somewhat compensated for by the surplus in the portfolio accounts. Foreigners bought $2,000 more of long-term financial instruments from the U.S. than the U.S. bought from other countries. When the balance in the long-term capital accounts is added to the current accounts balance, the result is called the *basic balance*. The U.S. basic balance is a positive $500. In the private short-term capital accounts, foreigners bought $3,800 worth of short-term securities issued by the U.S., while the U.S. invested $4,000 in short-term securities issued by foreign countries. The sum of the private short-term capital accounts and the basic balance produces another subtotal, often referred to as the *overall balance*. In the U.S., the overall balance produces a surplus of $300—a net source of external purchasing power for the U.S.

By definition, the net change in official reserves must be equal to the overall balance. Given the double-entry system of accounting in the balance of payments, the net of the accounts included in any balance must equal the net of the remaining accounts. In the U.S., the surplus in the overall balance of $300 equals the increase in official reserves (a debit or minus entry) of $300. Alternatively, we can say that the total of all the entries in the U.S. balance of payments is 0.

BALANCE OF PAYMENTS ADJUSTMENT

Balance of payments adjustment is the automatic response of an economy to a country's payments imbalances (payments deficits or surpluses). An adjustment is often necessary to correct an imbalance (a disequilibrium) of payments. Theoretically, if foreign exchange rates are freely floating, the market will automatically adjust for deficits through foreign exchange values and for surpluses through higher values. With fixed exchange rates, central banks must finance deficits, allow a devaluation, or use trade restrictions to restore equilibrium. Adjustment measures that can be taken to correct the imbalances include: (1) the use of fiscal and monetary policies to vary the prices of domestically produced goods and services vis-à-vis those made by other countries so as to make exports relatively cheaper (or more expensive) and imports more expensive (or cheaper) in foreign currency terms; and (2) the use of tariffs, quotas, controls, and the like to affect the price and availability of goods and services.

BALANCE OF TRADE

Also called *merchandise trade balance* or *visible trade*, the balance of trade is merchandise exports minus imports. Thus, if exports of goods exceed imports the trade balance is said to be "favorable" or to have a trade surplus, while an excess of imports over exports yields an "unfavorable" trade balance or a trade deficit. The balance of trade is an important item in calculating balance of payments.

See also BALANCE OF PAYMENTS.

BALANCE ON CURRENT ACCOUNT

See CURRENT ACCOUNT BALANCE.

BALANCE SHEET EXPOSURE

See TRANSLATION EXPOSURE.

BALANCE SHEET HEDGING

Balance sheet hedging is the MNC strategy of using hedges (such as forward contracts) to avoid currency risk (i.e., *translation exposure, transaction exposure*, and/or *economic exposure*) that would potentially adversely affect the company's balance sheet. This strategy involves bringing exposed assets equal to exposed liabilities. If the goal is protection against *translation exposure*, the procedure is to have monetary assets in a specific currency equal monetary liabilities in that currency. If the goal is to reduce *transaction* or *economic exposure*, the strategy is to denominate debt in a currency whose change in value will offset the change in value of future cash receipts.

BANKER'S ACCEPTANCE

Banker's acceptance (BA) is a *time draft* drawn on by a business firm and accepted by a bank to be paid at maturity. A bank creates a BA by approving a line of credit for a customer. It is an important source of financing in international trade, when the exporter of goods can be certain that the importer's *draft* will actually have funds behind it. Banker's acceptances are short-term, money-market instruments actively traded in the secondary market. Depending on the bank's creditworthiness, the acceptance becomes a financial instrument which can be discounted. In addition to the discount, an acceptance fee (usually 1.5% of the value of the draft) is charged to customers seeking acceptances.

See also DRAFT; LETTERS OF CREDIT; TRADE CREDIT INSTRUMENTS.

BANK FOR INTERNATIONAL SETTLEMENTS (BIS)

Bank for International Settlements (http://www.bis.org), established in 1930, promotes cooperation among central banks in international financial settlements. Members include: Australia, Austria, Belgium, Bulgaria, Canada, Czechoslovakia, Denmark, Finland, France, Germany, Greece, Hungary, Iceland, and Ireland.

BANK LETTER OF CREDIT POLICY

See EXPORT–IMPORT BANK.

BANK SWAPS

1. A *swap* between banks (commercial or central) of two or more countries for the purpose of acquiring temporarily needed foreign exchange.
2. A swap in which a bank in a *soft-currency* country will lend to an MNC subsidiary there, to avoid currency exchange problems. The MNC or its bank will make currency available to the lending bank outside the soft-currency country.

BARTER

Barter is international trade conducted by the direct exchange of goods or services between two parties without a cash transaction.

BASIC BALANCE

The basic balance is a *balance of payments* that measures all of the current account items and the net exports of long-term capital during a specified time period. It stresses the long-term trends in the balance of payments.

BASIS POINT

A basis point is a unit of measure for the change in interest rates for fixed income securities such as bonds and notes. One basis point is equal to 1/100th of a percent, that is, 0.01%. Thus, 100 basis points equal 1%. For example, an increase in a bond's yield from 6.0% to 6.5% is a rise of 50 basis points. A basis point should not be confused with a "point," which represents one percent.

B/E

See BILL OF EXCHANGE.

BEARER BOND

A bearer bond is a corporate or governmental bond that is not registered to any owner. Custody of the bond implies ownership, and interest is obtained by clipping a coupon attached to the bond. The benefit of the bearer form is easy transfer at the time of a sale, easy use as collateral for a debt, and what some cynics call "taxpayer anonymity," signifying that governments find it hard to trace interest payments in order to collect income taxes. Bearer bonds are common in Europe, but are seldom issued any more in the United States. The alternate form to a bearer bond is a *registered bond.*

BETA

Also called *beta coefficients*, beta (β), the second letter of Greek alphabet, is used as a statistical measure of risk in the *Capital Asset Pricing Model* (CAPM). It measures a security's (or mutual fund's) volatility relative to an average security (or market portfolio). Put another way, it is a measure of a security's return over time to that of the overall market. For example, if ABC's beta is 1.5, it means that if the stock market goes up 10%, ABC's common stock goes up 15%; if the market goes down 10%, ABC goes down 15%. Here is a guide for how to read betas:

Beta	What It Means
0	The security's return is independent of the market. An example is a risk-free security such as a T-bill.
0.5	The security is only half as responsive as the market.

1.0	The security has the same reponsive or rik as the market (i.e., average risk). This is the beta value of the market portfolio such as Standard & Poor's 500.
2.0	The security is twice as responsive, or risky, as the market.

Beta of a particular stock is useful in predicting how much the security will go up or down, provided that investors know which way the market will go. Beta helps to figure out risk and expected (required) return.

Expected (required) return = risk-free rate + beta × (market return − risk-free rate)

The higher the beta for a security, the greater the return expected (or demanded) by the investor.

EXAMPLE 24

XYZ stock actually returned 9%. Assume that the risk-free rate (for example, return on a T-bill) = 5%, market return (for example, return on the S&P 500) = 10%, and XYZ's beta =1.5. Then the return on XYZ stock required by investors would be

$$\begin{aligned}\text{Expected (required) return} &= 5\% + 1.5(10\% - 5\%) \\ &= 5\% + 7.5\% \\ &= 12.5\%\end{aligned}$$

Since the actual return (9%) is less than the required return (12.5%), you would not be willing to buy the stock.

Betas for stocks (and mutual funds) are widely available in many investment newsletters and directories. Exhibit 18 presents some betas for some selected MNCs:

EXHIBIT 18
Betas for Some Selected Multinational Corporations

Company	Ticker Symbol	Beta
Microsoft	MSFT	1.49
Pfizer	PFE	0.89
Dow	DOW	0.90
Wal-Mart	WMT	1.20
McDonald's	MCD	0.93
Honda	HMC	0.87
Nokia	NOT	1.91
IBM	IBM	1.07

Source: AOL Personal Finance Channel and MSN Money Central Investor. (http://moneycentral.msn.com/investor/home.asp), May 22, 2000.

Note: Beta is also used to determine a foreign direct investment project's cost of financing. See also FOREIGN DIRECT INVESTMENT.

BID

Also called a *quotation* or *quote*, bid is the price which a dealer is willing to pay for (i.e., buy) foreign exchange or a security.

BID–ASK SPREAD

The bid–ask spread is the spread between the bid (to buy) and the ask (to sell or offer) price and represents a transaction cost. It is based on the breadth and depth of the market for that currency as well as on the currency's volatility.

EXAMPLE 25

A Swiss francs quote of, say, \$0.7957–60 means that the bid rate is \$0.7957 and the ask rate is \$0.7960. The bid–ask spread is usually stated in terms of a percentage cost and calculated as follows:

$$\text{Percent spread} = \frac{\text{Ask price} - \text{Bid price}}{\text{Ask price}} \times 100$$

EXAMPLE 26

On Monday, February 21, 2000, the bid–ask spread on the Japanese yen was ¥110.886 – 936. (Source: Olsen and Associates; http://www.oanda.com.) Then the percent spread is: [(¥110.936 – ¥110.886)/¥110.936] × 100 = 0.004507%.

The bid–ask spread is the discount in the bid price as a percentage of the ask price. A spread of less than 1/10 of 1% is normal in the market for major traded currencies. *Note*: When quotations in American terms are converted to European terms, bid and ask reverse. The reciprocal of the bid becomes the ask, and the reciprocal of the ask becomes the bid.

See also CURRENCY QUOTATIONS.

BID PRICE

See BID RATE.

BID RATE

Also called the *buying rate*, bid rate is the rate at which a bank buys foreign currency from a customer by paying in home currency.

See also ASKED RATE.

BIG BANG

1. Advocating drastic changes in the policies of a country or an MNC.
2. The liberalization of the London capital markets that transpired in the month of October 1986.

BILATERAL EXCHANGES

Currencies participating in the *European Economic and Monetary Union* (EMU) are units of the euro until January 1, 2002. To convert one currency to another, you must use the triangulation method: Convert the first currency to the euro and then convert that amount in euros to the second currency, using the fixed conversion rates adopted on January 1, 1999 (see Exhibit 19).

EXHIBIT 19
Fixed Euro Conversion Rates

Country	Currency	Currency Literacy Abbreviation	Rate
Euroland	euro	EUR	1
Austria	schilling	ATS	13.7603
Belgium	franc	BEF	40.3399
Finland	makka	FIM	5.94573
France	franc	FRF	6.55957
Germany	mark	DEM	1.95583
Ireland	punt	IEP	.787564
Italy	lira	ITL	1,936.27
Luxembourg	franc	LUF	40.3399
Netherlands	guilder	NLG	2.20371
Portugal	escudo	PTE	200.482
Spain	peseta	ESP	166.386

See also EURO.

BILATERAL NETTING

See MULTILATERAL NETTING.

BILL OF EXCHANGE

Also called a *draft*, a bill of exchange (B/E) is an unconditional written agreement between two parties, written by an exporter instructing an importer or an importer's agent such as a bank to pay a specified amount of money at a specified time. Examples are acceptances or the commercial bank check. The business initiating the bill of exchange is called the *maker*, while the party to whom the bill is presented is called the *drawee*.

BILL OF LADING

The bill of lading (B/L) is a receipt issued to the exporter by a common carrier that acknowledges possession of the goods described on the face of the bill. It serves as a contract between the exporter and the shipping company. If it is properly prepared, a bill of lading is also a document of title that follows the merchandise throughout the transport process. As a document of title, it can be used by the exporter either as collateral for loans prior to payment or as a means of obtaining payment (or acceptance of a time draft) before the goods are released to the importer. There are different types of B/L:

1. A *negotiable* or *shipper's order* B/L can be bought, sold, or traded while goods are in transit.
2. A *straight* B/L is nether negotiable nor transferable.
3. An order B/L is cosigned to the exporter who keeps title to the merchandise until the B/L is endorsed.
4. An *on-board* B/L certifies that the goods have been actually placed on board the ship.
5. A *received-for-shipment* B/L simply acknowledges that the goods have been received for shipment.

BILL OF SALE

A bill of sale is a written document which transfers goods, title, or other interests from a seller to a buyer and specifies the terms and conditions of the transaction.

BIS

See BANK FOR INTERNATIONAL SETTLEMENTS.

B/L

See BILL OF LADING.

BLACK MARKETS

Black markets are illegal markets in foreign exchange. Developing nations generally do not permit free markets in foreign exchange and impose many restrictions on foreign currency transactions. These restrictions take many forms, such as limiting the amounts of foreign currency that may be purchased or having government licensing requirements. As a result, illegal markets in foreign exchange develop to satisfy trader demand. In many countries such illegal markets exist openly, with little government intervention.

BLACK–SCHOLES OPTION PRICING MODEL (OPM)

An option pricing equation developed in 1973 by Fischer Black and Myron Scholes provides the relationship between call option value and the five factors that determine the premium of an option's market value over its expiration value:

1. Time to maturity—the longer the option period, the greater the value of the option;
2. Stock price volatility—the greater the volatility of the underlying stock's price, the greater its value;
3. Exercise price—the lower the exercise price, the greater the value;
4. Stock price—the higher the price of the underlying stock, the greater the value; and
5. Risk-free rate—the higher the risk-free rate, the higher the value.

The formula is:

$$V = P[N(d_1)] - PV(E)[N(d_2)]$$

where

V = Current value of a call option
P = current stock price
$PV(E)$ = present value of exercise or strike price of the option $E = E/e^{-rt}$
r = risk-free rate of return, continuously compounded for t time periods
e = 2.71828
t = percentage of year until the expiration date (for example, 3 months means $t = 3/12 = 3/4 = 0.25$)
$N(d)$ = probability that the normally distributed random variable Z is less than or equal to d
σ = standard deviation per period of (continuously compounded) rate of return on the stock
$d_1 = \ln[P/PV(E)]/\sigma\sqrt{t} + \sigma\sqrt{t}/2$
$d_2 = d_1 - \sigma\sqrt{t}$

The formula requires readily available input data, with the exception of σ^2, or volatility. P, X, r, and t are easily obtained. The implications of the option model are as follows:

1. The value of the option increases with the level of stock price relative to the exercise price [$P/PV(E)$], the time to expiration, and the time to expiration times the stock's variability ($\sigma\sqrt{t}$).
2. Other properties:
   a. The option price is always less than the stock price.
   b. The option price never falls below the payoff to immediate exercise ($P - E$ or zero, whichever is larger).
   c. If the stock is worthless, the option is worthless.
   d. As the stock price becomes very large, the option price approaches the stock price less the present value of the exercise price.

EXAMPLE 27

The current price of Sigma Corporation's common stock is $59.375 per share. A call option on this stock has a $55 exercise price. It has 3 months to expiration. If the standard deviation of continuously compounded rate of return on the stock is 0.2968 and the risk-free rate is 5% per year, the value of this call option is:

First, calculate the time until the option expires in years,

$$t \text{ in years} = 90 \text{ days}/365 \text{ days} = 0.0822$$

Second, calculate the values of the other variables:

$$PV(E) = E/e^{-rt} = \$55/e^{0.05 \times 0.0822} = \$54.774$$

$$d_1 = \ln P/[P/PV(E)]\sigma\sqrt{t} + \sigma\sqrt{t}/2 = \ln[\$59.375/\$54.774]/(0.2968 \times \sqrt{0.0822})$$
$$+ (0.2968 \times \sqrt{0.0822})/2 = 0.9904$$

$$d_2 = d_1 - \sigma\sqrt{t} = 0.9904 - 0.2968 \times \sqrt{0.0822} = 0.9053$$

Next, use a table for the standard normal distribution (See the Appendix) to determine $N(d_1)$ and $N(d_2)$:

$$N(d_1) = N(0.9904) = 0.8389$$
$$N(d_2) = N(0.9053) = 0.8173$$

Finally, use those values to find the option's value:

$$V = P[N(d_1)] - PV(E)[N(d_2)]$$
$$= \$59.375[0.8389] - \$54.774[0.8173]$$
$$= \$5.05$$

This call option is worth $5.05, a little more than its value if it is exercised immediately, $4.375 ($59.375 – $55), as one should expect.

EXAMPLE 28

You want to determine the value of another option on the same stock that has an exercise price of $50 and expires in 45 days. The time until the option expires in years is t in years = 45 days/365 days = 0.1233.

The values of the other variables are:

$$PV(E) = E/e^{-rt} = \$50/e^{0.05 \times 0.1233} = \$49.6927$$

$$d_1 = \ln[P/PV(E)]/\sigma\sqrt{t} + \sigma\sqrt{t}/2 = \ln[\$59.375/\$49.6927]/(0.2968 \times \sqrt{0.1233}) + (0.2968 \times \sqrt{0.1233})/2 = 1.7602$$

$$d_2 = d_1 - \sigma\sqrt{t} = 1.7602 - 0.2968 \times \sqrt{0.1233} = 1.6560$$

Next, use a table for the standard normal distribution (See the Appendix) to determine $N(d_1)$ and $N(d_2)$:

$$N(d_1) = N(1.7603) = 0.9608$$
$$N(d_2) = N(1.6561) = 0.9511$$

Finally, use those values to find the option's value:

$$V = P[N(d_1)] - PV(E)[N(d_2)] = \$59.375[0.9608] - \$49.6927[0.9511] = \$9.78$$

The call option is worth more than the other option ($9.78 versus $5.05), because it has a lower exercise price and a longer time until expiration.

BLOCKED FUNDS

Blocked funds are funds in one nation's currency that may not be exchanged freely due to *exchange controls* or other reasons.

BOLIVAR

Venezuela's currency.

BOLIVIANO

Bolivia's currency.

BRADY BONDS

Brady bonds are bonds issued by emerging countries under a debt-reduction plan and are named after a former U.S. Secretary of the Treasury. They are traded on the international bond market.

BREAK-EVEN ANALYSIS

Break-even analysis is used to determine the amount of currency change that will equate the cost of local currency financing with the cost of home currency (dollar) financing. In general, the break-even rate of currency change is found as follows:

$$d^* = \frac{k_L - k_H}{1 + k_L}$$

where d = expected local currency devaluation if positive or revaluation if negative, k_L = dollar cost of the local currency loan, and k_H = cost of home currency financing.

EXAMPLE 29

Suppose that a U.S. firm in Switzerland is given a one-year loan at the quoted interest rate of 8% locally. The firm could also have borrowed funds in the U.S. for one year at 12%. Then

$$d^* = (0.08 - 0.12)/1.08 = -0.037 = -3.7\%$$

Which means the Swiss franc appreciation should equal 3.7% before it becomes less expensive to borrow dollars at 12% than Swiss franc at 8%.

In making investments, investors should use the following rules:

(a) For an expected devaluation, if $d < d^*$, borrow dollars, and if $d > d^*$, borrow the foreign currency.
(b) For an expected revaluation, if $d > d^*$, borrow dollars, and if $d < d^*$, borrow the foreign currency.

BREAK FORWARD

See FORWARD WITH OPTION EXIT.

BRETTON WOODS AGREEMENT

Bretton Woods Agreement is an agreement, implemented at an international conference with representatives of 40 countries in Bretton Woods, New Hampshire, that established the international monetary system in effect from 1945 to 1971. Each member government pledged to maintain a fixed, or pegged, exchange rate for its currency with respect to the U.S. dollar or gold. These fixed exchange rates were intended to reduce the riskiness of international transactions, thus promoting growth in global trade.

BRETTON WOODS CONFERENCE

A conference in 1944 in which representatives of 40 nations gathered to map a new international monetary system. They established the *International Monetary Fund*, the *World Bank*, and an international monetary system at Bretton Woods, New Hampshire.

BRITISH POUND

The currency of one of the United States' top allies and trading partners, the British pound is one of the world's most important currencies. Its relationship to the U.S. dollar is a key to the global marketplace and is seen as a barometer of the United Kingdom's economic strength versus the business climate in the United States.

BROKERS' MARKET

The brokers' market is the market for exchange of financial instruments between any two parties using a broker as an intermediary or agent. Along with the *interbank market*, the broker's market provides another area of large-scale foreign exchange dealing in the United States. A good number of foreign exchange brokerage firms make markets for foreign currencies in New York (as well as in London and elsewhere), creating trading in many currencies similar to that in the interbank market. The key differences are that the brokers (1) seek to match buyers and sellers on any given transaction, without taking a position themselves; (2) deal simultaneously with many banks (and other firms); and (3) offer both buy and sell positions to clients (where a bank may wish to operate on only one side of the market at any particular time). Also, the brokers deal "blind," offering rate quotations without naming the potential seller/buyer until a deal has been negotiated.

See also INTERBANK MARKET.

BULLDOGS

Bulldogs are sterling-denominated bonds issued within the United Kingdom by a foreign borrower. They are *foreign bonds* sold in the United Kingdom.

BUNDESBANK

The Bundesbank is the German central bank equivalent to the Federal Reserve System of the U.S. Its primary goals are to (1) set the discount rate, known as the *Lombard rate*; (2) monitor the money supply; and (3) back economic (fiscal and monetary) policies.

BURN RATE

Also called *cash burn rate*, burn rate is how quickly a company uses up its capital to finance operations before generating positive cash flow from operations. This rate is a critical key to survival in the case of small, fast growing companies that need constant access to capital. Many technology and Internet companies are examples. It is not uncommon for enterprises to lose money in their early goings, but it is important for financial analysts and investors to assess how much money those firms are taking in and using up. The number to examine is *free cash flow*, which is the company's operating cash flows (before interest) minus cash outlays for capital spending. It is the amount available to finance planned expansion of operating capacity. Burn rate is generally used in terms of cash spent per month. A burn rate of 1 million would mean the company is spending 1 million per month. When the burn rate begins to exceed plan or revenue fails to meet expectations, the usual recourse is to reduce the burn rate. In order to stay afloat, the business will have to reduce the staff, cut spending (possibly resulting in slower growth), or raise new capital, probably by taking on debt (resulting in interest expense) or by selling additional equity stock (diluting existing shareholders' ownership stake).

EXAMPLE 30

Secure-payments provider CyberCash had $26.4 million in cash at the end of March, and at its current burn rate of $7 million to $8 million per quarter, only has enough cash to last it through next February. The company generated just $155,000 in revenue in its most recent quarter. Once you have a burn rate such as this and reserves insufficient to cover your cash needs for a year, you have problems.

C

CABLE

Cable is the U.S. dollar per British pound *cross rate*.

CALL MONEY

1. Money lent by banks to brokers on demand (payable at call).
2. Also called *demand money* or *day-to-day money*, interest-bearing deposits payable upon demand. An example is Eurodeposits.

CALL OPTION

See CURRENCY OPTION; OPTION.

CANADIAN VENTURE EXCHANGE

Canadian Venture Exchange (CDNX) is a product of the merger of the Vancouver and Alberta stock exchanges. The objective of CDNX is to provide venture companies with effective access to capital while protecting investors. This exchange basically contains small-cap Canadian stocks. The CDNX is home to many penny stocks.

CAPITAL ACCOUNT

A capital account is a *balance of payment* account that records transactions involving the purchase or sale of capital assets. Capital account transactions are classified (e.g., portfolio, direct, or short-term investment).

See also BALANCE OF PAYMENTS.

CAPITAL ASSET PRICING MODEL

The Capital Asset Pricing Model (CAPM) quantifies the relevant risk of an investment and establishes the trade-off between risk and return (i.e., the price of risk). The CAPM states that the expected return on a security is a function of: (1) the risk-free rate, (2) the security's systematic risk, and (3) the expected risk premium in the market. The basic message of the model is that risk is priced in a portfolio context. A security risk consists of two components—diversifiable risk and nondiversifiable (or systematic) risk. Diversifiable risk, sometimes called *controllable risk* or *unsystematic risk*, represents the portion of a security's risk that can be controlled through diversification. This type of risk is unique to a given security and thus is not priced. Business, liquidity, and default risks fall into this category. Nondiversifiable risk, sometimes referred to as *noncontrollable risk* or *systematic risk*, results from forces outside of the firm's control and is therefore not unique to the given security. This type of risk must be priced and hence affects the required return on a project. Purchasing power, interest rate, and market risks fall into this category. Nondiversifiable risk is assessed relative to the risk of a diversified portfolio of securities, or the market portfolio. This type of risk is measured by the *beta* coefficient.

The CAPM relates the risk measured by beta to the level of expected or required rate of return on a security. The model, also called the *security market line* (*SML*; see Exhibit 20), is given as follows:

$$r_j = r_f + b(r_m - r_f)$$

where r_j = the expected (or required) return on security i, r_f = the risk-free security (such as a T-bill), r_m = the expected return on the market portfolio (such as Standard & Poor's 500 Stock Composite Index), and b = beta, an index of nondiversifiable (noncontrollable, systematic) risk. In words, the CAPM (or SML) equation shows that the required (expected) rate of return on a given security (r_j) is equal to the return required for securities that have no risk (r_f) plus a risk premium required by investors for assuming a given level of risk. The higher the degree of systematic risk (b), the higher the return on a given security demanded by investors.

See also BETA.

EXHIBIT 20
Security Market Line

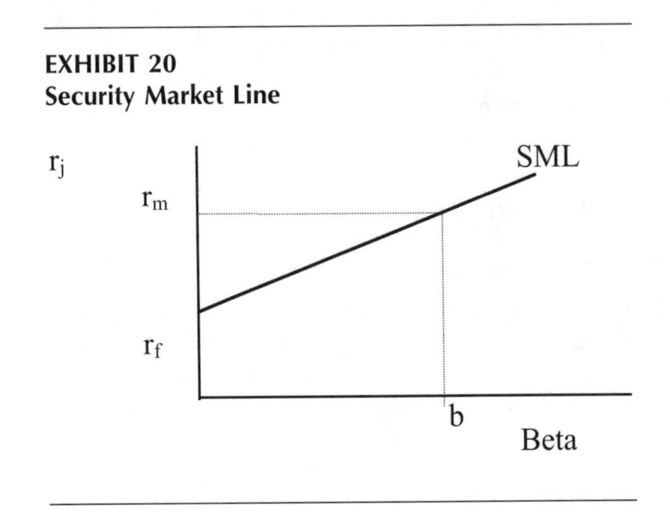

CAPITAL FLIGHT

Also called *capital exports*, capital flight is outflows of funds out of a country, typically for fears of *economic* or *political crisis*. This may be the result of political or financial crisis, tightening capital controls, tax increases, or fears of a domestic currency devaluation.

CAPITAL MARKETS

Capital markets are the markets for long-term debt and corporate equity issues. The *New York Stock Exchange (NYSE)*, which trades the stocks of many of the larger corporations, is a prime example of a capital market. The American Stock Exchange and the regional stock exchanges are also examples. In addition, securities are issued and traded through the thousands of brokers and dealers on the over-the-counter (OTC) market. The capital market is the major source of long-term financing for business and governments. It is an increasingly international one and in any country is not one institution but all those institutions that make up the supply and demand for long-term sources of capital, including the stock exchanges, underwriters, investment bankers, banks, and insurance companies.

CAPM

See CAPITAL ASSET PRICING MODEL.

CARIBBEAN COMMON MARKET

The Caribbean Common Market (CARICOM) consists of 14 sister-member countries of the Caribbean community: Antigua and Barbuda, Bahamas, Barbados, Belize, Dominica, Grenada, Guyana, Jamaica, Montserrat, St. Kitts and Nevis, St. Lucia, St. Vincent, Surinam, and Trinidad and Tobago. They have set as a goal that there will be a single market allowing for the free movement of labor. Conspicuous by their absence are the Cayman Islands and the British Virgin Islands, two major players in international banking and finance.

CARICOM

See CARIBBEAN COMMON MARKET.

CASH BURN RATE

See BURN RATE.

CASH FLOW RETURN ON INVESTMENT

Cash flow return on investment (CFROI), along with *economic value added (EVA)*, is a popular financial metric now embraced by MNCs. Unlike a traditional measure such as *return on investment (ROI)*, CFROI uses cash flow generated by the firm, rather than income or earnings derived from financial statements, as the "return." The capital base used is also *cash investment*. The formula is:

$$\text{CFROI} = \frac{\text{Profit from operations} - \text{Cash taxes} + \text{Depreciation}}{\text{Cash investment}}$$

See also ECONOMIC VALUE ADDED.

CENTRAL BANKS

Central banks are government agencies with authority over the size and growth of the nation's money supply. They commonly regulate commercial banks, act as the government's fiscal agent, and are architects of the nation's monetary policy. Central banks frequently intervene in the foreign exchange markets to smooth fluctuations. For example, the *Federal Reserve System* is the central bank in the U.S., while the *Bundesbank* is the central bank of Germany. For the monetary policies and economics performance of central banks, visit the Bank for International Settlements site (http://www.bis.org/cbanks.htm).

CENTRALIZED CASH MANAGEMENT

Centralized cash management is the management practice used by an MNC to place the responsibility for managing corporate cash balances from all affiliates under the custody of a single, central office (normally in New York or London). Centralized cash management offers the following potential gains to the MNC:

1. By pooling the cash holdings of affiliates where possible, it can hold a smaller total amount of cash, thus reducing its financing needs.
2. By centralizing cash management, it can have one group of people specialize in the performance of this task, thus achieving better decisions and economies of scale.
3. By reducing the amount of cash in any affiliate, it can reduce political risks as well as financial costs and enhance profitability.
4. It can net out intracompany accounts when there are multiple payables and receivables among affiliates, thus reducing the amount of money actually transferred among affiliates.
5. All decisions can be made using the overall corporate benefit as the criterion rather than the benefits of individual affiliates, when these might conflict.

The principal disadvantage of centralization is that it brings about both explicit and implicit costs. The explicit costs relate to the added management time and expanded communications required. Implicit costs might insclude that (1) affiliates might resent the tigher control necessary and (2) rigid centralization provides no incentive for local managers to cash in on specific opportunities of which only they may be aware.

CENTRALIZED CASH POOLING

See MULTICURRENCY CROSS-BORDER CASH POOLING.

CFROI

See CASH FLOW RETURN ON INVESTMENT.

CFTC

See COMMODITY FUTURES TRADING COMMISSION.

CHEAP DOLLAR

See STRONG DOLLAR.

CHIPS

See CLEARING HOUSE INTERBANK PAYMENT SYSTEM.

CLEAN FLOAT

Also called a *free float*, a clean float is a system of freely *floating exchange rates*.

CLEAN LETTER OF CREDIT

See NONDOCUMENTARY LETTER OF CREDIT.

CLEARING HOUSE INTERBANK PAYMENT SYSTEM

Clearing House Interbank Payment System (CHIPS) (http://www.chips.org) is a computerized clearing network developed by the New York Clearinghouse Association (http://www.theclearinghouse.org) for transfer of international dollar payments and settlement of interbank foreign exchange obligations, connecting some 150 depository institutions that have offices or affiliates in New York City. The transfers account for roughly 90% of all interbank transfers regarding international dollar payments.

COCKTAIL SWAP

The cocktail swap is a combination of currency and interest rate swaps, often involving a number of parties, currencies, and interest rates, which leaves the bank unexposed—as long as no default exists or as long as no contract is otherwise terminated prematurely.

COLLAR OPTION

As a form of hybrid option, a collar option involves the simultaneous purchase of a put option and sale of a call option, or vice versa. It is an option that limits the holder's exposure to a price band. The holder essentially has a call above the cap of that band and a put below its floor.

COMMERCIAL RISK

1. Also called *business risk*, the likelihood of loss in principal or variations in earnings (or cash flow) of a business due to change in demand and business operations.
2. In banking, the chance that a foreign debtor will fail to repay a loan because of adverse business conditions.

3. In connection with *Eximbank* guarantees, failure of repayment for reasons other than political factors, such as bankruptcy or insolvency.
4. Uncertainty that the exporter may not be paid for goods or services.

COMMODITY FUTURES TRADING COMMISSION

Commodity Futures Trading Commission (CFTC), as created by the Commodity Futures Trading Commission Act of 1974, is the government agency that currently regulates the nation's commodity futures industry.

COMMON MARKET

Common market is a group of countries seeking a high degree of *regional economic integration*. It involves: (1) eliminating internal trade barriers and establishing a common external tariff, (2) removing national restrictions on the movement of labor and capital among participating nations and the right of establishment for business firms, and (3) the pursuit of a common external trade policy. The best known is the *European Common Market*.

COMPANIES ACT OR ORDINANCE

This is legislation enacted by a tax haven to provide for the incorporation, registration, and operation of international business companies (IBCs). More commonly found in the Caribbean tax havens.

COMPARATIVE ADVANTAGE

Comparative advantage is a theory that everyone gains if each country specializes in the production of those goods that it produces relatively cheaply and more efficiently and imports those goods that other countries make relatively more efficiently. A country is said to have a *comparative advantage* in the production of such goods and services. This theory is the basis for free trade arguments.

COMPOUND AVERAGE RETURN

See ARITHEMATIC AVERAGE RETURN VS. COMPOUND (GEOMETRIC) AVERAGE RETURN.

CONFIRMED LETTER OF CREDIT

A *letter of credit* issued by one bank (a foreign bank) with its authenticity validated by another bank (a U.S. bank) obligating both banks to honor any drafts drawn in compliance. An exporter who requires a confirmed letter of credit from the importer is guaranteed payment from the U.S. bank even if the foreign buyer or bank defaults. It reads, for example, "we hereby confirm this credit and undertake to pay drafts drawn in accordance with the terms and conditions of the letter of credit."

CONSORTIUM BANK

A consortium bank, a special type of foreign affiliate, is a joint venture, incorporated separately and owned by two or more banks, usually of different nationalities.

See also FOREIGN SUBSIDIARIES AND AFFILIATES.

CONTINENTAL TERMS

Foreign exchange rate quotations in foreign currency; the foreign currency price of one U.S. dollar, also called *European terms*.

See also AMERICAN TERMS; EUROPEAN TERMS.

CONTINUITY OF CONTRACTS

This principle, backed by *European Council (EC)* regulation, says contracts cannot be canceled simply because they refer to a currency that is being replaced by the euro.

See also EURO.

CONTROLLED FOREIGN CORPORATION

Controlled Foreign Corporation (CFC) is an offshore company which, because of ownership or voting control of U.S. persons, is treated by the Internal Revenue Service as a U.S. tax reporting entity. Sections 951 and 957 of the Internal Revenue Code of 1986 collectively define the CFC as one in which a U.S. person owns 10% or more of a foreign corporation or in which 50% or more of the total voting stock is owned by U.S. shareholders collectively or 10% or more of the voting control is owned by U.S. persons.

CORRESPONDENT BANK

A bank in one country which, when so required, acts as an agent for a bank in another country, typically formalized by the holding of reciprocal bank accounts. Most major banks maintain correspondent banking relationships with local banks in market areas in which they wish to do business. International correspondent relationships provide international banking services. For example, Citibank of New York may have correspondent relationships with a bank in Cairo, Egypt. Correspondent services include accepting drafts, honoring letters of credit, furnishing credit information, collecting and disbursing international funds, and investing international money-markets funds.

Typically, correspondent services center around paying or collecting international funds because most transactions involve the importing or exporting of goods. In addition, the correspondent bank will provide introductions to local business people. Under a correspondent relationship, the U.S. bank usually will not maintain any permanent personnel in the foreign country; direct contact between the two banks may be limited to periodic conferences between their management to review services provided. In other cases, though, a correspondent relationship will be developed with a local institution even when the American bank has a branch or other presence in the country. This generally occurs when the U.S. bank either is precluded by local law from making, or does not want to make, the necessary investment to clear domestic currency payments in the country.

COST AND FREIGHT (CFR or C&F)

This is a delivery term in which the seller pays for transportation to the destination point. This could be combined with an "on board" instruction specifying when title to the goods changes hands (e.g., FOB).

COST, INSURANCE, AND FREIGHT (CIF)

A term used in the delivery of goods from the exporter to the importer. This is similar to *Cost and Freight (CFR* or *C&F)*, except the seller must also pay for insurance. Generally this means ocean marine insurance in case the ship sinks or the goods are otherwise damaged by an event other than war. War risk insurance may also be available and is generally paid for by the buyer.

COST OF CAPITAL

Also called a *hurdle rate, cutoff rate*, or *minimum required rate of return*, the cost of capital is the rate of return that is necessary to maintain market value (or stock price) of a firm. The firm's cost of capital is calculated as a weighted average of the costs of debt and equity funds.

It is often called a *weighted average of cost of capital (WACC)*. Equity funds include both capital stock (common stock and preferred stock) and retained earnings. These costs are expressed as annual percentage rates. For example, assume the following capital structure and the cost of each source of financing for a multinational company:

Source	Book Value	% of Total Weights	Cost
Debt	$20,000,000	40%	8%
Preferred stock	5,000,000	10	14
Common stock	20,000,000	40	16
Retained earnings	5,000,000	10	15
Totals	$50,000,000	100%	

The overall cost of capital is computed as follows: 8%(.4) + 14%(.1) + 16%(.4) + 15%(.1) = 12.5%. The cost of capital is used for *capital budgeting* purposes. Under the *net present value method*, the cost of capital is used as the discount rate to calculate the present value of future cash inflows. Under the *internal rate of return method*, it is used to make an accept-or-reject decision by comparing the cost of capital with the internal rate of return on a given project. A project is accepted when the internal rate exceeds the cost of capital.

COUNTERTRADE

Countertrade is an umbrella term for several sorts of trade in which the seller is required to accept the countervalue of its sale in local goods or services instead of in cash. Payment by a purchaser is entirely or partially in kind instead of *hard currencies* for products or technology from other countries. It is, therefore, a whole range of barterlike arrangements. More specifically, *countertrade* has evolved into a diverse set of activities that can be categorized as six distinct types of trading arrangements: (1) *barter*, (2) *counterpurchase*, (3) offset, (4) switch trading, (5) compensation or buyback, and (6) clearing account arrangements. Barter is a simple swap of one good for another. In counterpurchase, exporters agree to purchase a quantity of goods from a country in exchange for that country's purchase of the exporter's product. The goods being sold by each party are typically unrelated but may be equivalent in value. Offset occurs when the importing nation requires a portion of the materials, components, or subassemblies of a product to be procured in the local (importer's) market. Switch trading is a complicated form of barter, involving a chain of buyers and sellers in different markets. In a compensation or buy-back deal, exporters of heavy equipment, technology, or even entire facilities agree to purchase a certain percentage of the output of the facility. Clearing account arrangements are used to facilitate the exchange of products over a specified period of time. When the period ends, any balance outstanding must be cleared by the purchase of additional merchandise or settled by a cash payment.

COUNTRY RISK

See POLITICAL RISK.

COVERED INTEREST ARBITRAGE

Also called *covered investment arbitrage*, covered interest arbitrage is an investment in a second currency that is "covered" by a forward sale of that currency to protect against exchange rate fluctuations. More specifically, it involves buying or selling securities internationally and using the forward market to eliminate currency risk in order to take advantage of interest (return) differentials. It is a process whereby one earns a risk-free profit by (1) borrowing currency, (2) converting it into another currency where it is invested, (3) selling this other currency for future delivery against the initial original currency, and (4) using the proceeds of forward sale to repay the original loan. The profits in this transaction depend on interest rate differentials *minus* the discount or *plus* the premium on a forward sale.

EXAMPLE 31

Assume that

- The borrowing and lending rates are identical and the bid–ask spread in the spot and forward markets is zero.
- The interest rate on pounds sterling is 12% in London, and the interest rate on a comparable dollar investment in New York is 7%.
- The pound spot rate is $1.75 and the one-year forward rate is $1.68. These rates imply a forward discount on sterling of 4% [(1.68 – 1.75)/1.75] and a covered yield on sterling roughly equal to 8% (12% – 4%).

Because there is a covered interest differential in favor of London, funds will flow from New York to London. That is to say, covered interest arbitrage results.

The following are the steps the arbitrageur can take to profit from the discrepancy in rates based on a $1 million transaction. The *arbitrageur* will:

1. Borrow $1 million in New York at an interest rate of 7%. This means that at the end of one year, the arbitrageur must repay principal plus interest of $1,070,000.
2. Immediately convert the $1,000,000 to pounds at the spot rate of £1 = $1.75. This yields £571,428.57 ($1,000,000/$1.75) available for investment.
3. Invest the principal of £571,428.57 in London at 12% for one year. At the end of the year, the arbitrageur will have £640,000 (£571,428.57 × 1.12).
4. Simultaneously with the other transactions, sell the £640,000 in principal plus interest forward at a rate of £1 = $1.68 for delivery in one year. This transaction will yield $1,075,200 (£640,000 × $1.68) next year.
5. At the end of the year, collect the £640,000, deliver it to the bank's foreign exchange department in return for $1,075,200, and use $1,070,000 to repay the loan. The arbitrageur will earn $5,200 on this set of transactions.

The transactions associated with covered interest arbitrage will affect prices in both the money and foreign exchange markets. As pounds are bought spot and sold forward, boosting the spot rate and lowering the forward rate, the forward discount will tend to widen. Simultaneously, as money flows from New York, interest rates there will tend increase; at the same time, the inflow of funds to London will depress interest rates there. The process of covered interest arbitrage will continue until *interest parity* is attained, unless there is government interference.

To see if a covered interest opportunity exists, use the following steps:

Step 1: Use the *interest rate parity (IRP)* condition

$$\frac{F}{S} = \frac{(1 + r_h)}{(1 + r_f)}$$

where F = forward rate, S = spot rate, and r_h and r_f = home (domestic) and foreign interest rates, and rearrange it to show that the gross returns from investing at home and abroad are equal:

$$(1 + r_h) = (1 + r_f)(F/S)$$

Step 2: Substitute the interest rate and exchange rate data into the above equation. If they are not equal, a covered interest arbitrage opportunity exists. *Note:* Make sure to convert annualized interest rates into the rates for the time period compatible with forward rates given.

EXAMPLE 32

Suppose the current spot rate of the pound sterling is \$1.28, and the 90-day forward rate is \$1.30. The 3-month deposit rates in the U.S. and Great Britain are 3% and 4%, respectively. To see if a covered interest opportunity exists, check out the IRT condition.

$$(1 + r_h) = (1 + r_f)(F/S)$$
$$(1 + .03) = (1 + .04)(\$1.30/\$1.28)$$
$$1.030000 \neq 1.056250$$

A covered interest opportunity exists, the IRT condition does not hold (i.e., the gross returns from investing in the U.S. and in Great Britain are not the same). A profit is realized by

(a) Borrowing dollars in the U.S. for 3 months at 3% (or 12% per annum)—\$1.03
(b) Convert the dollars into pounds at the spot rate—1/\$1.28 = £0.78125
(c) Investing the pounds in a U.K. market for 3 months at 4% (or 16% per annum)—£0.78125 × (1 + 0.4) = £.8125
(d) Simultaneously covering the transactions with a 90-day forward contract to sell pounds at \$1.28 – £0.8125 × \$1.30/£ = \$1.05625. These transactions will result in a gain of \$0.02625 (\$1.05625 – \$1.03). This is equivalent to an annualized gain [or *annual percentage rate (APR)*] of 10.92%, calculated as

$$[(1 + 0.02625)^4 - 1] = 1.109207 - 1 = 0.109207 = 10.92\%$$

See also INTEREST RATE PARITY.

COVERING

Covering refers to buying or selling foreign currencies in amounts equivalent to future payments to be made or received. It is a way of protection against loss due to currency rate movements.

CREATION OF EURODOLLARS

How *Eurodollars* are created can be explained, step by step, by using a series of T-accounts to trace the movement of dollars into and through the Eurodollar market. Eurodollar creation involves a chain of deposits between the original depositor and the U.S. bank, and the transfer of control over the deposits from one Eurobank to another.

EXAMPLE 33

Assume that Henteleff and Co. in New York decides to shift \$1 million out of its NY bank time deposit and into a time deposit with a Eurobank in London. Transfer of the dollar deposit from the NY bank to the London bank creates a Eurodollar deposit.

1st stage:

First Eurobank	
Demand deposit	\$1M Eurodollar time deposit due to Henteleff

2nd Stage: If First Eurobank does not immediately have a commercial borrower or government to which it can loan the funds, First Eurobank will place the \$M in the Eurodollar interbank market.

First Eurobank		Second Eurobank	
Eurodollar $1M deposit in Second Eurobank	$1M Eurodollar deposit due to Henteleff	Demand deposit in the NY bank	$1M Eurodollar time deposit due to First Eurobank

3rd Stage: Eurobank needs the funds to lend Mr. Borrower.

Second Eurobank	
Loan to Mr. Borrower $1M	$1M Eurodollar time deposit due to Eurobank

Note: U.S. money supply is constant at $1 million, but the world's money supply is increased to $2 million (Henteleff and First Eurobank's $2 million deposits).

CREDIT RISK

Also called *default risk*, credit risk is the chance that a trading partner will default on a contract.

CREDIT SWAP

A credit swap is an exchange of currencies between an MNC and a bank (frequently a central bank) of a foreign country that is to be reversed by agreement at a later date. This is widely used, particularly where local credit is not available and where there is no forward exchange market. The basic advantage of the credit swap is the ability to minimize the risk and the cost of financing operations in a weak-currency country.

See also SWAPS.

CROSS-CURRENCY SWAP

Also called *circus swap*, the cross-currency swap is a currency swap combined with an *interest rate swap* (floating versus fixed rate), in the sense that the loans on which the service schedules are based differ by currency and type of interest payment.

CROSS INVESTMENT

Cross investment is a type of *foreign direct investment* made, as a defense measure, by oligopolistic companies in each other's home country.

CROSS RATE

The exchange rate between two currencies derived by dividing each currency's exchange rate with a third currency. For example, if dollars per pound is $1.5999/£ and yens per dollar is ¥110.66/$, the cross rate between Japanese yen and British pounds is

$$\text{Cross rate between yen and pound} = \frac{Dollars}{Pound} \times \frac{Yens}{Dollar} = \frac{Yens}{Pound}$$

$$= \$/£ \times ¥/\$ = ¥/£$$

$$= 1.5999 \text{ dollars per pound} \times 110.66 \text{ yens per dollar}$$

$$= 177.05 \text{ yens per pound}$$

Because most currencies are quoted against the dollar, it may be necessary to work out the cross rates for currencies other than the dollar. The cross rate is needed to consummate financial transactions between two countries.

KEY CURRENCY CROSS RATES
21 Feb 2000

	British	Euro	Japan	U.S.
British	—	0.6172	0.5653	0.6250
Euro	1.6203	—	0.9153	1.0129
Japan	177.05	109.25	—	110.66
U.S.	1.5999	0.9873	0.9037	—

Source: Bloomberg, L.P. (http://www.bloomberg.com).

Note: Cross currency table calculator can be accessed by http://www.xe.net/currency/table.htm.

CUMULATIVE TRANSLATION ADJUSTMENT ACCOUNT

Cumulative translation adjustment (CTA) account is an entry in a translated balance sheet in which gains and/or losses from currency translation have been accumulated over a period of years.

CURRENCY ARBITRATE

Currency arbitrate is a form of *arbitrage* that takes advantage of divergences in exchange rates in different money markets by buying a currency in one market and selling it in another.

CURRENCY BOARD

A currency board is a government institution that exchanges domestic currency for foreign currency at a fixed rate of exchange. Under a currency board system, there is no central bank. The board has no discretionary monetary policy. Instead, market forces alone determine the money supply. The board attempts, however, to promote price stability and follow a responsible fiscal policy.

CURRENCY CALL OPTION

See CURRENCY OPTION.

CURRENCY COCKTAIL BOND

A bond denominated in a blend (or cocktail) of currencies.

CURRENCY DIVERSIFICATION

Currency diversification is the practice aimed at slashing the impact of unforeseen currency fluctuations by engaging activities in a portfolio of different currencies. Exposure to a diversified currency portfolio normally results in less *currency risk* than if all of the exposure was in one foreign currency.

CURRENCY FORECASTING

The idea of cutting the impact of unforeseen currency fluctuations by engaging activities in a portfolio of different currencies.
See FOREIGN EXCHANGE RATE FORECASTING.

CURRENCY FUTURES CONTRACT

A currency futures contract is a contract for future delivery of a specific quantity of a given currency, with the exchange rate fixed at the time the contract is entered. *Futures contracts* are similar to *forward contracts* except that they are standardized and traded on the organized exchanges and the gains and losses on the contracts are settled each day.

See also FOREIGN CURRENCY FUTURES; FOREIGN CURRENCY FUTURES; FORWARD CONTRACTS.

CURRENCY INDEXES

Currency indexes are economic indicators that attempt to measure foreign currencies. Two popular currency indexes are:

- *Federal Reserve Trade-Weighted Dollar*: The index reflects the currency units of more than 50% of the U.S. purchase, principal trading countries.The index measures the currencies of ten foreign countries: the United Kingdom, Germany, Japan, Italy, Canada, France, Sweden, Switzerland, Belgium, and the Netherlands. The index is weighted by each currency's base exchange rate and then averaged on a geometric basis. This weighting process indicates relative significance in overseas markets. The base year was 1973. The index is published by the Federal Reserve System and is found in its Federal Reserve Bulletin or at various Federal Reserve Internet sites such as http://woodrow.mpls.frb.fed.us/economy. The MNC should examine the trend in this index to determine foreign exchange risk exposure associated with its investment portfolio and financial positions. Also, the Federal Reserve trade-weighted dollar is the basis for commodity futures on the New York Cotton Exchange.
- *J.P. Morgan Dollar Index*: The index measures the value of currency units versus dollars. The index is a weighted-average of 19 currencies including those of France, Italy, United Kingdom, Germany, Canada, and Japan. The weighting is based on the relative significance of the currencies in world markets. The base of 100 was established for 1980 through 1982. The index highlights the impact of foreign currency units in U.S. dollar terms. The MNC can see the effect of foreign currency conversion on U.S. dollar investment.

See also BRITISH POUND; DEUTSCHE MARK; YEN.

CURRENCY OPTION

Foreign currency options are financial contracts that give the buyer the right, but not the obligation, to buy (or sell) a specified number of units of foreign currency from the option seller at a fixed dollar price, up to the option's expiration date. In return for this right the buyer pays a premium to the seller of the option. They are similar to foreign currency futures, in that the contracts are for fixed quantities of currency to be exchanged at a fixed price in the future. The key difference is that the maturity date for an option is only the last day to carry out the currency exchange; the option may be "exercised," that is, presented for currency exchange, at any time between its issuance and the maturity date, or not at all. Currency options are used as a hedging tool and for speculative purposes.

EXAMPLE 34

The buyer of a call option on British pounds obtains the right to buy £50,000 at a fixed dollar price (i.e., the exercise price) at any time during the (typically) three-month life of the option. The seller of the same option faces a contingent liability in that the seller will have to deliver the British pounds at any time, if the buyer chooses to exercise the option. The market value of an option depends on its exercise price, the remaining time to its expiration, the exchange rate in the spot market, and expectations about the future exchange rate. An option may sell for a price near zero or for thousands of dollars, or anywhere in between. Notice that the buyer of a call option on British pounds may pay a small price to obtain the option but does not have to exercise the option if the actual exchange rate moves favorably. Thus, an option is superior to a forward contract having the same maturity and exercise price because it need not be used—and the cost is just its purchase price. However, the price of the option is generally greater than the expected cost of the forward contract; so the user of the option pays for the flexibility of the instrument.

A. Currency Option Terminology

Foreign currency option definitions are as follows.

1. The *amount* is how much of the underlying foreign currency involved.
2. The seller of the option is referred to as the *writer* or *grantor.*
3. A *call* is an option to buy foreign currency, and a *put* is an option to sell foreign currency.
4. The *exercise* or *strike* price is the specified exchange rate for the underlying currency at which the option can be exercised.
   - *At the money*—exercise price equal to the spot price of the underlying currency. An option that would be profitable if exercised immediately is said to be *in the money.*
   - *In the money*—exercise price below the current spot price of the underlying currency, while in-the-money puts have an exercise price above the current spot price of the underlying currency.
   - *Out of the money*—exercise price above the current spot price of the underlying currency, while out-of-the-money puts have an exercise price below the current spot price of the underlying currency. An option that would not be profitable if exercised immediately is referred to as *out of the money.*
5. There are broadly two *types of options*: *American* option can be exercised at any time between the date of writing and the expiration or maturity date and *European* option can be exercised only on its expiration date, not before.
6. The *premium* or *option price* is the cost of the option, usually paid in advance by the buyer to the seller. In the over-the-counter market, premiums are quoted as a percentage of the transaction amount. Premiums on exchange-traded options are quoted as a dollar (domestic currency) amount per unit of foreign currency.

B. Foreign Currency Options Markets

Foreign currency options can be purchased or sold in three different types of markets:

1. Options on the physical currency, purchased on the over-the-counter (interbank) market;
2. Options on the physical currency, purchased on an organized exchange such as the Philadelphia Stock Exchange; and
3. Options on futures contracts, purchased on the International Monetary Market (IMM).

B.1. Options on the Over-the-Counter Market

Over-the-counter (OTC) options are most frequently written by banks for U.S. dollars against British pounds, German marks, Swiss francs, Japanese yen, and Canadian dollars. They are usually written in round lots of $85 to $10 million in New York and $2 to 83 million in London. The main advantage of over-the-counter options is that they are tailored to the specific needs of the firm. Financial institutions are willing to write or buy options that vary by amount (national principal), strike price, and maturity. Although the over-the-counter markets were relatively illiquid in the early years, the market has grown to such proportions that liquidity is now considered quite good. On the other hand, the buyer must assess the writing bank's ability to fulfill the option contract. Termed *counterparty risk*, the financial risk associated with the counterparty is an increasing issue in international markets. Exchange-traded options are more the sphere of the financial institutions themselves. A firm wishing to purchase an option in the over-the-counter market normally places a call to the currency option desk of a major money center bank, specifies the currencies, maturity, strike rate(s), and asks for an *indication*, a bid-offer quote.

B.2. Options on Organized Exchanges

Options on the physical (underlying) currency are traded on a number of organized exchanges worldwide, including the Philadelphia Stock Exchange (PHLX) and the London International Financial Futures Exchange (LIFFE). Exchange-traded options are settled through a clearinghouse, so that buyers do not deal directly with sellers. The clearinghouse is the counterparty to every option contract and it guarantees fulfillment. Clearinghouse obligations are in turn the obligation of all members of the exchange, including a large number of banks. In the case of the Philadelphia Stock Exchange, clearinghouse services are provided by the Options Clearing Corporation (OCC).

The Philadelphia Exchange has long been the innovator in exchange-traded options and has in recent years added a number of unique features to its United Currency Options Market (UCOM) making exchange-traded options much more flexible—and more competitive—in meeting the needs of corporate clients. UCOM offers a variety of option products with standardized currency options on eight major currencies and two cross-rate pairs (non-U.S. dollar), with either American- or European-style pricing. The exchange also offers *customized currency options*, in which the user may choose exercise price, expiration date (up to two years), and premium quotation form (units of currency or percentage of underlying value). Cross-rate options are also available for the DM/¥ and £/DM. By taking the U.S. dollar out of the equation, cross-rate options allow one to hedge directly the currency risk that arises in dealing with nondollar currencies. Contract specifications are shown in Exhibit 21. The PHLX trades both American-style and European-style currency options. It also trades month-end options (listed as EOM, or end of month), which ensures the availability of a short-term (at most, a two- or sometimes three-week) currency option at all times and long-term options, which extend the available expiration months on PHLX dollar-based and cross-rate contracts providing for 18- and 24-month European-style options. In 1994, the PHLX introduced a new option contract, called the *Virtual Currency Option*, which is settled in U.S. dollars rather than in the underlying currency.

EXHIBIT 21
Philadelphia Stock Exchange Currency Option Specifications

	Austrian Dollar	British Pound	Canadian Dollar	Deutsche Mark	Swiss Franc	Euro	Japanese Yen
Symbol							
American	XAD	XBP	XCD	XDM	SXF	XEU	XJY
European	CAD	CBP	CCD	CDM	CSF	ECU	CJY
Contract size	A$50,000	£31,250	C$50,000	DM 62,500	SFr 62,500	€62,500	¥6,250,000
Exercise Price Intervals	1¢	2.5¢	0.5¢	1¢[1]	1¢[1]	2¢	0.01¢[1]
Premium Quotations	Cents per unit	Cents per unit	Cents per unit	Cents per unit	Cents per unit	Cents per unit	Hundredths of a cent per unit
Minimum Price Change	$0.(00)01	$0.(00)01	$0.(00)01	$0.(00)01	$0.(00)01	$0.(00)02	$0.(00)01
Minimum Contract Price Change	$5.00	$3.125	$5.00	$6.25	$6.25	$6.25	$6.25
Expiration Months	March, June, September, and December + two near-term months						
Exercise Notice	No automatic exercise of in-the-money options						
Expiration Date	Friday before third Wednesday of the month (Friday is also the last trading day)						
Expiration Settlement Date	Third Wednesday of month						
Daily Price Limits	None						
Issuer & Guarantor	Options Clearing Corporation (OCC)						
Margin for Uncovered Writer	Option premium plus 4% of the underlying contract value less out-of-money amount, if any, to a minimum of the option premium plus ¾% of the underlying contract value. Contract value equal spot price times unit of currency per contract.						
Position & Exercise Limits	100,000 contracts						
Trading Hours	2:30 A.M.–2:30 P.M. Philadelphia time, Monday through Friday[2]						
Taxation	Any gain or loss: 60% long-term/40% short-term						

[1] Half-point strike prices (0.5¢) for SFr (0.5¢), and ¥ (0.005¢) in the three near-term months only.
[2] Trading hours for the Canadian dollar are 7:00 A.M.–2:30 P.M. Philadelphia time, Monday through Friday.

Source: Adapted from *Standardized Currency Options Specifications*, Philadelphia Stock Exchange, May 2000. (http://www.phlx.com/products/standard.html)

B.3. Currency Option Quotations and Prices

Some recent currency option prices from the Philadelphia Stock Exchange are presented in Exhibit 22. Quotations are usually available for more combinations of strike prices and expiration dates than were actually traded and thus reported in the newspaper such as the *Wall Street Journal*. Exhibit 22 illustrates the three different prices that characterize any foreign currency option. *Note*: Currency option strike prices and premiums on the U.S. dollar are quoted here as direct quotations ($/DM, $/¥, etc.) as opposed to the more common usage of indirect quotations used throughout the book. This approach is standard practice with option prices as quoted on major option exchanges like the Philadelphia Stock Exchange.

The three prices that characterize an "August 48 1/2 call option" are the following:

EXHIBIT 22
Foreign Currency Option Quotations
(Philadelphia Stock Exchange)

Option and Underlying	Strike Price	Calls—Last Aug.	Calls—Last Sept.	Calls—Last Dec.	Puts—Last Aug.	Puts—Last Sept.	Puts—Last Dec.
62.500 German marks	Cents per unit						
48.51	46	—	—	2.76	0.04	0.22	1.16
48.51	46 1/2	—	—	—	0.06	0.30	—
48.51	47	1.13	—	1.74	0.10	0.38	1.27
48.51	47 1/2	0.75	—	—	0.17	0.55	—
48.51	48	0.71	1.05	1.28	0.27	0.89	1.81
48.51	48 1/2	0.50	—	—	0.50	0.99	—
48.51	49	0.30	0.66	1.21	0.90	1.36	—
48.51	49 1/2	0.15	0.40	—	2.32	—	—
48.51	50	—	0.31	—	2.32	2.62	3.30

1. *Spot rate.* In Exhibit 22, "option and underlying" means that 48.51 cents, or $0.4851, was the spot dollar price of one German mark at the close of trading on the preceding day.
2. *Exercise price.* The exercise price or "strike price" listed in Exhibit 22 means the price per mark that must be paid if the option is exercised. The August call option on marks of 48 1/2 means $0.4850/DM. Exhibit 22 lists nine different strike prices, ranging from $0.4600/DM to $0.5000/DM, although more were available on that date than are listed here.
3. *Premium.* The premium is the cost or price of the option. The price of the August 48 1/2 call option on German marks was 0.50 U.S. cents per mark, or $0.0050/DM. There was no trading of the September and December 48 1/2 call on that day. The premium is the market value of the option. The terms *premium, cost, price*, and *value* are all interchangeable when referring to an option. All option premiums are expressed in cents per unit of foreign currency on the Philadelphia Stock Exchange except for the French franc, which is expressed in tenths of a cent per franc, and the Japanese yen, which is expressed in hundredths of a cent per yen.

The August 48 1/2 call option premium was 0.50 cents per mark, and in this case, the August 48 1/2 put premium was also 0.50 cents per mark. As one option contract on the Philadelphia Stock Exchange consists of 62,500 marks, the total cost of one option contract for the call (or put in this case) is DM62,500 × $0.0050/DM = $312.50.

B.4. Speculating in Option Markets

Options differ from all other types of financial instruments in the patterns of risk they produce. The option owner has the choice of exercising the option or allowing it to expire unused. The owner will exercise it only when exercising is profitable, which means when the option is in the money. In the case of a call option, as the spot price of the underlying currency moves up, the holder has the possibility of unlimited profit. On the downside, however, the holder can abandon the option and walk away with a loss never greater than the premium paid.

C. Buyer of a Call

To see how currency options might be used, consider a U.S. importer, called MYK Corporation with a DM 62,500 payment to make to a German exporter in two months (see Exhibit 23). MYK could purchase a European call option to have the DMs delivered to him at a specified exchange rate (the exercise price) on the due date. Assume that the option premium is $0.005/DM, and the strike price is 48 1/2 ($0.4850/DM). MYK has paid $312.50 for a DM 48 1/2 call option, which gives him the right to buy DM 62,500 at a price of $0.4850 per mark at the end of two months. Exhibit 24 illustrates the importer's gains and losses on the call option. The vertical axis measures profit or loss for the option buyer, at each of several different spot prices for the mark up to the time of maturity.

At all spot rates *below* (out-of-the-money) the strike price of $0.485, MYK would choose not to exercise its option. This decision is obvious, since at a spot rate of $0.485, for example, MYK would prefer to buy a German mark for $0.480 on the spot market rather than exercise his option to buy a mark at $0.485. If the spot rate remains below $0.480 until August when the option expires, he would not exercise the option. His total loss would be limited to only what he paid for the option, the $0.005/DM purchase price. At any lower price for the mark, his loss would similarly be limited to the original $0.005/DM cost.

Alternatively, at all spot rates *above* (in-the-money) the strike price of $0.485, MYK would exercise the option, paying only the strike price for each German mark. For example, if the spot rate were $0.495 cents per mark at maturity, he would exercise his call option, buying German marks for $0.485 each instead of purchasing them on the spot market at $0.495 each. The German marks could be sold immediately in the spot market for $0.495 each, with MYK pocketing a gross profit of $0.0010/DM, or a net profit of $0.005/DM after deducting the original cost of the option of $0.005/DM for a total profit of $312.50 ($0.005/DM × 62,500 DM). The profit to MYK, if the spot rate is greater than the strike price, with a strike price of $0.485, a premium of $0.005, and a spot rate of $0.495, is

Profit = Spot Rate – (Strike Price + Premium)
= $0.495/DM – ($0.485/DM + $0.005/DM)
= $0.005/DM or a total of $312.50 (S0.005/DM × 62,500 DM)

More likely, MYK would realize the profit by executing an offsetting contract on the options exchange rather than taking delivery of the currency. Because the dollar price of a mark could rise to an infinite level (off the upper right-hand side of Exhibit 24), maximum profit is unlimited. The buyer of a call option thus possesses an attractive combination of outcomes: limited loss and unlimited profit potential.

The *break-even price* at which the gain on the option just equals the option premium is $0.490/DM. The premium cost of $0.005, combined with the cost of exercising the option of $0.485, is exactly equal to the proceeds from selling the marks in the spot market at $0.490. Note that MYK will still exercise the call option at the break-even price. By exercising it MYK at least recovers the premium paid for the option. At any spot price above the exercise price but below the break-even price, the gross profit earned on exercising the option and selling the underlying currency covers part (but not all) of the premium cost.

D. Writer of a Call

The position of the writer (seller) of the same call option is illustrated in the bottom half of Exhibit 23. Because this is a *zero-sum game*, the profit from selling a call, shown in Exhibit 23, is the mirror image of the profit from buying the call. If the option expires when the spot price of the underlying currency is below the exercise price of \$0.485, the holder does not exercise the option. What the holder loses, the writer gains. The writer keeps as profit the entire premium paid of \$0.005/DM. Above the exercise price of \$0.485, the writer of the call must deliver the underlying currency for \$0.485/DM at a time when the value of the mark is above \$0.485. If the writer wrote the option naked—that is, without owning the currency—that seller will now have to buy the currency at spot and take the loss. The amount of such a loss is unlimited and increases as the price of the underlying currency rises. Once again, what the holder gains, the writer loses, and vice versa. Even if the writer already owns the currency, the writer will experience an opportunity loss, surrendering against the option the same currency that could have been sold for more in the open market.

For example, the loss to the writer of a call option with a strike price of \$0.485, a premium of \$0.005, and a spot rate of \$0.495/DM is

$$\begin{aligned} \text{Profit} &= \text{Premium} - (\text{Spot Rate} - \text{Strike Price}) \\ &= \$0.005/\text{DM} - (\$0.495/\text{DM} - \$0.485/\text{DM}) \\ &= -\$0.005/\text{DM or a total of } \$312.50\ (-\$0.005/\text{DM} \times 62{,}500\ \text{DM}) \end{aligned}$$

but only if the spot rate is greater than or equal to the strike rate. At spot rates less than the strike price, the option will expire worthless and the writer of the call option will keep the premium earned. The maximum profit that the writer of the call option can make is limited to the premium. The writer of a call option would have a rather unattractive combination of potential outcomes: limited profit potential and unlimited loss potential. Such losses can be limited through other techniques.

EXHIBIT 23
Profit or Loss For Buyer and Seller of a Call Option

Contract size:	62,500	DM
Expiration date:	2	months
Exercise, or strike price:	0.4850	\$/DM
Premium, or option price:	0.0050	\$/DM

Profit or Loss for Buyer of a Call Option

Ending Spot Rate (\$/DM)	0.475	0.480	0.485	0.490	0.495	0.500
Payments:						
Premium	(313)	(313)	(313)	(313)	(313)	(313)
Exercise cost	0	0	0	(30,313)	(30,313)	(30,313)
Receipts:						
Spot sale of DM	0	0	0	30,625	30,938	31,250
Net (\$):	(313)	(313)	(313)	0	313	625

Profit or Loss for Seller of a Call Option

The writer of an option profits when the buyer of the option suffers losses, i.e., a zero-sum game. The net position of the writer is, therefore, the negative of the position of the holder.

Net (\$):	313	313	313	0	(313)	(625)

EXHIBIT 24
German Mark Call Option
(Profit or Loss Per Option)

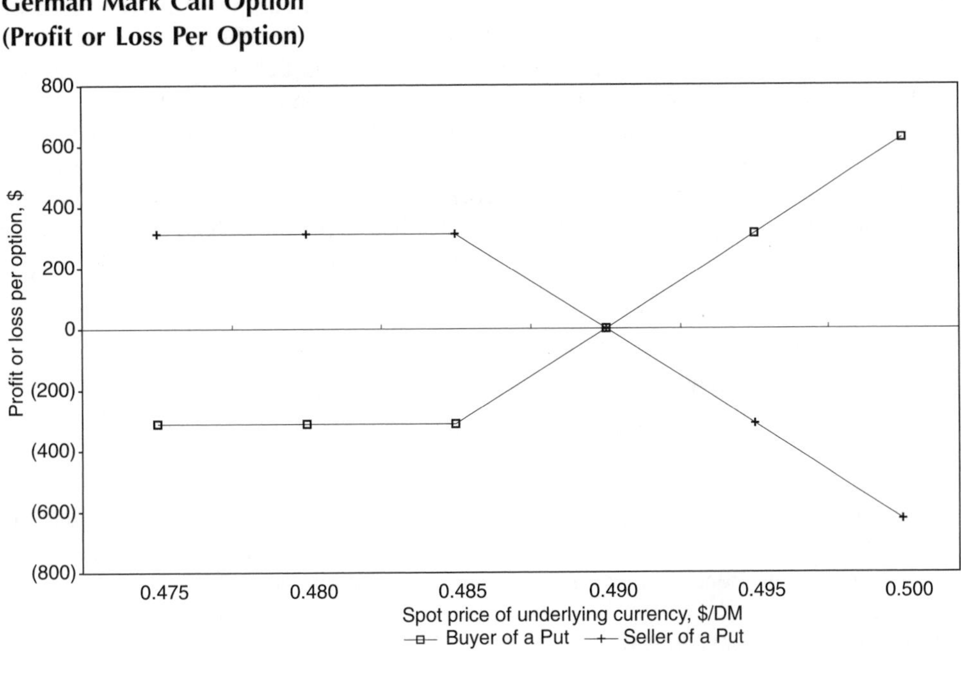

E. Buyer of a Put

The position of MYK as buyer of a put is illustrated in Exhibit 25. The basic terms of this put are similar to those just used to illustrate a call. The buyer of a put option, however, wants to be able to sell the underlying currency at the exercise price when the market price of that currency drops (not rises as in the case of a call option). If the spot price of a mark drops to, say, \$0.475/DM, MYK will deliver marks to the writer and receive \$0.485/DM. Because the marks can now be purchased on the spot market for \$0.475 each and the cost of the option was \$0.005/DM, he will have a net gain of \$0.005/DM. Explicitly, the profit to the holder of a put option if the spot rate is less than the strike price, with a strike price of \$0.485/DM, a premium of \$0.005/DM, and a spot rate of \$0.475/DM is

$$
\begin{aligned}
\text{Profit} &= \text{Strike Price} - (\text{Spot Rate} + \text{Premium}) \\
&= \$0.485/\text{DM} - (\$0.475/\text{DM} + \$0.005/\text{DM}) \\
&= \$0.005/\text{DM or a total of } \$312.50\ (\$0.005/\text{DM} \times 62{,}500\ \text{DM})
\end{aligned}
$$

The break-even price for the put option is the strike price less the premium, or \$0.480/DM in this case. As the spot rate falls further below the strike price, the profit potential would increase, and MYK's profit could be unlimited (up to a maximum of \$0.480/DM, when the price of a DM would be zero). At any exchange rate above the strike price of \$0.485, MYK would not exercise the option, and so would have lost only the \$0.005/DM premium paid for the put option. The buyer of a put option has an almost unlimited profit potential with a limited loss potential. Like the buyer of a call, the buyer of a put can never lose more than the premium paid up front.

F. Writer of a Put

The position of the writer of the put sold to MYK is shown in the lower portion of Exhibit 25. Note the symmetry of profit/loss, strike price, and break-even prices between the buyer and the writer of the put, as was the case of the call option. If the spot price of marks drops below $0.485 per mark, the option will be exercised by MYK. Below a price of $0.480 per mark, the writer will lose more than the premium received from writing the option ($0.005/DM), falling below break-even. Between $0.480/DM and $0.485/DM the writer will lose part, but not all, of the premium received. If the spot price is above $0.485/DM, the option will not be exercised, and the option writer pockets the entire premium of $0.005/DM. The loss incurred by the writer of a $0.485 strike price put, premium $0.005, at a spot rate of $0.475, is

$$\begin{aligned}\text{Loss} &= \text{Premium} - (\text{Strike Price} - \text{Spot Rate})\\ &= \$0.005/\text{DM} - (\$0.0485/\text{DM} - \$0.475/\text{DM} - \$0.005/\text{DM})\\ &= -\$0.005/\text{DM or a total of }\$312.50\ (-\$0.005/\text{DM} \times 62{,}500\ \text{DM})\end{aligned}$$

but only for spot rates that are less than or equal to the strike price. At spot rates that are greater than the strike price, the option expires out-of-the-money and the writer keeps the premium earned up-front. The writer of the put option has the same basic combination of outcomes available to the writer of a call: limited profit potential and unlimited loss potential up to a maximum of $0.480/DM.

EXHIBIT 25
Profit or Loss for Buyer and Seller of a Put Option

Contract size:	62,500	DM
Expiration date:	2	months
Exercise, or strike price:	0.4850	$/DM
Premium, or option price:	0.0050	$/DM

Profit or Loss for Buyer of a Put Option

Ending Spot Rate ($/DM)	0.470	0.475	0.480	0.485	0.490	0.495	0.500
Payments:							
Premium	(313)	(313)	(313)	(313)	(313)	(313)	(313)
Spot Purchase of DM	(29.375)	(29,688)	(30,000)	0	0	0	0
Receipts:							
Exercise of option	30,313	30,313	30,313	0	0	0	0
Net ($):	625	313	0	(313)	(313)	(313)	(313)

Profit or Loss for Seller of a Put Option

The writer of an option profits when the holder of the option suffers losses, i.e., a zero-sum game. The net position of the writer is, therefore, the negative of the position of the holder.

Net ($):	(625)	(313)	0	313	313	313	313

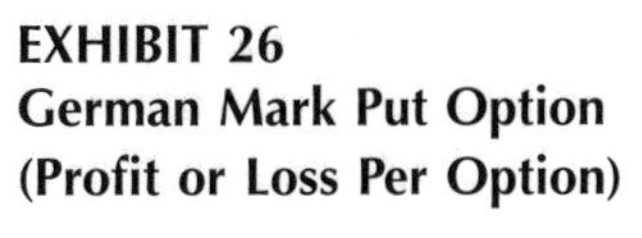

EXHIBIT 26
German Mark Put Option
(Profit or Loss Per Option)

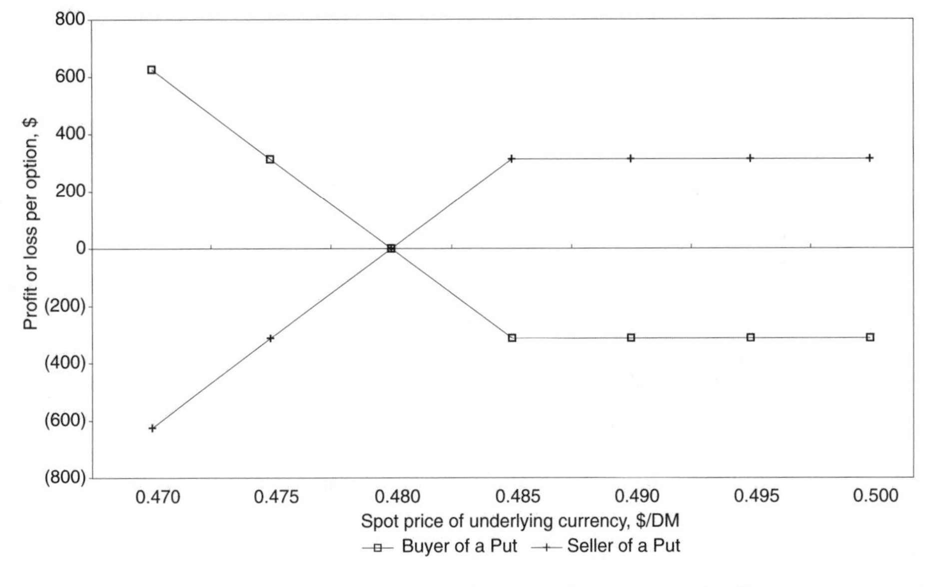

G. Option Pricing and Valuation

Exhibit 27 illustrates the profit/loss profile of a European-style call option on British pounds. The call option allows the holder to buy British pounds (£) at a strike price of $1.70/£. The value of this call option is actually the sum of two components:

$$\text{Total Value (Premium)} = \text{Intrinsic Value} + \text{Time Value}$$

Intrinsic value is the financial gain if the option is exercised immediately. It is shown by the solid line in Exhibit 28, which is zero until reaching the strike price, then rises linearly (1 cent for each 1 cent increase in the spot rate). Intrinsic value will be zero when the option is out-of-the-money—that is, when the strike price is above the market price—as no gain can be derived from exercising the option. When the spot price rises above the strike price, the intrinsic value becomes positive because the option is always worth at least this value if exercised. The time value of an option exists since the price of the underlying currency, the spot rate, can potentially move further in-the-money between the present time and the option's expiration date.

EXHIBIT 27
Intrinsic Value, Time Value, Total Value of a Call Option on British Pounds

Spot($/£) (1)	Strike Price (2)	Intrinsic Value of Option (1) − (2) = (3)	Time Value of Option (4)	Total Value (3) + (4) = (5)
1.65	1.70	0.00	1.37	1.37
1.66	1.70	0.00	1.67	1.67
1.67	1.70	0.00	2.01	2.01
1.68	1.70	0.00	2.39	2.39
1.69	1.70	0.00	2.82	2.82
1.70	1.70	0.00	3.30	3.30
1.71	1.70	1.00	2.82	3.82
1.72	1.70	2.00	2.39	4.39
1.73	1.70	3.00	2.01	5.01
1.74	1.70	4.00	1.67	5.67
1.75	1.70	5.00	1.37	6.37

Note from Exhibit 28 that the time value of a call option varies with option contract periods.

EXHIBIT 28
Intrinsic Value, Time Value, Total Value of a Call Option on British Pounds

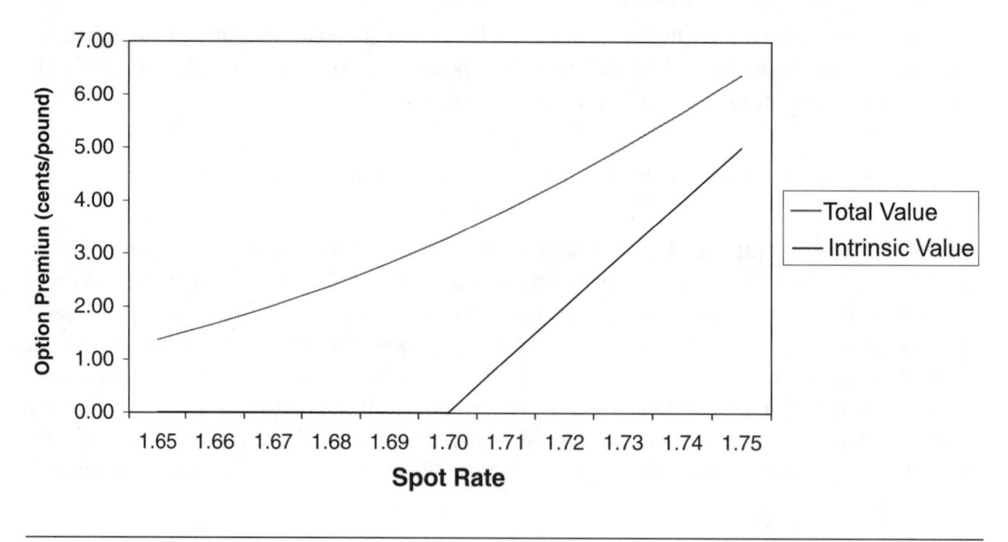

Note from Exhibit 29 that the time value of a call option varies with option contract periods.

EXHIBIT 29
The Value of a Currency Call Option before Maturity

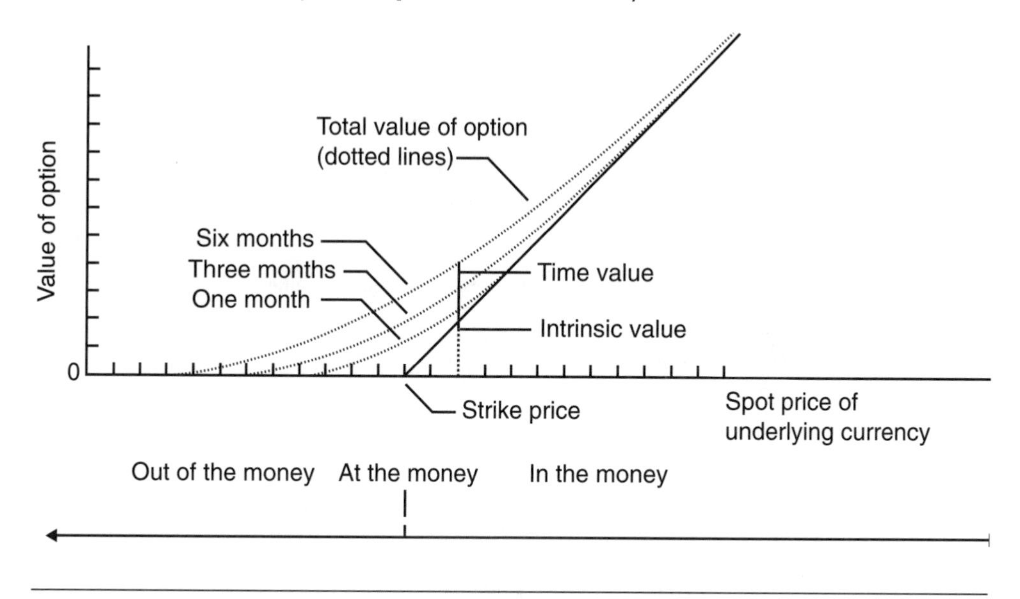

See also CURRENCY OPTION PRICING.

CURRENCY OPTION PRICING

Based on the work of *Black* and *Scholes* and others, the model yields the option premium. The basic theoretical model for the pricing of a European call option is:

$$V = e_f^{-rt}S[N(d_1)] - e_d^{-rt}E[N(d_2)]$$

where

V = Premium on a European call
e = 2.71828
S = spot exchange rate (in direct quote)
E = exercise or strike rate
r_f = foreign interest rate
r_d = domestic interest rate
t = number of time periods until the expiration date (For example, 90 days means $t = 90/365 = 0.25$)
$N(d)$ = probability that the normally distributed random variable Z is less than or equal to d
σ = standard deviation per period of (continuously compounded) rate of return

The two density functions, d_1 and d_2, and the formula are determined as follows:

$$V = [FN(d_1) - E[N(d_2)]e_d^{-rt}$$

$$d_1 = \ln[F/E]/\sigma\sqrt{t} + \sigma\sqrt{t}/2$$

$$d_2 = d_1 - \sigma\sqrt{t}$$

Note: In the final derivations, the spot rate (S) and foreign interest rate (r_f) have been replaced with the forward rate (F).

The premium for a European put option is similarly derived:

$$V = \{F[N(d_1) - 1] - E[N(d_2) - 1]\}e_d^{-rt}$$

EXAMPLE 35

Given the following data on basic exchange rate and interest rate values:

Data	Symbols	Numerical values
Spot rate	S	\$1.7/£
90-day forward	F	\$1.7/£
Exercise or strike rate	E	\$1.7/£
U.S. interest rate	r_d	0.08 = 8%
British pound interest rate	r_f	0.08 = 8%
Time	t	90/365
Standard deviation	σ	0.01 = 10%

$$d_1 = \ln[F/E]/\sigma\sqrt{t} + \sigma\sqrt{t}/2$$

$$d_1 = \ln[1.7/1.7]/(\sqrt{90/365}) + (.1)\sqrt{90/365}/2 = 0.025$$

$$d_2 = d_1 - \sigma\sqrt{t} = 0.025 - (.1)\sqrt{90/365} = -0.025$$

The values of d_1 and d_2 are found from the normal distribution table (see Table 5 in the Appendix).

$$N(d_1) = 0.51; \quad N(d_2) = 0.49$$

Substituting these values into the option premium formula yields:

$$V = [FN(d_1) - E[N(d_2)]e_d^{-rt} = [(1.7)(0.51) - (1.7)(0.49)]2.71827^{-0.08(90/365)}$$
$$= \$0.033/£$$

See also BLACK-SCHOLES OPTION PRICING MODEL.

CURRENCY OPTION PRICING SENSITIVITY

If *currency options* are to be used effectively for hedging or speculative purposes, it is important to know how option prices (values or premiums) react to their various components. Four key variables that impact option pricing are: (1) changing spot rates, (2) time to maturity, (3) changing volatility, and (4) changing interest differentials.

The corresponding measures of sensitivity are:

1. *Delta*—The sensitivity of option premium to a small change in the spot exchange rate.
2. *Theta*—The sensitivity of option premium with respect to the time to expiration
3. *Lambda*—The sensitivity of option premium with respect to volatility.
4. *Rho* and *Phi*—The sensitivity of option premium with respect to the interest rate differentials.

Exhibit 30 describes how these sensitivity measures are interpreted.

EXHIBIT 30
Interpretations of Option Pricing Sensitivity Measures

Sensitivity Measures	Interpretation	Reasoning
Delta	The higher the delta, the greater the chance of the option expiring in-the-money.	Deltas of .7 or up are considered high.
Theta	Premiums are relatively insensitive until the last 30 or so days.	Longer maturity options are more highly valued. This gives a trader the ability to alter an option position without incurring significant time value deterioration.
Lambda	Premiums rise with increases in volatility.	Low volatility may cause options to sell. A trader is hoping to buy back for a profit immediately after volatility falls, causing option premiums to drop.
Rho	Increases in home interest rates cause call option premiums to increase.	A trader is willing to buy a call option on foreign currency before the home interest rate rises (interest rate for the home currency), which will allow the trader to buy the option before its price increases.
Phi	Increases in foreign interest rates cause call option premiums to decrease.	A trader is willing to sell a call option on foreign currency before the foreign interest rate rises (interest rate for the foreign currency), which will allow the trader to sell the option before its price decreases.

See also CURRENCY OPTION; CURRENCY OPTION PRICING.

CURRENCY PUT OPTION

See CURRENCY OPTION.

CURRENCY QUOTATIONS

Currency quotes are always given in pairs because a dealing bank usually does not know whether a prospective customer is in the market to buy or to sell a foreign currency. The first rate is the bid, or buy rate; the second is the sell, ask, or offer rate.

EXAMPLE 36

Suppose the pound sterling is quoted at $1.5918–29. This quote means that banks are willing to buy pounds at $l.5918 and sell them at $1.5929. Note that the banks will always buy low and sell high. In practice, however, they quote only the last two digits of the decimal. Thus, sterling would be quoted at 18–19 in this example.

Note that when *American terms* are converted to *European terms* or direct quotations are converted to indirect quotations, bid and ask quotes are reversed; that is, the reciprocal of the American (direct) bid becomes the European (indirect) ask and the reciprocal of the American (direct) ask becomes the European (indirect) bid.

EXAMPLE 37

So, in Example 1, the reciprocal of the American bid of \$1.5918/£ becomes the European ask of £0.6282 and the reciprocal of the American ask of \$1.5929/£ equals the European bid of £0.6278/\$ resulting in a direct quote for the dollar in London of £0.6278–82. Exhibit 31 summarizes this result.

EXHIBIT 31
Direct Versus Indirect Currency Quotations

Direct (American)	Indirect (European)
\$1.5918–29	£0.6278–82

See also BID–ASK SPREAD; DIRECT QUOTE; INDIRECT QUOTE.

CURRENCY REVALUATION

Also called *appreciation* or *strengthening*, revaluation of a currency refers to a rise in the value of a currency that is pegged to gold or to another currency. The opposite of revaluation is weakening, deteriorating, devaluation, or depreciation. Revaluation can be achieved by raising the supply of foreign currencies via restriction of imports and promotion of exports.

See also DEVALUATION.

CURRENCY RISK

Also called *foreign exchange risk, exchange rate risk*, or *exchange risk*, currency risk is the risk that tomorrow's exchange rate will differ from today's rate. In financial activities involving two or more currencies, it reflects the risk that a change (gain or loss) in an entity's economic value can occur as a result of a change in exchange rates. Currency risk applies to all types of multinational businesses—international trade contracts, international *portfolio investments*, and *foreign direct investments* (*FDIs*). Currency risk exists when the contract is written in terms of the foreign currency or denominated in foreign currency. Also, when you invest in a foreign market, the return on the foreign investment in terms of the U.S. dollar depends not only on the return on the foreign market in terms of local currency but also on the change in the exchange rate between the local currency and U.S. dollar.

The idea of exchange risk in trade contracts is illustrated in the following example.

EXAMPLE 38

Case I. An American automobile distributor agrees to buy a car from the manufacturer in Detroit. The distributor agrees to pay \$25,000 upon delivery of the car, which is expected to be 30 days from today. The car is delivered on the thirtieth day and the distributor pays \$25,000. Notice that, from the day this contract was written until the day the car was delivered, the *buyer* knew the *exact dollar amount* of his liability. There was, in other words, *no uncertainty* about the value of the contract.

Case II. An American automobile distributor enters into a contract with a British supplier to buy a car from the United Kingdom for 8,000 pounds. The amount is payable on the delivery of the car, 30 days from today. Suppose, the range of spot rates that we believe can occur on the date the contract is consummated is \$2 to \$2.10. On the thirtieth day, the American importer will pay some amount in the range of 8,000 × \$2.00 = \$16,000 to 8,000 × 2.10 = \$16,800 for the car. As of today, the American firm is uncertain regarding its future dollar outflow 30 days hence. That is, the *dollar value of the contract is uncertain.*

These two examples help illustrate the idea of foreign exchange risk in international trade contracts. In the case of the domestic trade contract, given as Case I, the exact dollar amount of the future dollar payment is known today with certainty. In the case of the international trade contract given in Case II, where the *contract is written in the foreign currency*, the exact dollar amount of the contract is not known. The variability of the exchange rate induces variability in the future cash flow. This is the risk of exchange-rate changes, *exchange risk*, or *currency risk*. Currency risk exists when the contract is written in terms of the foreign currency or *denominated* in foreign currency. There is no exchange risk if the international trade contract is written in terms of the domestic currency. That is, in Case II, if the contract were written in dollars, the American importer would face *no* exchange risk. With the contract written in dollars, the British exporter would bear *all* the exchange risk, because the British exporter's future pound receipts would be uncertain. That is, he would receive payment in dollars, which would have to be converted into pounds at an unknown (as of today) pound–dollar exchange rate. In international trade contracts of the type discussed here, at least one of the two parties bears the exchange risk. Certain types of international trade contracts are denominated in a third currency, different from either the importer's or the exporter's domestic currency. In Case II, the contract might have been denominated in the Deutsche mark. With a DM contract, both the importer and the exporter would be subject to exchange-rate risk.

Exchange risk is not limited to the two-party trade contracts; it exists also in foreign direct or portfolio investments. The next example illustrates how a change in the dollar affects the return on a foreign investment.

EXAMPLE 39

You purchased bonds of a Japanese firm paying 12% interest. You will earn that rate, assuming interest is paid in marks. What if you are paid in dollars? As Exhibit 32 shows, you must then convert yens to dollars before the payout has any value to you. Suppose that the dollar appreciated 10% against the yen during the year after purchase. (A currency appreciates when acquiring one of its units requires more units of a foreign currency.) In this example, 1 yen required 0.01 dollars, and later, 1 yen required only 0.0091 dollars; at the new exchange rate it would take 1.099 (0.01/0.0091) yens to acquire 0.01 dollars. Thus, the dollar has appreciated while the yen has depreciated. Now, your return realized in dollars is only 10.92%. The adverse movement in the foreign exchange rate—the dollar's appreciation—reduced your actual yield.

EXHIBIT 32
Exchange Risk and Foreign Investment Yield

Transaction	Yens	Exchange Rate: No. of Dollars per 1 Yen	Dollars
On 1/1/20X1			
Purchased one German bond with a 12% coupon rate	500	\$0.01*	\$5.00
On 12/31/20X1			
Expected interest received	60	0.01	0.60
Expected yield	12%		12%

(*Continued*)

On 12/31/20X1			
Actual interest received	60	0.0091**	0.546
Realized yield	12%		10.92%***

* For illustrative purposes assume that the direct quote is $0.01 per yen.
** $0.01/(1 + .1) = $0.01/1.1 = $0.0091.
*** $0.546/$5.00 = .1092 = 10.92%.

Note, however, that currency swings work both ways. A weak dollar would boost foreign returns of U.S. investors. Exhibit 33 is a quick reference to judge how currency swings affect your foreign returns.

EXHIBIT 33
Currency Changes vs. Foreign Returns in U.S. Dollars

	Change in Foreign Currency against the Dollar Return				
Foreign	**20%**	**10%**	**0%**	**−10%**	**−20%**
20%	44%	32	20	8	−4
10	32	21	10	−1	−12
0	20	10	0	−10	−20
−10	8	−1	−10	−19	−28
−20	−4	−12	−20	−28	−36

CURRENCY RISK MANAGEMENT

Foreign exchange rate risk exists when the contract is written in terms of the foreign currency or denominated in the foreign currency. The exchange rate fluctuations increase the riskiness of the investment and incur cash losses. The financial manager must not only seek the highest return on temporary investments but must also be concerned about changing values of the currencies invested. You do not necessarily eliminate foreign exchange risk. You may only try to contain it. In countries where currency values are likely to drop, financial managers of the subsidiaries should:

- Avoid paying advances on purchase orders unless the seller pays interest on the advances sufficient to cover the loss of purchasing power.
- Not have excess idle cash. Excess cash can be used to buy inventory or other real assets.
- Buy materials and supplies on credit in the country in which the foreign subsidiary is operating, extending the final payment date as long as possible.
- Avoid giving excessive trade credit. If accounts receivable balances are outstanding for an extended time period, interest should be charged to absorb the loss in purchasing power.
- Borrow local currency funds when the interest rate charged does not exceed U.S. rates after taking into account expected devaluation in the foreign country.

A. Ways to Neutralize Foreign Exchange Risk

Foreign exchange risk can be neutralized or hedged by a change in the asset and liability position in the foreign currency. Here are some ways to control exchange risk.

A.1. Entering a Money-Market Hedge

Here the exposed position in a foreign currency is offset by borrowing or lending in the money market.

EXAMPLE 40

XYZ, an American importer enters into a contract with a British supplier to buy merchandise for 4,000 pounds. The amount is payable on the delivery of the good, 30 days from today. The company knows the exact amount of its pound liability in 30 days. However, it does not know the payable in dollars. Assume that the 30-day money-market rates for both lending and borrowing in the U.S. and U.K. are .5% and 1%, respectively. Assume further that today's foreign exchange rate is $1.735 per pound.

In a money-market hedge, XYZ can take the following steps:

Step 1. Buy a one-month U.K. money-market security, worth 4,000/(1 + .005) =3,980 pounds. This investment will compound to exactly 4,000 pounds in one month.

Step 2. Exchange dollars on today's spot (cash) market to obtain the 3,980 pounds. The dollar amount needed today is 3,980 pounds × $1.7350 per pound = $6,905.30.

Step 3. If XYZ does not have this amount, it can borrow it from the U.S. money market at the going rate of 1%. In 30 days XYZ will need to repay $6,905.30 × (1 + .1) = $7,595.83.

Note: XYZ need not wait for the future exchange rate to be available. On today's date, the future dollar amount of the contract is known with certainty. The British supplier will receive 4,000 pounds, and the cost of XYZ to make the payment is $7,595.83.

A.2. Hedging by Purchasing Forward (or Futures) Exchange Contracts

A forward exchange contract is a commitment to buy or sell, at a specified future date, one currency for a specified amount of another currency (at a specified exchange rate). This can be a hedge against changes in exchange rates during a period of contract or exposure to risk from such changes. More specifically, do the following: (1) Buy foreign exchange forward contracts to cover payables denominated in a foreign currency and (2) sell foreign exchange forward contracts to cover receivables denominated in a foreign currency. This way, any gain or loss on the foreign receivables or payables due to changes in exchange rates is offset by the gain or loss on the forward exchange contract.

EXAMPLE 41

In the previous example, assume that the 30-day forward exchange rate is $1.6153. XYZ may take the following steps to cover its payable.

Step 1. Buy a forward contract today to purchase 4,000 pounds in 30 days.

Step 2. On the 30th day pay the foreign exchange dealer 4,000 pounds × $1.6153 per pound = $6,461.20 and collect 4,000 pounds. Pay this amount to the British supplier.

Note: Using the forward contract XYZ knows the exact worth of the future payment in dollars ($6,461.20).

Note: The basic difference between futures contracts and forward contracts is that futures contracts are for specified amounts and maturities, whereas forward contracts are for any size and maturity.

A.3. Hedging by Foreign Currency Options

Foreign currency options can be purchased or sold in three different types of markets: (1) options on the physical currency, purchased on the over-the counter (interbank) market;

(2) options on the physical currency, purchased on organized exchanges such as the Philadelphia Stock Exchange and the Chicago Mercantile Exchange; and (3) options on futures contracts, purchased on the International Monetary Market (IMM) of the Chicago Mercantile Exchange.

A.4. Using Currency Swaps

Currency swaps are temporary exchanges of funds between two parties—central banks or the central bank and MNC—that do not go through the foreign exchange market. Suppose a U.S. MNC wants to inject capital into its Ghanan subsidiary. The U.S. company signs a swap contract with the central Ghanan bank, then deposits dollars at the bank. The bank then makes a loan in Ghanan currency to the subsidiary firm. At the end of the loan period, the subsidiary pays off the loan to the bank, which returns the original dollar deposit to the U.S. MNC. Usually, the central bank does not pay interest on the foreign currency deposit it receives but does charge interest on the loan it makes. Therefore, the cost of the swap includes *two* interest components: the interest on the loan and the foregone interest on the deposit. In recent years, MNCs have made *direct* swaps with each other. In the late 1970s some British and U.S. companies were swapping currency, typically for about 10 years. Because British interest rates were higher, the U.S. firm paid a 2% fee to the British firm. To protect against movements in U.S.–U.K. exchange rates, many swap contracts often had a *top-off provision*, calling for renegotiation at settlement time if the exchange rate moved over 10%.

A.5. Repositioning Cash by Leading and Lagging the Time at Which an MNC Makes Operational or Financial Payments

Often, money- and forward-market hedges are not available to eliminate exchange risk. Under such circumstances, leading (accelerating) and lagging (decelerating) may be used to *reduce* risk.

A.6. Maintaining Balance between Receivables and Payables Denominated in a Foreign Currency

MNCs typically set up *multilateral netting* centers as a special department to settle the outstanding balances of affiliates of an MNC with each other on a net basis. These act as a clearing house for payments by the firm's affiliates. If there are amounts due among affiliates they are offset insofar as possible. The net amount would then be paid in the currency of the transaction; thus, a much lower quantity of the currency must be acquired.

A.7. Maintaining Monetary Balance

Monetary balance refers to minimizing *accounting exposure*. If a company has net positive exposure (more monetary assets than liabilities), it can use more financing from foreign monetary sources to balance things. MNCs with assets and liabilities in more than one foreign currency may try to reduce risk by balancing off exposure in the different countries. Often, the monetary balance is practiced across *several* countries simultaneously.

A.8. Positioning of Funds through Transfer Pricing

A transfer price is the price at which an MNC sells goods and services to its foreign affiliates or, alternatively, the price at which an affiliate sells to the parent. For example, a parent that wishes to transfer funds from an affiliate in a depreciating-currency country may charge a higher price on the goods and services sold to this affiliate by the parent or by affiliates from strong-currency countries. Transfer pricing affects not only transfer of funds from one entity to another but also the income taxes paid by both entities.

CURRENCY SPREAD

A currency spread involves buying an option at one strike price and selling a similar option at a different strike price. Thus, the currency spread limits the option holder's downside risk on the currency bet but at the cost of limiting the position's upside potential as well. There are two types of currency spreads:

- A *bull spread*, which is designed to bet on a currency's appreciation, involves buying a call at one strike price and selling another call at a higher strike price.
- A *bear spread*, which is designed to bet on a currency's decline, involves buying a put at one strike price and selling another put at a lower strike price.

CURRENCY SWAP

Currency swaps are temporary exchanges of monies between two parties that do not go through the foreign exchange market. In *official* swaps, the two parties are central banks. *Private* swaps are between central banks and MNCs. Currency swaps are often used to minimize *currency risk*. See CURRENCY RISK MANAGEMENT; SWAPS.

CURRENCY TRANSLATION METHODS

Accountants are concerned with the appropriate way to translate foreign currency-denominated items on financial statements into their home currency values. If currency values change, translation gains or losses may result. A foreign currency asset or liability is said to be *exposed* if it must be translated at the current exchange rate. Regardless of the translation method selected, measuring accounting exposure is conceptually the same. It involves determining which foreign currency-denominated assets and liabilities will be translated at the current (postchange) exchange rate and which will be translated at the historical (prechange) exchange rate. The former items are considered to be exposed, while the latter items are regarded as not exposed. Translation exposure is the difference between exposed assets and exposed liabilities.

There are various alternatives available to measure *translation (accounting) exposure*. The basic translation methods are the *current-rate method*, *current/noncurrent method*, *monetary/nonmonetary method*, and *temporal method*. The current-rate method treats all assets and liabilities as exposed. The current/noncurrent method treats only current assets and liabilities as being exposed. The monetary/nonmonetary method treats only monetary assets and liabilities as being exposed. The temporal method translates financial assets and all liabilities valued at current cost as exposed and historical cost assets and liabilities as unexposed. Exhibit 33 summarizes these four currency translation methods.

EXHIBIT 34
Four Currency Translation Methods

	Items Translated at	
	Current Rate	**Historical Rate**
Current rate	All assets and all liabilities and common stock	—
Current/noncurrent	Current assets and current liabilities	Fixed assets and long-term liabilities Common stock
Monetary/nonmonetary	Monetary assets and all liabilities	Physical assets Common stock
Temporal	Financial assets and all liabilities and physical assets valued at current price	Physical assets valued at historical cost Common stock

EXAMPLE 42

G&G France, the French subsidiary of a U.S. company, G&G, Inc., has the following balance sheet expressed in French francs:

Assets (FFr thousands)		Liabilities (FFr thousands)	
Cash, marketable securities	7,000	Accounts payable	14,000
Accounts receivable	18,000	Short-term debt	8,000
Inventory	31,000	Long-term debt	45,000
Net fixed assets	63,000	Equity	52,000
	FFr 119,000		FFr 119,000

(1) Suppose the current spot rate is \$0.21/FFr. G&G's translation exposure would be calculated as follows:

Under the current rate method, G&G France's exposure is its equity of FFr 52 million, or \$10.92 million (0.21 × 52 million). Under the current/noncurrent method, G&G France's accounting exposure is FFr 34 million (7 + 18 + 31 – 14 – 8, in millions), or \$7.14 million (0.21 × 34 million). Its monetary/nonmonetary method accounting exposure is –FFr 42 million (7 + 18 – 14 – 8 – 45, in millions), or –\$8.82 million (0.21 × –42 million). G&G's temporal method exposure is the same as its current/noncurrent method exposure. The calculations assume that all assets and liabilities are denominated in francs.

(2) Suppose the French Franc depreciates to \$0.17. The balance sheets for G&G France at the new exchange rate are shown below.

Current Rate Method

Assets (U.S. \$ thousands)		Liabilities (U.S. \$ thousands)	
Cash, marketable securities	1,190	Accounts payable	2,380
Accounts receivable	3,060	Short-term debt	1,360
Inventory	5,270	Long-term debt	7,650
Net fixed assets	10,710	Equity	8,840
	\$20,230		\$20,230

Current/noncurrent Method

Assets (U.S. \$ thousands)		Liabilities (U.S. \$ thousands)	
Cash, marketable securities	1,190	Accounts payable	2,380
Accounts receivable	3,060	Short-term debt	1,360
Inventory	5,270	Long-term debt	9,450
Net fixed assets	13,230	Equity	9,560
	\$22,750		\$22,750

Monetary/nonmonetary Method

Assets (U.S. \$ thousands)		Liabilities (U.S. \$ thousands)	
Cash, marketable securities	1,190	Accounts payable	2,380
Accounts receivable	3,060	Short-term debt	1,360
Inventory	6,510	Long-term debt	7,650
Net fixed assets	13,230	Equity	12,600
	\$23,990		\$23,990

Temporal Method

Cash, marketable securities	1,190	Accounts payable	2,380
Accounts receivable	3,060	Short-term debt	1,360
Inventory	5,270	Long-term debt	9,450
Net fixed assets	13,230	Equity	9,560
	$22,750		$22,750

The translation gain (loss) equals the franc exposure multiplied by the –$0.04 change in the exchange rate. These translation gains (losses) are as follows: current rate method—loss of $2.08 million (–0.04 × 52 million); current/noncurrent method—loss of $1.36 million (–0.04 × 34 million); monetary/nonmonetary method—gain of $1.68 million (–0.04 × –42 million); temporal method—loss of $1.36 million (–0.04 × 34 million). These gains (losses) show up on the equity account and equal the difference in equity values calculated at the new exchange rate of $0.17/FFr and the old exchange rate of $0.21/FFr.

CURRENT ACCOUNT

The current account in the *balance of payments*, analogous to the revenues and expenses of a business, is the sum of the merchandise, services, investment income, and unilateral transfer accounts. When combined, they provide important insights into a country's international economic performance, just as a firm's profit and loss statement conveys vital information about its performance.

See also BALANCE OF PAYMENTS; OFFICIAL SETTLEMENTS BALANCE.

CURRENT ACCOUNT BALANCE

A *balance of payments* that measures a nation's merchandise *trade balance* plus its net receipts of unilateral transfers during a specified time period.

See also BALANCE OF PAYMENTS.

CURRENT/NONCURRENT METHOD

Also called *net current asset* or *net working capital method*, under the current/noncurrent method, all current accounts (assets and liabilities) are translated at the current rate of foreign exchange, and all noncurrent accounts at their historical exchange rates.

See also CURRENT RATE METHOD; MONETARY/NONMONETARY METHOD; TEMPORAL METHOD.

CURRENT RATE METHOD

Under the current rate method, the exchange rate at the balance sheet date is used to translate the financial statement of a foreign subsidiary into the home currency of the MNC. Under the current rate method: (1) All balance sheet assets and liabilities are translated at the current rate of exchange in effect on the balance sheet date; (2) income statement items are usually translated at an average exchange rate for the reporting period; (3) all equity accounts are translated at the historical exchange rates that were in effect at the time the accounts first entered the balance sheet; and (4) translation gains and losses are reported as a separate item in the stockholders' equity section of the balance sheet. Translation gains and losses are only included in net income when there is a sale or liquidation of the entire investment in a foreign entity. Although this method may seem a logical choice, it is incompatible with the historic cost principle, which is a *generally accepted accounting principle* (GAAP) in many countries, including the U.S.

EXAMPLE 43

Consider the case of a U.S. firm that invests $100,000 in a French subsidiary. Assume the exchange rate at the time is $1 = FFr 5. The subsidiary converts the $100,000 into francs, which yields it FFr 500,000. It then goes out and purchases some land with this money. Subsequently, the dollar depreciates against the franc, so that by year-end $1 = FFr 4. If this exchange rate is used to convert the value of the land back into U.S. dollars for the purpose of preparing consolidated accounts, the land will be valued at $125,000. The piece of land would appear to have increased in value by $25,000, although in reality the increase would be simply a function of an exchange rate change. Thus the consolidated accounts would present a somewhat misleading picture.

See also CURRENCY TRANSLATION METHODS; CURRENT/NONCURRENT METHOD; MONETARY/NONMONETARY METHOD; TEMPORAL METHOD.

D

DEBENTURE

A long-term debt instrument that is not collateralized. Because it is unsecured debt, it is issued usually by large, financially strong companies with excellent bond ratings.

DEBT SWAP

Also called a *debt-equity swap*, a debt swap is a set of transactions in which an MNC buys a country's dollar bank debt at a discount and swaps this debt with the central bank for local currency that it can use to acquire local equity.

DEFAULT RISK

Default risk is the risk that a borrower will be unable to make interest payments or principal repayments on debt. For example, there is a great amount of default risk inherent in the bonds of a company experiencing financial difficulty. Exhibit 35 presents the degree of default risk for some investment instruments.

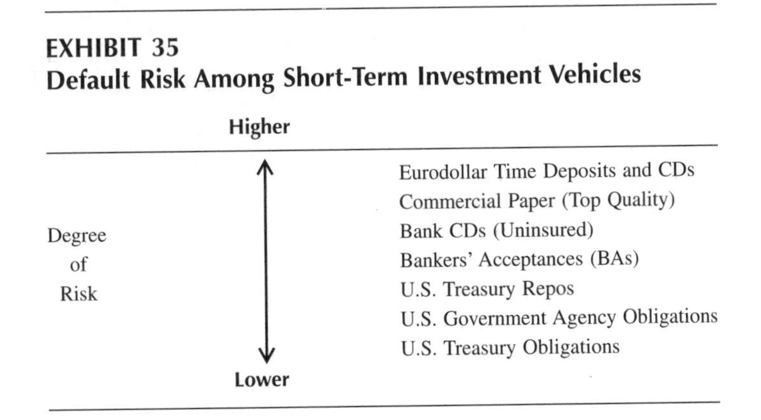

EXHIBIT 35
Default Risk Among Short-Term Investment Vehicles

See also RISK.

DELTA

In *option*, delta is the ratio of change of the option price to a small change in the price of the underlying asset. Denoted with δ, it is also equal to the derivative of the option price to the security price.

See also CURRENCY OPTION; CURRENCY OPTION PRICING SENSITIVITY; OPTION.

DELTA HEDGE

A powerful hedging strategy using options with steady adjustment of the number of options used, as a function of the *delta* of the option.

DEMAND MONEY

See CALL MONEY.

DEPRECIATION

1. A drop in the foreign exchange value of a floating currency. The opposite of depreciation is *appreciation*. This term contrasts with *devaluation*, which is a drop in the foreign exchange value of a currency that is pegged to gold or to another currency. In other words, the par value is reduced.

   See also APPRECIATION OF THE DOLLAR; DEPRECIATION OF THE DOLLAR.
2. The decline in economic potential of limited life assets originating from wear and tear, natural deterioration through interaction of the elements, and technical obsolescence.
3. The spreading out of the original cost over the estimated life of the fixed assets such as plant and equipment.

DEPRECIATION OF THE DOLLAR

Also called *cheap dollar, weak dollar, deterioration of the dollar*, or *devaluation of the dollar*, depreciation of the dollar refers to a drop in the foreign exchange value of the dollar relative to other currencies.

See also APPRECIATION OF THE DOLLAR.

DERIVATIVES

Derivatives are leveraged instruments that are linked either to specific financial instruments or indicators (such as foreign currencies, government bonds, stock price indices, or interest rates) or to particular commodities (such as gold, sugar, or coffee) that may be purchased or sold at a future date. Derivatives may also be linked to a future exchange, according to contractual arrangement, of one asset for another. The instrument, which is a contract, may be tradable and have a market value. Among derivative instruments are options (on currencies, interest rates, commodities, or indices), traded financial futures, and arrangements such as currency and interest rate swaps. Firms use derivative instruments to hedge their risks from swings in securities prices or currency exchange rates. They also can be used for speculative purposes, that is, to make risk bets on market movements.

See also FINANCIAL DERIVATIVE.

DEUTSCHE MARK

Germany's currency.

DEVALUATION

The process of officially dropping the value of a country's currency relative to other foreign currencies.

See DEPRECIATION OF THE DOLLAR.

DFI

See FOREIGN DIRECT INVESTMENT.

DINAR

Monetary unit of Abu Dhabi, Aden, Algeria, Bahrain, Iraq, Jordan, Kuwait, Libya, South Yemen, Tunisia, and Yugoslavia.

DIRECT FOREIGN INVESTMENT

See FOREIGN DIRECT INVESTMENT.

DIRECT QUOTE

The price of a unit of foreign currency expressed in the home country's currency: dollars per pound, dollars per yen, etc. For example, the direct quote for Japanese yen is $.00909 per yen if the home country is the U.S. Direct and *indirect* quotations are reciprocals.

$$\text{Direct quote} = \frac{1}{\textit{Indirect quote}} = \frac{1}{110 \text{ yens}} = \$.00909 \text{ per yen}$$

Spot (cash) rates, both direct and indirect, are presented in Exhibit 36.

EXHIBIT 36
Foreign Exchange Rates

Tuesday, March 13, 2001

EXCHANGE RATES

The New York foreign exchange mid-range rates below apply to trading among banks in amounts of $1 million and more, as quoted at 4 p.m. Eastern time by Reuters and other sources. Retail transactions provide fewer units of foreign currency per dollar. Rates for the 12 Euro currency countries are derived from the latest dollar-euro rate using the exchange ratios set 1/1/99.

	U.S. $ EQUIV.		CURRENCY PER U.S. $	
COUNTRY	Tue	Mon	Tue	Mon
Argentina (Peso)	1.0001	1.0001	.9999	.9999
Australia (Dollar)	.5053	.5092	1.9792	1.9637
Austria (Schilling)	.06648	.06751	15.043	14.813
Bahrain (Dinar)	2.6525	2.6525	.3770	.3770
Belgium (Franc)	.0227	.0230	44.0994	43.4253
Brazil (Real)	.4850	.4840	2.0620	2.0660
Britain (Pound)	1.4490	1.4645	.6901	.6828
1-month forward	1.4486	1.4641	.6903	.6830
3-months forward	1.4474	1.4628	.6909	.6836
6-months forward	1.4454	1.4608	.6919	.6846
Canada (Dollar)	.6475	.6445	1.5443	1.5517
1-month forward	.6476	.6445	1.5441	1.5515
3-months forward	.6479	.6448	1.5434	1.5508
6-months forward	.6483	.6452	1.5426	1.5500
Chile (Peso)	.001703	.001687	587.25	592.75
China (Renminbi)	.1208	.1208	8.2783	8.2784
Colombia (Peso)	.0004389	.0004388	2278.30	2278.75
Czech. Rep. (Koruna)				
Commercial rate	.02647	.02681	37.776	37.296
Denmark (Krone)	.1224	.1243	8.1688	8.0471
Ecuador (US Dollar)-e	1.0000	1.0000	1.0000	1.0000
Finland (Markka)	.1539	.1562	6.4998	6.4005
France (Franc)	.1395	.1416	7.1709	7.0613
1-month forward	.1395	.1417	7.1688	7.0590
3-months forward	.1395	.1417	7.1664	7.0571
6-months forward	.1396	.1418	7.1628	7.0536
Germany (Mark)	.4677	.4750	2.1381	2.1054
1-month forward	.4678	.4751	2.1375	2.1047
3-months forward	.4680	.4752	2.1368	2.1042
6-months forward	.4682	.4755	2.1357	2.1031
Greece (Drachma)	.002685	.002726	372.51	366.81
Hong Kong (Dollar)	.1282	.1282	7.7999	7.8000
Hungary (Forint)	.003434	.003489	291.17	286.61
India (Rupee)	.02144	.02149	46.640	46.540
Indonesia (Rupiah)	.0000972	.0000952	10287.50	10500.00
Ireland (Punt)	1.1614	1.1795	.8610	.8478
Israel (Shekel)	.2420	.2420	4.1330	4.1330
Italy (Lira)	.0004724	.0004798	2116.72	2084.36
Japan (Yen)	.008351	.008306	119.74	120.39
1-month forward	.008387	.008343	119.23	119.86
3-months forward	.008454	.008408	118.29	118.94
6-months forward	.008554	.008505	116.90	117.58
Jordan (Dinar)	1.4085	1.4065	.7100	.7110
Kuwait (Dinar)	3.2584	3.2616	.3069	.3066
Lebanon (Pound)	.0006604	.0006604	1514.25	1514.25
Malaysia (Ringgit)-b	.2632	.2632	3.8001	3.8000
Malta (Lira)	2.2553	2.2789	.4434	.4388
Mexico (Peso)				
Floating rate	.1044	.1035	9.5810	9.6600
Netherland (Guilder)	.4151	.4215	2.4091	2.3723
New Zealand (Dollar)	.4180	.4203	2.3923	2.3793
Norway (Krone)	.1113	.1132	8.9819	8.8376
Pakistan (Rupee)	.01668	.01689	59.950	59.200
Peru (new Sol)	.2843	.2843	3.5178	3.5176
Philippines (Peso)	.02077	.02078	48.150	48.125
Poland (Zloty)-d	.2455	.2479	4.0725	4.0338
Portugal (Escudo)	.004563	.004634	219.17	215.82
Russia (Ruble)-a	.03487	.03489	28.677	28.661
Saudi Arabia (Riyal)	.2666	.2666	3.7506	3.7505
Singapore (Dollar)	.5683	.5677	1.7595	1.7614
Slovak Rep. (Koruna)	.02094	.02126	47.758	47.032
South Africa (Rand)	.1277	.1285	7.8300	7.7800
South Korea (Won)	.0007843	.0007819	1275.00	1279.00
Spain (Peseta)	.005498	.005583	181.89	179.11
Sweden (Krona)	.0998	.1013	10.0170	9.8670
Switzerland (Franc)	.5943	.6036	1.6827	1.6567
1-month forward	.5952	.6046	1.6801	1.6540
3-months forward	.5967	.6061	1.6759	1.6500
6-months forward	.5990	.6083	1.6694	1.6439
Taiwan (Dollar)	.03075	.03084	32.520	32.430
Thailand (Baht)	.02295	.02283	43.570	43.800
Turkey (Lira)	.00000104	.00000109	963000.00	920000.00
United Arab (Dirham)	.2723	.2723	3.6729	3.6730
Uruguay (New Peso)				
Financial	.07871	.07874	12.705	12.700
Venezuela (Bolivar)	.001420	.001420	704.00	704.45
SDR	1.2878	1.2907	.7765	.7748
Euro	.9148	.9290	1.0931	1.0764

Special Drawing Rights (SDR) are based on exchange rates for the U.S., German, British, French, and Japanese currencies. Source: International Monetary Fund.

a-Russian Central Bank rate. b-Government rate. d-Floating rate; trading band suspended on 4/11/00. e-Adopted U.S. dollar as of 9/11/00. Foreign Exchange rates are available from Readers' Reference Service (413) 592-3600.

See also INDIRECT QUOTE; AMERICAN TERMS; EUROPEAN TERMS.

DIRHAM

Monetary unit of the United Arab Republic and Morocco.

DIRTY FLOAT

Also called *managed float*, dirty float is a *flexible* (*market-determined*) *exchange rate system* in which central banks intervene directly in foreign exchange markets from time to time to manipulate short-term swings in exchange rates in a direction perceived to be in the national interest. A country floats "dirty" if it attempts to hold down a strong currency in order to keep its exports competitive, when all objective analysis would suggest that the currency is undervalued.

See also MANAGED FLOAT.

DIRTY PRICE

The total price of a bond, including accrued interest.

DISCOUNTING

1. Deducting anticipated undesirable economic or company's negative earnings, news from security prices prior to their announcement. For example, a highly anticipated interest rate hike by the Federal Reserve Bank of the U.S. may have already been factored in stock prices a few weeks ahead.
2. Process of determining the present worth of expected future cash flows generated by a project—either a single sum or a series of cash sums to be paid or received. This process is used to evaluate the value of a *foreign direct investment* (FDI) project.

   See also ADJUSTED PRESENT VALUE; INTERNAL RATE OF RETURN; NET PRESENT VALUE.
3. In connection with forward foreign exchange rates, the percentage amount by which the forward rate is less than the spot rate.

   See also FORWARD PREMIUM OR DISCOUNT.
4. Interest deducted in advance from a loan, *forfaiting*, or *factoring*. For example, the discount rate for forfaiting is set at a fixed rate, typically about 1.25% above the local borrowing rate or the *London Interbank Offer Rate (LIBOR)*.

   See also FACTORING; FORFAITING.

DISINVESTMENT

1. Also called *divestiture,* the selling off or closing down of all or part of a foreign direct investment (e.g., foreign subsidiaries) for economic or other reasons.
2. Pulling out of the capital invested in a foreign country.

DIVERSIFIABLE RISK

Also called *controllable risk, company-specific risk*, or *unsystematic* risk, diversifiable risk is that part of the total risk of a security associated with such random events as lawsuits, strikes, winning or losing a major contract, and other events that are unique to a particular company. This type of risk can be diversified away and hence is not priced in a portfolio.

DIVERSIFICATION

1. Entry into a different business activity outside of the firm's traditional business. This may involve a different product, stage of the production process, or country. Some companies wish to diversify their operations by getting into various industries. This can be a long-term, strategic decision on the part of management.
2. Allocation of investments among different companies, different industries, or different regions in order to ease risk. Diversification exists by owning securities of companies having negative or no correlation.

   See also PORTFOLIO DIVERSIFICATION; PORTFOLIO THEORY.

DOBRA

Monetary unit of San Tome and Principe.

DOCUMENTARY COLLECTION

A method of payment for a foreign trade transaction that adopts a *draft* and other important documents but *not* a *letter of credit.*

DOCUMENTARY DRAFT

A *sight* or *time draft* that is accompanied by documents such as invoices, *bills of lading*, inspection certificates, and insurance papers.

DOCUMENTARY LETTER OF CREDIT

Documentary letter of credit (L/C) is a *letter of credit* for which the issuing bank provides that specified documents must be attached to the draft. The documents are to guarantee the importer that the goods have been sent and that title to the merchandise has been duly transferred to the importer. Most L/Cs in commercial transactions are documentary.

DOLLAR INDEXES

Also called *currency indexes*, dollar indexes measure the value of the dollar and are provided by the Federal Reserve Board (FRB), Morgan Guaranty Trust Company of New York, and Federal Reserve Bank of Dallas. They show different movements because they include different countries and are based on different concepts and methodologies. The data are provided in nominal values (market exchange rates) and in real values (purchasing power corrected for inflation). The FRB index is published in a press release and in the monthly *Federal Reserve Bulletin*; the Morgan index is published in the bimonthly *World Financial Markets*; and the FRB Dallas index is published monthly in *Trade-Weighted Value of the Dollar.* The FRB and Morgan indexes include 10 and 18 industrial nations, respectively, and the FRB Dallas index includes all of the 131 U.S. trading partners.

See also CURRENCY INDEXES.

DOW JONES GLOBAL STOCK INDEXES

Dow Jones Global Stock Indexes is a grouping of indexes tracking stocks around the globe. The biggest index, the Dow World, tracks shares in 33 countries. The index is based on an equal weighted average of commodity prices. A 100 base value was assigned to the U.S. index on June 30, 1982; a 100 base went to the rest of the world indexes for Dec. 31, 1991. Indexes are tracked in both local currency and U.S. dollars, though the dollar tracking contains far more analytical data, such as 52-week high and low and year-to-date percentage performance. The index appears daily in *The Wall Street Journal.* The index can be used to examine the difference between performances in various stock markets and exchanges around the globe. That performance can be used to determine if shares in those countries are good or poor values. In addition, stock indexes often hint how a country's economy is performing, because rising stock markets typically appear in healthy economies.

Note: Currency swings play an important role in calculating investment performance from foreign markets. So if one country's stock market is performing well but its currency is weak versus the dollar, a U.S. investor may still suffer. Conversely, a country with a strong currency and weak stock market may produce profits for U.S. investors despite the rocky climate for equities.

DRACHMA

Greece's currency.

DRAFT

Also called a *bill of exchange*, a draft is an unconditional order to pay. It is the instrument normally used in foreign trade to effect payment. It is simply an order written by an exporter (seller) requesting an importer (buyer) or its agent to pay a specified amount of money at a specified time. There are three parties involved: the drawer, the payee, and the drawee. The person or business initiating the draft is known as the *drawer, maker*, or *originator*. The *payee* is the party receiving payment. The party (usually an importer's bank) to whom the draft is addressed is the *drawee*. There are different types of drafts: (1) A *sight draft* is payable upon presentation of documents; (2) a *time draft* is payable at some stated future date called *usance* or *tenor*; (3) when a time draft is drawn on and accepted by a commercial business, it is called a *trade acceptance*; (4) when a time draft is drawn on and accepted by a bank, it is called a *banker's acceptance*; (5) a *clean draft* is not accompanied by any necessary documents; and (6) a *documentary draft* is accompanied by all required papers.

See also DOCUMENTARY DRAFT; LETTERS OF CREDIT; TRADE CREDIT INSTRUMENTS.

DRAGON BOND

A U.S. dollar-denominated bond issued in the so-called "Dragon" economies of Asia, such as Hong Kong, Singapore, and Taiwan.

DUAL CURRENCY BOND

A dual currency bond is a bond denominated in one currency but paying interest in another currency at a specified exchange rate. This type of bond can also redeem proceeds in a different currency than the currency of denomination.

E

EAFE (EUROPE, AUSTRALIA, FAR EAST) INDEX

More exactly known as the *Morgan Stanley Capital International Europe, Australia, Far East Index*, the EAFE Index reflects the performance of all major stock markets outside North America. It is a market-weighted index composed on 1,041 companies representing the stock markets of Europe, New Zealand, and the Far East. It is considered the key "rest-of-the-world" index for U.S. investors, much as is the *Dow Jones Industrial Average* for the American market. It is used as a guide to see how U.S. shares are faring against other markets around the globe. It also serves as a performance benchmark for international mutual funds that hold non-U.S. assets. Morgan Stanley has created its own indexes for 18 major foreign markets. To make the EAFE Index, those country indexes are weighted to reflect the total market capitalization of each country's markets as a share in the world market. The index's base of 100 is for January 1, 1970. The index is quoted two ways: one in local currencies and a second in the U.S. dollar. This shows how American investors would fare addressing both share price and currency fluctuations. The EAFE Index can be found in newspapers such as *Barron's*. When the EAFE Index is performing better than the U.S. markets, it may be time for investors to shift money overseas. Conversely, when U.S. market indexes are doing better than the EAFE Index, a shift away from foreign assets may be in order. *A Word of Caution*: Currency fluctuations can play a major part of any overseas investment. A rising EAFE may be more a reflection of a weak U.S. dollar than improving foreign economies or strong opportunities in overseas stocks.

EASDAQ

European Association of Securities Dealers Automated Quotation (http://www.easdaq.com), the only pan-European stock market, offers international growth companies and investors seamless cross-border trading, clearing, and settlement within a unified market infrastructure. It is a fully regulated market, independent of any national exchange. Trading takes place through member firms in 15 countries.

EC

See EUROPEAN COMMUNITY.

ECONOMIC AND FINANCE COUNCIL

The Economic and Finance Council (ECOFIN), made up of the finance ministers of *Euroland's* 11 member-states, has the key legal and political responsibilities for managing the euro. See also EURO.

ECONOMIC EXPOSURE

Also called *operating exposure*, economic exposure is the extent to which an MNC's future international cash flows and its market value are affected by an unexpected change in exchange rates. Economic exposure involves the long-run effect of changes in exchange rates on future prices, sales, and costs. It differs from *translation exposure* and *transaction exposure* in that it is subjective and thus not easily quantified. The best strategy to control economic exposure is to internationally diversify operations and financing. Exhibit 37 compares economic exposure with translation (accounting) exposure.

EXHIBIT 37
Translation Exposure vs Economic Exposure

Translation (Accounting) Exposure	Economic (Operating) Exposure
Occurs only when an MNC has foreign subsidiaries	Occurs for any type of foreign operations
A backward looking concept, reflecting past decisions as reported in the subsidiary's financial statements	A forward looking concept, focusing on future cash flows
Deals with accounting values due to translation	Deals with cash flows (not just accounting values) and market value of the MNC
Subject to accounting rules and regulations	Subject to economic facts such as outstanding commitments denominated in foreign currency and operating exposure
Looks only at items on the financial statements; ignores off-balance-sheet contracts and future operations	Incorporates all cash flows and sources of value

In order to see the difference between the economic and translation exposure, consider the following example.

EXAMPLE 44

A U.S. MNC has a subsidiary in France. On December 31 of each year the parent company consolidates the balance sheet and income statement of the subsidiary with its U.S. operations. If the exchange rate prevailing on December 31 of any year is used to translate the assets and liabilities of the French subsidiary, the total book or accounting value of the subsidiary, measured in dollars, will fluctuate from year to year, even if the total franc-denominated assets and liabilities do not. This fluctuation in the value of the subsidiary is due to the accounting or translation exposure of the subsidiary. Economically, it makes little sense to say that the value of the French plant has changed over the years, while nothing physical has changed. The change appears in the books only—due to an arbitrary accounting rule. What should matter, however, is the way in which the earnings (or cash flow) stream of the French operations changes because of changes in the franc–dollar rate. The measurement of the responsiveness of the future earnings stream of the French subsidiary to changes in the dollar–franc exchange rate is captured in the concept of economic (or operating) exposure. Economic exposure is, conceptually, a sound way of capturing the effects of exchange-rate changes on the *value* of the firm, although it is difficult to measure this exposure.

See also TRANSACTION EXPOSURE; TRANSLATION EXPOSURE.

ECONOMIC RISK

1. The long-run effect of changes in exchange rates on future prices, sales, and costs.
2. The likelihood that events, including economic mismanagement, will cause drastic changes in a country's business environment that adversely affect the profit and other goals of a particular business enterprise.
3. The chance of loss or uncertain variations in earnings of a business due to economic conditions, including foreign exchange risk, inflation risk, and interest rate risk.

ECONOMIC VALUE ADDED

Economic value added (EVA), a registered trademark of Stern & Steward & Company, is one of the two well-known *financial metrics* (measures of performance) of an MNC or its affiliates. The other is *cash flow return on investment* (*CFROI*). Also called *residual income*,

EVA is the operating profit which an MNC or its division is able to earn above some minimum rate of return on its assets. EVA is the value created by a company in excess of the cost of capital for the investment base. It attempts to determine whether management has, in fact, added value to the entity over and above what the providers of capital (both credit and equity holders) to the firm *require*. The formula is:

EVA = Net operating profit after taxes – (weighted average cost of capital × capital employed)

Improving EVA can be achieved in three ways: (1) invest capital in high-performing projects, (2) use less capital, and (3) increase profit without using more capital.
See also CASH FLOW RETURN ON INVESTMENT.

ECU

See EUROPEAN CURRENCY UNIT.

EDGE ACT AND AGREEMENT CORPORATION

Also referred to as an *Edge Act Corporation.* An Edge Act and Agreement Corporation is a subsidiary, located in the United States, of a U.S. commercial bank incorporated under federal law to be allowed to engage in various international banking, investment, and financing activities, including equity participations. The Edge Act subsidiary may be located in a state other than that of the parent bank. The Edge subsidiary, operating abroad, is free of restraints of U.S. law and may perform whatever services and functions are legal in the countries where it operates. The Edge Act was proposed by Senator Walter E. Edge of New Jersey and enacted in 1919 by the Congress.

EFFECTIVE INTEREST RATE

1. The *effective annual yield* is better known as the *annual percentage rate* (*APR*). Different types of investments use different compounding periods. For example, most bonds pay interest semi-annually. Some banks pay interest quarterly. If an investor wishes to compare investments with different compounding periods, he/she needs to put them on a common basis.

   See also ANNUAL PERCENTAGE RATE.
2. *Real interest rate* on a loan. It is the nominal interest rate divided by the actual proceeds of the loan. For example, assume you took out a $10,000, one year, 10% discounted loan. The effective interest rate equals: $1,000/($10,000 – $1,000) = $1,000/$9,000 = 11%. In this discount loan, the actual proceeds are only $9,000, which effectively increases the cost of the loan.
3. Yield to maturity.

   See YIELD; YIELD TO MATURITY.

EFFICIENT MARKET

An efficient market is one in which all available and relevant information is already reflected in the prices of traded securities. The term is most frequently applied to foreign exchange markets and securities markets. It is very difficult for investors to outperform the market. If competition exists and transaction costs are low, prices tend to respond rapidly to new information, and speculation opportunities quickly dissipate. The efficient market hypotheses have been subjected to numerous empirical tests, but with mixed results. For example, the more recent studies seriously challenge the view of unbiased forward exchange rates.

EFFICIENT PORTFOLIO

Efficient portfolio is the central gist of Markowitz's *portfolio theory*. Efficient portfolio theory claims that rational investors behave in a way reflecting their aversion to taking increased risk without being compensated by an adequate increase in expected return. Also, for any given expected return, most investors will prefer a lower risk and, for any given level of risk, prefer a higher return to a lower return. In Exhibit 38, an efficient set of portfolios that lie along the ABC line, called "efficient frontier," is noted. Along this frontier, the investor can receive a maximum return for a given level of risk or a minimum risk for a given level of return. Specifically, comparing three portfolios A, B, and D, portfolios A and B are clearly more efficient than D, because portfolio A could produce the same expected return but at a lower risk level, while portfolio B would have the same degree of risk as portfolio D but would afford a higher return. Investors try to find the optimum portfolio by using the indifference curve, which shows the investor's trade-off between risk and return. By matching the indifference curve showing the risk-return trade-off with the best investments available in the market as represented by points on the efficient frontier, investors are able to find an optimum portfolio.

EXHIBIT 38
Efficient Portfolio

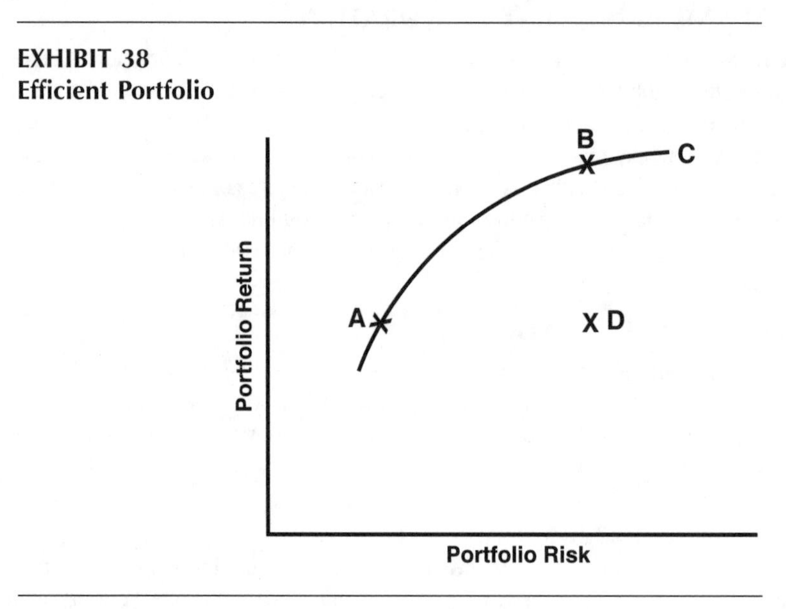

See also PORTFOLIO THEORY.

EMBEDDED DERIVATIVE

An embedded derivative is not itself a derivative, but a derivative implied by a contract such as a bond that is contingent on an underlying commodity.

EQUILIBRIUM EXCHANGE RATE

Exchange rate at which demand for a currency exactly matches the supply of the currency for sale.

EQUITY METHOD

The equity method is an accounting method used in preparing a parent company's nonconsolidated financial reports that treats income at the foreign subsidiary's level as being received by the parent company when it is earned by the foreign entity. This method requires that the value of the investment in the parent company's balance sheet be increased or decreased to recognize the parent's share of profits or losses since the acquisition. To compute this profit or loss, however, a foreign subsidiary's financial statements must be translated into U.S currency, with the translation profit or loss specified.

ESCUDO

Monetary unit of Azores, Cape Verde Islands, Guinea-Bissau, Madeira, Mozambique, Portugal, Portuguese East Africa, and Timor.

EURO

The euro is the European currency that made its debut on January 1, 1999 and that unites many European economies. The symbol is €, and the ISO code is EUR. The history behind the creation of the euro is as follows. The European Union Treaty (Maastricht Treaty) of 1993 created the European Currency Unit (ECU), a basket of currencies which includes 15 currencies of the European Union (EU) countries. The ECU took two different forms: the official ECU and the private ECU. The official ECU did not trade in the foreign exchange market and was used between central banks and international financial institutions. The private ECU was freely traded in the foreign exchange market, and its exchange rate resulted from the supply and demand.

Beginning with 1999, as provided in the Maastricht Treaty, the ECU was replaced by the euro. It is based on fixed conversion rates between the currencies of the different countries which constitute the EMU, thus there is no longer trading in currency rates between ECU members that agreed to use the euro.

According to the Maastricht criteria, the following 11 countries qualified for entry and joined EMU in January 1999: Germany, France, Italy, Spain, Portugal, Belgium, Luxembourg, the Netherlands, Finland, Austria, and Ireland. Of the four remaining EU members, Greece is scheduled to join in 2001, and Britain, Sweden, and Denmark have opted out for now. Foreign exchange rates are quoted daily in business dailies as well as major newspapers, on computer services such as America Online, and on financial TV networks and specialty shows. The introduction of the euro in 1999 is only for operations carried out in the money, foreign exchange, and financial markets. For most retail transactions, the changeover to the euro will start only after the date of the physical introduction of euro banknotes and coins denominated in euros, by 2002 at the latest. When the euro conversion is complete, perhaps by 2002, investors will notice these changes in Western European investments:

- Stocks will be priced and settled in euro only.
- Government debt will be quoted in euro only.
- Stock and bond deals, such as mergers, will be stated in euro.
- Financial statements from companies and governments will be in euro.

Exhibit 39 presents a series of events leading to the debut of the euro.
See also EUROPEAN CURRENCY UNIT; MAASTRICHT TREATY.

EXHIBIT 39
Europe Moves Toward a Single Market

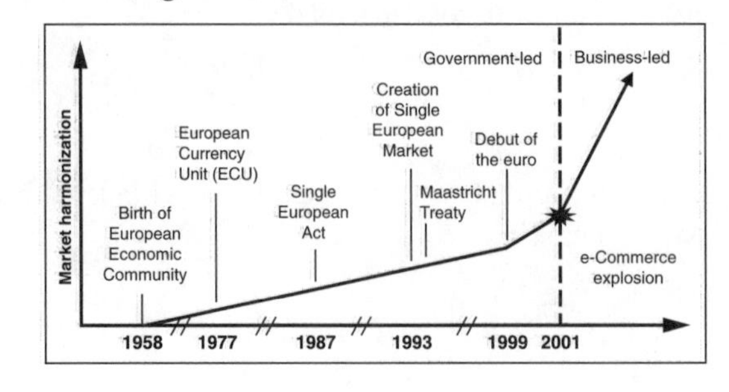

EUROBANKS

Eurobanks are those banks that accept deposits and make loans in foreign currencies.

EUROBILL OF EXCHANGE

A Eurobill of exchange is a bill of exchange drawn and accepted in the ordinary manner but denominated in foreign currency and approved as being payable outside the country in whose currency it is denominated.

EUROBOND MARKET

The Eurobond market is an international market for long-term debt, whereas the foreign bond market is a domestic market issued by a foreign borrower. A Eurobond market is the market for bonds in any country denominated in any currency other than the local one. A bond originally offered outside the country in whose currency it is denominated, Eurobonds are typically dollar-denominated bonds originally offered for sale to investors outside of the United States.

EUROBONDS

A Eurobond is a bond that is sold simultaneously in a number of countries by an international syndicate. It is a bond sold in a country other than the one in whose currency the bond is denominated. Examples include a General Motors issue denominated in dollars and sold in Japan and a German firm's sale of pound-denominated bonds in Switzerland. Eurobonds are underwritten by an *international underwriting syndicate* of banks and other securities firms. For example, a bond issued by a U.S. corporation, denominated in U.S. dollars, but sold to investors in Europe and Japan (not to investors in the United States), would be a Eurobond. Eurobonds are issued by MNCs, large domestic corporations, governments, governmental agencies, and international institutions. They are offered simultaneously in a number of different national capital markets, but not in the capital market of the country, nor to residents of the country, in whose currency the bond is denominated. Eurobonds appeal to investors for several reasons: (1) They are generally issued in bearer form rather than as registered bonds. So investors who desire anonymity, whether for privacy reasons or for tax avoidance, prefer Eurobonds. (2) Most governments do not withhold taxes on interest payments associated with Eurobonds. While depositors in the short-term *Eurocurrency market* are primarily corporations, potential buyers of Eurobonds are often private individuals.

EUROCHEQUE

A check from a European bank that can be cashed at over 200,000 banks around the world displaying the "European Union" pinnacle. It is similar to an American traveler's check.

EURO-CLEAR

Telecommunications network that notifies all traders regarding outstanding issues of Eurobonds for sale.

EURO-COMMERCIAL PAPER

Euro-commercial papers (Euro-CP or ECP) are short-term notes of an MNC or bank, sold on a discount basis in the *Eurocurrency market.* The proceeds of the issuance of Euro-commercial papers at a discount by borrows is computed as follows:

$$\text{Market price} = \frac{\text{Face value}}{1 + \left[\left(\frac{N}{360}\right) \times \left(\frac{N}{100}\right)\right]}$$

where Y = yield per annum and N = days remaining until maturity.

EXAMPLE 45

The proceeds from the sale of a $10,000 face value, 90-day issue Euro-CP priced to yield 8% per annum (reflecting current market yields on similar debt securities for comparable credit ratings) would be:

$$\text{Market price} = \$10{,}000/\{1 + [(90/360) \times (8/100)]\} = \$9{,}803.90$$

EUROCREDIT LOANS

Eurocredit loans are loans of one year or longer made by *Eurobanks.*

EUROCREDIT MARKET

Eurocredit market is the group of banks that accept deposits and extend loans in large denominations and a variety of currencies. *Eurobanks* are major players in this and the *Eurocurrency market.* The Eurocredit loans are longer than so-called Eurocurrency loans.

EUROCURRENCY

Eurocurrency is a dollar or other freely convertible currency outside the country of the currency in which funds are denominated. A U.S. dollar in dollar-denominated loans, deposits, and bonds in Europe is called a *Eurodollar.* There are Eurosterling (British pounds deposited in banks outside the U.K.), Euromarks (Deutsche marks deposited outside Germany), and Euroyen (Japanese yen deposited outside Japan).

EUROCURRENCY BANKING

Eurocurrency banking is not subject to domestic banking regulations, such as reserve requirements and interest-rate restrictions. This enables *Eurobanks* to operate more efficiently, cheaply, and competitively than their domestic counterparts and to attract intermediation business out of the domestic and into the external market. Eurocurrency banking is a *wholesale* rather than a retail business. The customers are corporations and governments—not individuals. They do not want checking accounts; they want to earn interest on their deposits. Therefore, they lend on a short-term time deposits or they buy somewhat larger longer-term certificates of deposits. They borrow anything from overnight call money to 8-year term loans. Interest rates in the *Eurocurrency market* may be fixed or floating. Floating rates are usually tied to the rate at which the banks lend to one another.

EUROCURRENCY MARKET

Also called a *Eurodollar market* or a *Euromarket*, a Eurocurrency market is a market for a currency deposited in a bank outside the country of its origin, say, the United States, which is based primarily in Europe and engaged in the lending and borrowing of U. S. dollars and other major currencies outside their countries of origin to finance international trade and investment. The Eurocurrency market then consists of those banks (*Eurobanks*) that accept deposits and make loans in foreign currencies. The term *Eurocurrency markets* is misleading for two reasons: (1) they are not currency markets where foreign exchange is traded, rather they are money markets for short-term deposits and loans; and (2) the prefix *euro-* is no longer accurate since there are important offshore markets in the Middle East and the Far East.

EURODEPOSIT

A eurodeposit or *Eurodollar deposit*, is a dollar-denominated deposit held in banks outside of the U.S.

EXAMPLE 46

A Swedish investor may deposit U.S. dollars with a bank outside the U.S., perhaps in Stockholm or in London. This deposit is then considered a eurodeposit.

See also EURODOLLAR.

EURODOLLAR

A Eurodollar is not some strange banknote. It is simply a U.S. dollar deposited in a bank outside the United States. Eurodollars are so called because they originated in Europe, but Eurodollars are really any dollars deposited in any part of the world outside the United States. They represent claims held by foreigners for U.S. dollars. Typically, these are time deposits ranging from a few days up to one year. These deposit accounts are extensively used abroad for financial transactions such as short-term loans, the purchase of dollar certificates of deposit, or the purchase of dollar bonds (*called Eurobonds*) often issued by U.S. firms for the benefit of their overseas operations. In effect, Eurodollars are an international currency. See also CREATION OF EURODOLLARS.

EUROLAND

See EUROPEAN ECONOMIC AND MONETARY UNION.

EUROMARKET DEPOSITS

Also called *eurodeposits*, Euromarket deposits are dollars deposited outside of the United States. Other important Eurocurrency deposits include the Euroyen, the Euromark, the Eurofranc, and the Eurosterling.

EUROMARKETS

Also called *Eurocurrency markets*, Euromarkets are offshore money and capital markets in which the currency of denomination is not the official currency of the country where the transaction takes place. They are the international markets that are engaged in the lending and borrowing of U.S. dollars and other major currencies outside their countries of origin to finance international trade and investment. The main financial instrument used in the Eurocurrency market for long-term investment purposes is the *Eurobond*. Despite its name, the market is not restricted to European currencies or financial centers. It began as the Eurodollar market in the late 1950s.

EURO.NM ALL SHARE INDEX

The EURO.NM all share index (http://www.euronm.com) is a pan-European grouping of regulated stock markets dedicated to high growth companies. EURO.NM member markets are Le Nouveau Marche (Paris Stock Exchange), Neuer Market (Deutsche Borse), EURO.NM Amsterdam (Amsterdam Exchanges), EURO.NM Belgium (Brussels Exchange), and Nuovo Mercata (Italian Exchange).

EURONOTE

Short- to medium-term unsecured debt security issued by MNCs outside the country of the currency it is denominated in.

EUROPEAN CENTRAL BANK

The European Central Bank (http://www.ecb.int/) is a new, fully independent institution, located in Frankfurt, Germany, created by the *European Economic and Monetary Union* that is charged with ensuring economic stability related to the *euro*. It is directed by a governing council made up of six members of the bank's executive board and governors from the central banks of the 11 countries participating in the euro.

See also EURO.

EUROPEAN COMMISSION

The European Commission (EC) (http://europa.euint/euro/) has exclusive responsibility for all legal and regulatory proposals governing the *European Economic and Monetary Union*. It also is in charge of monitoring economic developments in the European Union and of making policy recommendations to the *Economic and Finance Council* when necessary.

EUROPEAN COMMUNITY

Also called *European Economic Community (EEC)* or *common market*, European Community (EC) is the association of Western European countries formed in 1958 that has reduced costly political and economic rivalries, eliminated most tariffs among member nations, harmonized some fiscal and monetary policies, and broadly attempted to increase economic integration among them.

EUROPEAN CURRENCY UNIT

The European Currency Unit (ECU) was a basket of the currencies of the 15 members of the European Economic Community (EEC). It is weighted by the economic importance of each member country. Created by the *European Monetary System*, it serves a reserve currency numeraire. The weighting is based on the foreign currency in the ECU on a percentage relationship to the equivalent U.S. dollar. The objective is to keep a stable relationship in European currencies among members. It may be used as the numeraire for denomination of a number of financial instruments. International contracts, bank accounts, Eurobonds, and even traveler's checks are being denominated in ECUs. The ECU was replaced by the euro, which is being used by only the 11 nations that joined the European Economic and Monetary Union; the rate is 1 ECU to 1 euro.

EUROPEAN ECONOMIC AND MONETARY UNION

The European Economic and Monetary Union (EMU or Euroland), is the group of 11 countries that fixed their currencies to the *euro* (Austria, Belgium, Finland, France, Germany, Ireland, Italy, Luxembourg, the Netherlands, Portugal, and Spain).

EUROPEAN MONETARY SYSTEM

European Monetary System (EMS) is a mini-*IMF* system, formed in 1979 by 12 European countries, under which they agreed to maintain their exchange rates within an established range about fixed central rates in relation to one another. These central exchange rates are denominated in currency units per *European Currency Unit* (*ECU*). The EMS observes exchange rate fluctuations between member-nation currencies, controls inflation, and makes loans to member governments, primarily to serve the goal of *balance of payments* stability.

EUROPEAN PARLIAMENT

The European Parliament is the body that supervises the *European Union*. The citizens of all 15 member-states elect parliament members.

EUROPEAN TERMS

Foreign exchange quotations for the U.S. dollar, expressed as foreign currency price of one U.S. dollar. For example, 1.50 DM/$. This may also be called "German terms."

See also AMERICAN TERMS.

EUROPEAN UNION

The European Union (EU) is a group of 15 member countries (Austria, Belgium, Britain, Denmark, Finland, France, Germany, Greece, Ireland, Italy, Luxembourg, the Netherlands, Portugal, Spain, and Sweden), 11 of which are in the *European Economic and Monetary Union*. The EU has its own flag and anthem and celebrates Europe Day on May 9.

EVA

See ECONOMIC VALUE ADDED.

EXCHANGE AGIO

See FORWARD PREMIUM OR DISCOUNT.

EXCHANGE CONTROLS

Government regulations that limit outflows of funds from a country. These restrictions relate to access to foreign currency at the central bank and multiple exchange rates for different users.

EXCHANGE FUNDS

See SWAP FUNDS.

EXCHANGE RATE

See CURRENCY RISK; FOREIGN EXCHANGE RATE.

EXCHANGE RATE FORECASTING

See FOREIGN EXCHANGE RATE FORECASTING.

EXCHANGE RISK

See CURRENCY RISK.

EXCHANGE RISK ADAPTATION

Exchange risk adaptation is the strategy of structuring the MNC's activities to lessen the potential impact of unexpected changes in foreign exchange rate. This strategy includes all methods of *hedging* against exchange rate changes. In the extreme, exchange risk calls for protecting all liabilities denominated in foreign currency with equal-value, equal-maturity assets denominated in that foreign currency.

EXCHANGE RISK AVOIDANCE

Exchange risk avoidance is an MNC's strategy of attempting to escape foreign currency transactions. It includes: (1) eliminating dealings or activities that involve high currency risk and (2) charging higher prices when exchange risk seems to be greater.

EXCHANGE RISK TRANSFER

Exchange risk transfer is the strategy of transferring exchange risk to others. This strategy involves the use of an insurance policy or guarantee.

EX DOCK

A term used in delivery. The seller is responsible for all costs required to deliver the goods at the port of destination. Title to the goods passes to the buyer at the dock of the port of importation.

EXERCISE PRICE

Also called *strike price*, the price at which an *option* may be used to buy/sell foreign exchange.

EX FACTORY

A term used in delivery of goods. The goods are transferred to the buyer at the point of origin, the seller's factory. The seller is responsible for all costs of making the goods available at the factory. The buyer assumes all further expenses.

EX-IM BANK

See EXPORT-IMPORT BANK.

EXPATRIATES

See EXPROPRIATION.

EXPECTATIONS THEORY

The expectations theory of exchange rates states that the percentage difference between the forward rate and today's spot rate is equal to the expected change in the spot rate.

See also FORWARD RATES AS UNBIASED PREDICTORS OF FUTURE SPOT RATES.

EXPIRATION DATE

1. The last day that an *option* may be exercised into the underlying futures contract upon the exercise of the option.
2. The last day of trading for a futures contract.

EXPORT-IMPORT BANK

Also called as *EX-IM bank* or *Eximbank*, the Export-Import Bank (http://www.exim.gov) is a U.S. government agency that finances and facilitates for U.S. exports through credit risk protection and funding programs. The EX-IM bank was established in 1934 with the original intention to facilitate Soviet–American trade. It operates as an independent agency of the U.S. government and, as such, carries the full faith and credit of the United States. The EX-IM bank provides fixed-rate financing for U.S. export sales facing competition from foreign export financing agencies. Other programs provided make international factoring more feasible because they offer credit assurance alternatives that promise funding sources the security they need to agree to a deal. When the EX-IM bank is involved, the payor must be a foreign company buying from a U.S. company. Just as the EX-IM bank makes international commerce a realistic alternative for wary U.S. companies, it helps make international factoring as feasible as domestic factoring.

The EX-IM bank has nothing to do with imports, in spite of the name, but it plays a key role in determining the competitiveness of the U.S. among its trading partners because of the buyer credit programs, which is often a major component of an overseas customer's ability to finance and, therefore, to buy American products. EX-IM bank's willingness and ability to insure foreign private or sovereign buyers in any corner of the world often determines whether a U.S. supplier can offer competitive or acceptable terms to the foreign buyer. EX-IM bank states that its responsibilities are: (1) to assume most of the risks inherent in financing the production and sale of exports when the private sector is unwilling to assume such risks, (2) to provide funding to foreign buyers of U.S. goods and services when such funding is not available from the private sector, and (3) to help U.S. exporters meet officially supported and/or subsidized foreign credit competition. These roles fit into four functional categories: foreign loan guarantees, supplier credit working capital guarantees, direct loans to foreign buyers, and export credit insurance.

A. Guarantee Programs

The two most widely used guarantee programs are the following:

- *The Working Capital Guarantee Program.* This program encourages commercial banks to extend short-term export financing to eligible exporters. By providing a comprehensive guarantee that covers 90 to 100% of the loan's principal and interest, EX-IM bank's guarantee protects the lender against the risk of default by the exporter. It does not protect the exporter against the risk of nonpayment by the foreign buyer. The loans are fully collateralized by export receivables and export inventory and require the payment of guarantee fees to EX-IM bank. The export receivables are usually supported with export credit insurance or a letter of credit.
- *The Guarantee Program.* This program encourages commercial lenders to finance the sale of U.S. capital equipment and services to approved foreign buyers. The EX-IM bank guarantees 100% of the loan's principal and interest. The financed amount cannot exceed 85% of the contract price. This program is designed to finance products sold on a medium-term basis, with repayment terms generally between one and five years. The guarantee fees paid to EX-IM bank are determined by the repayment terms and the buyer's risk. EX-IM bank now offers a leasing program to finance capital equipment and related services.

B. Loan Programs

Two of the most popular loan programs are the following:

- *The Direct Loan Program.* Under the program, EX-IM bank offers fixed-rate loans directly to the foreign buyer to purchase U.S. capital equipment and services on a medium- or long-term basis. The total financed amount cannot exceed 85% of the contract price. Repayment terms depend upon the amount but are typically one to five years for medium-term transactions and seven to ten years for long-term transactions. EX-IM bank's lending rates are generally below market rates.
- *The Project Finance Loan Program.* The program allows banks, EX-IM bank, or a combination of each to extend long-term financing for capital equipment and related services for major projects. These are typically large infrastructure projects, such as power generation, whose repayment depends on project cash flows. Major U.S. corporations are often involved in these types of projects. The program typically requires a 15% cash payment by the foreign buyer and allows for guarantees of up to 85% of the contract amount. The fees and interest rates will vary depending on project risk.

C. Bank Insurance Programs

EX-IM bank offers several insurance policies to banks.

- *The Bank Letter of Credit Policy.* This policy enables banks to confirm letters of credit issued by foreign banks supporting a purchase of U.S. exports. Without this insurance, some banks would not be willing to assume the underlying commercial and political risk associated with confirming a letter of credit. The banks are insured up to 100% for sovereign (government) banks and 95% for all other banks. The premium is based on the type of buyer, repayment term, and country.
- *The Financial Institution Buyer Credit Policy.* Issued in the name of the bank, this policy provides insurance coverage for loans by banks to foreign buyers on a short-term basis. A variety of short-term and medium-term insurance policies are available to exporters, banks, and other eligible applicants. Basically, all the policies provide insurance protection against the risk of nonpayment by foreign buyers. If the foreign buyer fails to pay the exporter because of commercial reasons such as cash flow problems or insolvency, EX-IM bank will reimburse the exporter between 90 and 100% of the insured amount, depending upon the type of policy and buyer. If the loss is due to political factors, such as foreign exchange controls or war, EX-IM bank will reimburse the exporter for 100% of the insured amount. The insurance policies can be used by exporters as a marketing tool by enabling them to offer more competitive terms while protecting them against the risk of nonpayment. The exporter can also use the insurance policy as a financing tool by assigning the proceeds of the policy to a bank as collateral. Certain restrictions may apply to particular countries, depending upon EX-IM bank's experience, as well as the existing economic and political conditions.
- *The Small Business Policy.* This policy provides enhanced coverage to new exporters and small businesses. Firms with very few export credit sales are eligible for this policy. The policy will insure short-term credit sales (under 180 days) to approved foreign buyers. In addition to providing 95% coverage against commercial risk defaults and 100% against political risk, the policy offers lower premiums and no annual commercial risk loss deductible. The exporter can assign the policy to a bank as collateral.

- *The Umbrella Policy.* Issued to an "administrator," such as a bank, trading company, insurance broker, or government agency, the policy is administerd for multiple exporters and relieves the exporters of the administrative responsibilities associated with the policy. The short-term insurance protection is similar to the Small Business Policy and does not have a commercial risk deductible. The proceeds of the policy may be assigned to a bank for financing purposes.
- *The Multi-Buyer Policy.* Used primarily by the experienced exporter, the policy provides insurance coverage on short-term export sales to many different buyers. Premiums are based on an exporter's sales profile, credit history, terms of repayment, country, and other factors. Based upon the exporter's experience and the buyer's creditworthiness, EX-IM bank may grant the exporter authority to preapprove specific buyers up to a certain limit.
- *The Single-Buyer Policy.* This policy allows an exporter to selectively insure certain short-term transactions to preapproved buyers. Premiums are based on repayment term and transaction risk. There is also a medium-term policy to cover sales to a single buyer for terms between one and five years.

EX-IM bank, in addition to other federal support programs for export finance and promotion, can be viewed as a competitive weapon provided by the U.S. to help match export marketing advantages with those extended by foreign governments on behalf of their exporters and U.S. firms' foreign competition. Another advantage is that the EX-IM bank has a wealth of information on foreign buyers as a result of its insurance, guarantee, and lending activities. Information that has been given in confidence to the EX-IM bank will not be divulged; however, general information about the repayment habits of buyers insured or funded by EX-IM bank is available. You can call or fax Credit Services at EX-IM bank for further information. EX-IM bank's Washington headquarters are at 811 Vermont Avenue NW, Washington, D.C. 20571, and its toll-free number for general information is 1-800-565-3946, fax (202) 565-3380. There are five regional offices in New York, Miami, Chicago, Houston, and Los Angeles.

EXPOSURE NETTING

Exposure netting is the acceptance of open positions in two or more currencies that are considered to balance one another and therefore require no further internal or external *hedging*. Thus, exposures in one currency are offset with exposures in the same or another currency.

An *open* position exists when the firm has greater assets than liabilities (or greater liabilities than assets) in one currency. A *closed*, or *covered*, position exists when assets and liabilities in a currency are identical.

EXPROPRIATION

Expropriation is the forced seizure or takeover of the host government of property rights or assets owned by a foreigner or foreign corporation without compensation (or with inadequate compensation). Such an action is not in violation of international law if it is followed by prompt, adequate, and effective compensation. If not, it is called *confiscation*.

EXTRACTIVE FDI

A form of *foreign direct investment* (*FDI*) adopted by the MNC for the sole purpose of securing raw materials such as oil, copper, or other materials.

F

FACTOR

1. The basis on which a shipping charge is based such as a rate per mile the cargo is transported.
2. A firm that buys accounts receivable from exporters using short-term maturities of no longer than a year and then assumes responsibility for collecting the receivables. This usually involves no recourse, which means the factor must bear the risk of collection. Some banks and commercial finance companies factor (buy) accounts receivable. The purchase is made at a discount from the account's value. Customers remit either directly to the factor (notification basis) or indirectly through the seller.

FACTORING

Discounting without recourse an account receivable by an intermediate company called a *factor.* The exporter receives immediate (discounted) payment, and the factor receives eventual payment from the importer.

FADE-OUT

Fade-out is a host government policy toward *foreign direct investment (FDI)* that calls for progressive divestment of foreign ownership over time, ending with either complete local ownership or limited foreign ownership share. For example, a *joint venture* may have served the goal of helping a firm acquire local experience in the initial entry state but no longer serves this need at a later stage.

FAIR VALUE

1. The theoretical value of a security based on current market conditions. The fair value is such that no *arbitrage* opportunities exist.
2. Price negotiated at *arm's-length* between a willing buyer and a willing seller, each acting in his or her own best interest.
3. The *fair market value* of a multinational company's activities that is used as a basis to determine tax.
4. The "proper" value of the spread between the Standard & Poor's 500 futures and the actual S&P Index that makes no economic difference to investors whether they own the futures or the actual stocks that make up the S&P 500. Their buy and sell decisions will be driven by other factors. Through a complex formula using current short-term interest rates and the amount of time left until the futures contract expires, one can determine what the spread between the S&P futures and the cash "should be." The formula for determining the fair value

$$F = S[1 + (i - d)t/360]$$

where F = break-even futures price, S = spot index price, i = interest rate (expressed as a money-market yield), d = dividend rate (expressed as a money-market yield), and t = number of days from today's spot value date to the value date of the futures contract.

FAS

See FREE ALONGSIDE.

FASB NO. 8

See STATEMENT OF FINANCIAL ACCOUNTING STANDARDS NO. 8.

FASB NO. 52

See STATEMENT OF FINANCIAL ACCOUNTING STANDARDS NO. 52.

FDI

See FOREIGN DIRECT INVESTMENT.

FIAT MONEY

Fiat money is nonconvertible paper money backed only by full faith that the monetary authorities will not cheat (by issuing more money).

FINANCE DIRECTOR

The finance director is the financial executive responsible for the finance function of the company. The finance director may be a controller, treasurer, or Chief Financial Officer (CFO). A group finance director of an MNC typically reports to the Chief Executive Officer (CEO) and the member of the main board of directors with responsibility for leading the finance function and contributing actively to the overall strategy and development of the business. This involves a blend of hands-on operational involvement and high level influencing/negotiating with banks, venture capitals, and strategic alliance partners/suppliers.

FINANCIAL DERIVATIVE

A transaction, or contract, whose value depends on or, as the name implies, derives from the value of underlying assets such as stocks, bonds, mortgages, market indexes, or foreign currencies. One party with exposure to unwanted risk can pass some or all of that risk to a second party. The first party can assume a different risk from the second party, pay the second party to assume the risk, or, as is often the case, create a combination. The participants in derivatives activity can be divided into two broad types—dealers and end-users. Dealers include investment banks, commercial banks, merchant banks, and independent brokers. In contrast, the number of end-users is large and growing as more organizations are involved in international financial transactions. End-users include businesses; banks; securities firms; insurance companies; governmental units at the local, state, and federal levels; "supernational" organizations such as the World Bank; mutual funds; and both private and public pension funds. The objectives of end-users may vary. A common reason to use derivatives is so that the risk of financial operations can be controlled. Derivatives can be used to manage foreign exchange exposure, especially unfavorable exchange rate movements. Speculators and arbitrageurs can seek profits from general price changes or simultaneous price differences in different markets, respectively. Others use derivatives to *hedge* their position; that is, to set up two financial assets so that any unfavorable price movement in one asset is offset by favorable price movement in the other asset. There are five common types of derivatives: *options, futures, forward contracts, swaps*, and *hybrids*. The general characteristics of each are summarized in Exhibit 40. An important feature of derivatives is that the types of risk are not unique to derivatives and can be found in many other financial activities. The risks for derivatives are especially difficult to manage for two principal reasons: (1) the derivative products are complex, and (2) there are very real difficulties in measuring the risks associated derivatives.

It is imperative for financial officers of a firm to know how to manage the risks from the use of derivatives. Exhibit 40 compares major types of financial derivatives.

EXHIBIT 40
General Characteristics of Major Types of Financial Derivatives

Type	Market	Contract	Definition
Option	OTC or Organized Exchange	Custom* or Standard	Gives the buyer the right but *not* the obligation to buy or sell a specific amount at a specified price within a specified period
Futures	Organized Exchange	Standard	*Obligates* the holder to buy or sell at a specified price on a specified date
Forward	OTC	Custom	Same as futures
Swap	OTC	Custom	Agreement between the parties to make periodic payments to each other during the swap period
Hybrid	OTC	Custom	Incorporates various provisions of other types of derivatives

* Custom contracts vary and are negotiated between the parties with respect to their value, period, and other terms.

See also CURRENCY OPTION; FORWARD CONTRACT; FUTURES; OPTION; SWAPS.

FINANCIAL FUTURES

Financial futures are types of futures contracts in which the underlying commodities are financial assets. Examples are debt securities, foreign currencies, and market baskets of common stocks.

FINANCIAL MARKETS

The financial markets are composed of *money markets* and *capital markets*. Money markets, also called *credit markets*, are the markets for debt securities that mature in the short term (usually less than one year). Examples of money-market securities include U.S. Treasury bills, government agency securities, bankers' acceptances, commercial paper, and negotiable certificates of deposit issued by government, business, and financial institutions. The money-market securities are characterized by their highly liquid nature and a relatively low default risk. Capital markets are the markets in which long-term securities issued by the government and corporations are traded. Unlike the money market, both debt instruments (bonds) and equity share (common and preferred stocks) are traded. Relative to money-market instruments, those of the capital market often carry greater default and market risks but return a relatively high yield in compensation for the higher risks. The New York Stock Exchange, which handles the stock of many of the larger corporations, is a prime example of a capital market. The American Stock Exchange and the regional stock exchanges are yet another example. These exchanges are organized markets. In addition, securities are traded through the thousands of brokers and dealers on the over-the-counter (or unlisted) market, a term used to denote an informal system of telephone contacts among brokers and dealers. There are other markets including (1) the foreign exchange market, which involves international financial transactions between the U.S. and other countries; (2) the commodity markets which handle various commodity futures; (3) the mortgage market that handles various home loans; and (4) the insurance, shipping, and other markets handling short-term credit accommodations in their operations. A primary market refers to the market for new issues, while a secondary market is a market in which previously issued, "secondhand" securities are exchanged. The New York Stock Exchange is an example of a secondary market.

FISHER EFFECT

The Fisher effect, named after Irving Fisher, states that that nominal interest rates (r) in each country equal the required real rate of return (R) *plus* a premium for expected inflation (I) over the period of time for which the funds are to be lent (i.e., $r = R + I$). To be precise,

$$\begin{aligned} 1 + \text{Nominal rate} &= (1 + \text{Real rate})(1 + \text{Expected inflation rate}) \\ 1 + r &= (1 + R)(1 + I) \end{aligned}$$

or

$$r = R + I + RI$$

which is approximated as $r = R + I$. The theory implies that countries with higher rates of inflation have higher interest rates than countries with lower rates of inflation. *Note*: The equation requires a forecast of the future rate of inflation, not what inflation has been.

EXAMPLE 47

If you have $100 today and loan it to your friend for a year at a nominal rate of interest of 11.3%, you will be paid $111.30 in one year. But if during the year inflation (prices of goods and services) goes up by 5%, it will take $105 at year end to buy the same goods and services that $100 purchased at the start of the year. Then the increase in your purchasing power over the year can be quickly found by using the approximation $r = R + \text{I}$:

$$\begin{aligned} 11.3\% &= R + 5\% \text{ or } r = 11.3\% - 5\% = 6.3\% \\ \text{To be precise, use } r &= R + I + RI: \\ 11.3\% &= R + 0.05 + 0.05R \\ R &= 0.06 = 6\% \end{aligned}$$

In other words, at the new higher prices, your purchasing power will have increased by only 6%, although you have $11.30 more than you had at the beginning of the year. To see why, suppose that at the start of the year, one unit of the market basket of goods and services cost $1, so you could buy 100 units with your $100. At year-end, you have $11.30 more, but each unit now costs $1.05 (with the 5% rate of inflation). This means that you can purchase only 106 units ($111.30/$1.05), representing a 6% increase in real purchasing power.

The generalized version of the Fisher effect claims that real returns are equalized across countries through arbitrage—that is, r_h and r_f where the subscripts h and f are home and foreign real rates. If expected real returns were higher in one country than another, capital would flow from the second to the first currency. This process of arbitrage would continue, in the absence of government intervention, until expected real returns were equalized. In equilibrium, then, with no government interference, it should follow that nominal *interest rate differential* will approximately equal the anticipated inflation rate differential, or

$$\frac{1 + r_h}{1 + r_f} = \frac{1 + I_h}{1 + I_h} \qquad \text{(Equation 1)}$$

where r_h and r_f are the nominal home and foreign currency interest rates, respectively. If these rates are relatively small, then this exact relationship can be approximated by

$$r_h - r_f = I_h - I_f \qquad \text{(Equation 2)}$$

Note: Equation 1 can be converted into Equation 2 by subtracting 1 from both sides and assuming that r_h and r_f are relatively small. This generalized version of the Fisher effect says that currencies with high rates of inflation should bear higher interest rates than currencies with lower rates of inflation.

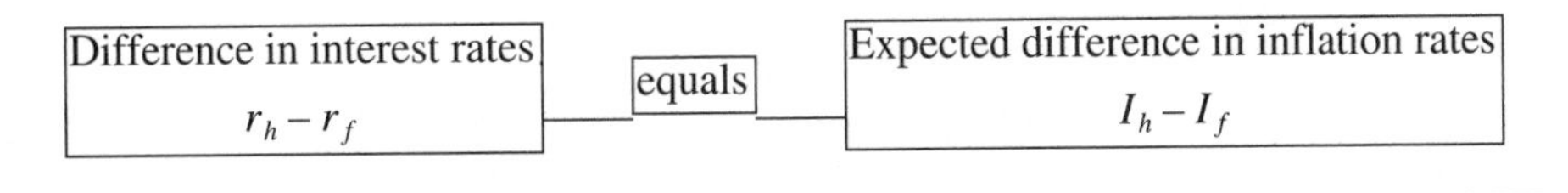

EXAMPLE 48

If inflation rates in the United States and the United Kingdom are 4% and 7%, respectively, the Fisher effect says that nominal interest rates should be about 3% higher in the United Kingdom than in the United States.

A graph of Equation 2 is shown in Exhibit 41. The horizontal axis shows the expected difference in inflation rates between the home country and the foreign country, and the vertical axis shows the interest differential between the two countries for the same time period. The parity line shows all points for which $r_h - r_f = I_h - I_f$. Point A, for example, is a position of equilibrium, as the 2% higher rate of inflation in the foreign country ($r_h - r_f = -2\%$) is just offset by the 2% lower home currency interest rate ($I_h - I_f = -2\%$). At point B, however, where the real rate of return in the home country is 1% higher than in the foreign country (an inflation differential of 3% versus an interest differential of only 2%), funds should flow from the foreign country to the home country to take advantage of the real differential. This flow will continue until expected real returns are equal.

EXHIBIT 41
The Fisher Effect

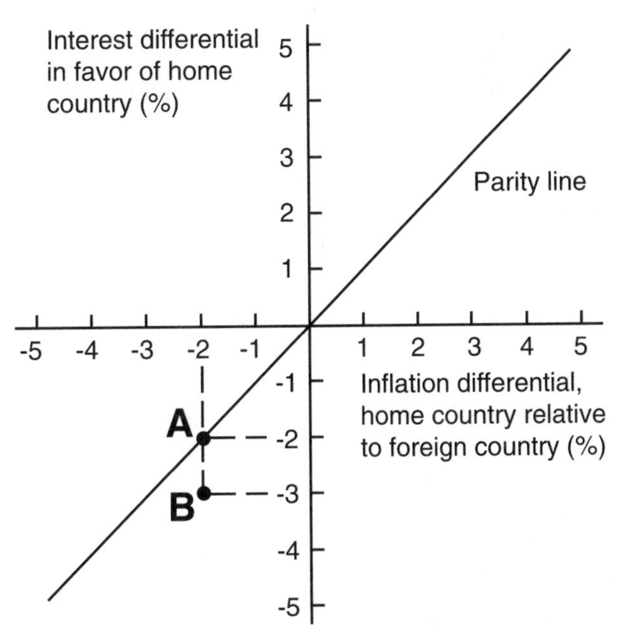

FIXED EXCHANGE RATES

An international financial arrangement under which the values of currencies in terms of other currencies are fixed by the governments involved and by governmental intervention in the foreign exchange markets.

See also FOREIGN EXCHANGE RATE.

FLEXIBLE EXCHANGE RATES

See FLOATING EXCHANGE RATES.

FLOATING EXCHANGE RATES

Also called *flexible exchange rates*, floating exchange rates are a system in which the values of currencies in terms of other currencies are determined by the supply of and demand for the currencies in foreign exchange markets. Arrangements may vary from *free float*, i.e., absolutely no government intervention, to *managed float*, i.e., limited but sometimes aggressive government intervention, in the foreign exchange market.

See also FOREIGN EXCHANGE RATE.

FOB

See FREE ON BOARD.

FOREIGN BOND

A foreign bond is a bond issued by a foreign borrower on a foreign capital market just like any domestic (local) firm. The bond must of course be denominated in a local currency—the currency of the country in which the issue is sold. The terms must conform to local custom and regulations. A foreign bond is the simplest way for an MNC to raise long-term debt for its foreign expansion. A bond issued by a German corporation, denominated in dollars, and sold in the U.S. in accordance with SEC and applicable state regulations, to U.S. investors by U.S. investment bankers, would be a foreign bond. Except for the foreign origin of the borrower, this bond will be no different from those issued by equivalent U.S. corporations. Foreign bonds have nicknames: foreign bonds sold in the U.S. are *Yankee bonds*, foreign bonds sold in Japan are *Samurai bonds*, and foreign bonds sold in the United Kingdom are *Bulldogs*. Exhibit 42 below specifically reclassifies foreign bonds from a U.S. investor's perspective.

EXHIBIT 42
Foreign Bonds to U.S. Investors

	Sales	
Issuer	**In the U.S.**	**In Foreign Countries**
Domestic	Domestic bonds	Eurobonds
Foreign	Yankee bonds	Foreign bonds Eurobonds

FOREIGN BOND MARKET

The foreign bond market is the market for long-term loans to be raised by MNCs for their foreign expansion. It is that portion of the *domestic* market for bond issues floated by foreign companies or governments. In contrast, the *Eurobond market* is an international market for long-term debt.

See also EUROBONDS; FOREIGN BOND.

FOREIGN BRANCHES

A foreign branch is a legal and operational part of the parent bank. Creditors of the branch have full legal claims on the bank's assets as a whole and, in turn, creditors of the parent bank have claims on its branches' assets. Deposits of both foreign branches and domestic branches are considered total deposits of the bank, and reserve requirements are tied to these total deposits. Foreign branches are subject to two sets of banking regulations. First, as part of the parent bank, they are subject to all legal limitations that exist for U.S. banks. Second, they are subject to the regulation of the host country. Domestically, the OCC is the overseas regulator and examiner of national banks, whereas state banking authorities and the Federal Reserve Board share the authority for state-chartered member banks. Granting power to open a branch overseas resides with the Board of Governors of the Federal Reserve System. As a practical matter, the Federal Reserve System and the OCC dominate the regulation of foreign branches. The attitudes of host countries toward establishing and regulating branches of U.S. banks vary widely.

Typically, countries that need capital and expertise in lending and investment welcome the establishment of U.S. bank branches and allow them to operate freely within their borders. Other countries allow the establishment of U.S. bank branches but limit their activities relative to domestic banks because of competitive factors. Some foreign governments may fear that branches of large U.S. banks might hamper the growth of their country's domestic banking industry. As a result, growing nationalism and a desire for locally controlled credit have slowed the expansion of American banks abroad in recent years. The major advantage of foreign branches is a worldwide name identification with the parent bank. Customers of the foreign branch have access to the full range of services of the parent bank, and the value of these services is based on the worldwide value of the client relationship rather than only the local office relationship. Furthermore, deposits are more secure, having their ultimate claim against the much larger parent bank and not merely the local office. Similarly, legal loan limits are a function of the size of the parent bank and not of the branch. The major disadvantages of foreign branches are the cost of establishing them and the legal limits placed on the activities in which they may engage.

FOREIGN CREDIT INSURANCE ASSOCIATION

The Foreign Credit Insurance Association (FCIA) is a private U.S. insurance association that insures exporters in conjunction with the *Eximbank*. It offers a broad range of short-term and medium-term insurance policies to protect losses from political and commercial risks. For providing the insurance, the FCIA charges premiums based on the types of buyers and countries and the terms of payment.

FOREIGN CURRENCY FUTURES

A futures contract promises to deliver a specified amount of foreign currency by some given future date. Foreign currency futures differ from forward contracts in a number of significant ways, although both are used for trading, hedging, and speculative purposes. Participants include MNCs with assets and liabilities denominated in foreign currency, exporters and importers, speculators, and banks. Foreign currency futures are contracts for future delivery of a specific quantity of a given currency, with the exchange rate fixed at the time the contract is entered. Futures contracts are similar to *forward contracts* except that they are traded on organized futures exchanges and the gains and losses on the contracts are settled each day.

Like *forward contracts*, a foreign currency futures contract is an agreement calling for future delivery of a standard amount of foreign exchange at a fixed time, place, and price. It is similar to futures contracts that exist for commodities (e.g., hogs, cattle, lumber), for interest-bearing deposits, and for gold. Unlike forward contracts, futures are traded on organized exchanges with specific rules about the terms of the contracts and with an active secondary market.

A. Futures Markets

In the United States the most important marketplace for foreign currency futures is the International Monetary Market (IMM) of Chicago, organized in 1972 as a division of the Chicago Mercantile Exchange. Since 1985, contracts traded on the IMM have been interchangeable with those traded on the Singapore International Monetary Exchange (SIMEX).

Most major money centers have established foreign currency futures markets during the past decade, notably in New York (New York Futures Exchange, a subsidiary of the New York Stock Exchange), London (London International Financial Futures Exchange), Canada, Australia, and Singapore. So far, however, none of these rivals has come close to duplicating the trading volume of the IMM.

B. Contract Specifications

Contract specifications are defined by the exchange on which they are traded. The major features that must be standardized are the following:

- *A specific sized contract.* A German mark contract is for DM125,000. Consequently, trading can be done only in multiples of DM125,000.
- *A standard method of stating exchange rates.* American terms are used; that is, quotations are the dollar cost of foreign currency units.
- *A standard maturity date.* Contracts mature on the third Wednesday of January, March, April, June, July, September, October, or December. However, not all of these maturities are available for all currencies at any given time. "Spot month" contracts are also traded. These are not spot contracts as that term is used in the interbank foreign exchange market, but are rather short-term futures contracts that mature on the next following third Wednesday, that is, on the next following standard maturity date.
- *A specified last trading day.* Contracts may be traded through the second business day prior to the Wednesday on which they mature. Therefore, unless holidays interfere, the last trading day is the Monday preceding the maturity date.
- *Collateral.* The purchaser must deposit a sum as an *initial margin* or collateral. This is similar to requiring a *performance bond* and can be met by a letter of credit from a bank, Treasury bills, or cash. In addition, a *maintenance margin* is required. The value of the contract is *marked-to-market* daily, and all changes in value are paid in cash daily. The amount to be paid is called the *variation margin.*

EXAMPLE 49

The initial margin on a £62,500 contract may be US$ 3,000, and the maintenance margin US$ 2,400. The initial US$ 3,000 margin is the initial equity in your account. The buyer's equity increases (decreases) when prices rise (fall). As long as the investor's losses do not exceed US$ 600 (that is, as long as the investor's equity does not fall below the maintenance margin, US$ 2,400), no margin call will be issued to him or her. If his or her equity, however, falls below US$ 2,400, he or she must add variation margin that restores his or her equity to US$ 3,000 by the next morning.

- *Settlement.* Only about 5% of all futures contracts are settled by the physical delivery of foreign exchange between the buyer and seller. Most often, buyers and sellers offset their original position prior to delivery date by taking an opposite position. That is, if one had bought a futures contract, that position would be closed out by selling a futures contract for the same delivery date. The complete buy/sell or sell/buy is called a *round turn*.

- *Commissions*. Customers pay a commission to their broker to execute a round turn and only a single price is quoted. This practice differs from that of the interbank market, where dealers quote a bid and an offer and do not charge a commission.
- *Clearing house as counterparty*. All contracts are agreements between the client and the exchange clearing house, rather than between the two clients. Consequently, clients need not worry that a specific counterparty in the market will fail to honor an agreement.

Currency futures contracts are currently available in over ten currencies including the British pound, Canadian dollar, Deutsche mark, Swiss franc, Japanese yen, and Australian dollar. The IMM is continually experimenting with new contracts. Those that meet the minimum volume requirements are added and those that do not are dropped. The number of contracts outstanding at any one time is called the *open interest*. Contract sizes are standardized according to amount of foreign currency—for example, £62,500, C$100,000, SFr 125,000. Exhibit 43 shows contract specifications. Leverage is high; margin requirements average less than 2% of the value of the futures contract. The leverage assures that investors' fortunes will be decided by tiny swings in exchange rates. The contracts have minimum price moves, which generally translate into about S10 to S12 per contract. At the same time, most exchanges set daily price limits on their contracts that restrict the maximum price daily move. When these limits are reached, additional margin requirements are imposed and trading may be halted for a short time. Instead of using the bid–ask spreads found in the interbank market, traders charge commissions. Though commissions will vary, a *round trip*—that is, one buy and one sell—costs as little as $15. The low cost, coupled with the high degree of leverage, has provided a major incentive for speculators to participate in the market.

EXHIBIT 43
Chicago Mercantile Exchange Foreign Currency Futures Specifications*

	Austrian Dollar	British Pound	Canadian Dollar	Deutsche Mark	Swiss Franc	French Franc	Japanese Yen
Symbol	AD	BP	CD	DM	SF	FR	JY
Contract size	A$100,000	£62,500	C$100,000	DM125,000	SFr125,000	FFr500,000	¥12,500,000
Margin requirements							
Initial	$1,148	$1,485	$608	$1,755	$2,565	$1,755	$4,590
Maintenance	$850	$1,100	$450	$1,300	$1,900	$1,300	$3,400
Minimum Price Change	0.00001 (1 pt.)	0.0002 (2 pts.)	0.0001 (1 pt.)	0.0001 (1 pt.)	0.0001 (1 pt.)	0.00002 (2 pts.)	0.000001 (1 pt.)
Value of 1 point	$10.00	$6.25	$10.00	$12.50	$12.50	$12.50	$10.00
Months traded	January, March, April, June, July, September, October, December, and spot month						
Last day of trading	The second business day immediately preceding the third Wednesday of the delivery month						
Trading Hours	7:20 A.M.–2:00 P.M. (Central Time)						

Note: Contract specifications are also available for currencies such as Brazilian real, Mexican peso, Russian ruble, and New Zealand dollar.

Source: Adapted from *Contract Specifications for Currency Futures and Options, Chicago Mercantile Exchange*, May 2000. (http://www.cme.com/clearing/spex/cscurrency.htm).

C. Futures Contracts on Euros

The Chicago Mercantile Exchange (CME) has developed futures contracts on euros so that MNCs can easily hedge their positions in euros, as shown in Exhibit 44. U.S.-based MNCs commonly consider the use of the futures contract on the euro with respect to the dollar (column 2). However, there are also futures contracts available on cross-rates between the euro and the British pound (column 3), the euro and the Japanese yen (column 4), and the euro and the Swiss franc (column 5). Settlement dates on all of these contracts are available in March, June, September, and December. The futures contracts on cross-rates allow for easy hedging by foreign subsidiaries that wish to exchange euros for widely used currencies other than the dollar.

EXHIBIT 44
Futures Contracts on Euros

	Euro/U.S.$	Euro/Pound	Euro/Yen	Euro/Swiss franc
Ticker Symbol	EC	RP	RY	RF
Trading Unit	125,000 euros	125,000 euros	125,000 euros	125,000 euros
Quotation	$ per euro	Pounds per euro	Yen per euro	SF per euro
Last Day of Trading	Second business day before third Wednesday of the contract month			

D. Reading Newspaper Quotations

Futures trading on the IMM in Japanese yens for a Tuesday was reported as shown in Exhibit 45.

EXHIBIT 45
Foreign Currency Futures Quotations

						Lifetime		Open
	Open	High	Low	Settle	Change	High	Low	Interest
	JAPAN YEN (CME)—12.5 million yen; $ per yen (.00)							
Sept	.9458	.9466	.9386	.9389	−.0046	.9540	.7945	73,221
Dec	.9425	.9470	.9393	.9396	−.0049	.9529	.7970	3,455
Mar	. ..	. ...	... .	.9417	−.0051	.9490	.8700	318
	Est vol 28,844; vol Wed 36,595; open int 77,028 + 1.820							

The head, JAPAN YEN, shows the size of the contract (12.5M yen) and states that the prices are stated in US$ cents. The June 2000 contract had expired more than a month previously, so the three contracts being traded on July 29, 2000, are the September 2000, December 2000, and the March 2001 contracts. Detailed descriptions of the quotations follow:

1. In each row, the first four prices relate to trading on Thursday, July 29—the price at the start of trading (open), the highest and lowest transaction price during the day, and the settlement price ("Settle"), which is representative of the transaction prices around the close. The settlement (or closing) price is the basis of marking to market.
2. The column "Change" contains the change of today's settlement price relative to yesterday's. For instance, on Thursday, July 29, the settlement price of the September contract dropped by 0.0046 cents, implying that a holder of a purchase contract

has lost 12.5m × (0.0046/100) = US$ 575 per contract and that a seller has made US$ 575 per contract.

3. The next two columns show the extreme (highest and lowest) prices that have been observed over the contract's trading life. For the March contract, the "High-Low" range is narrower than for the older contracts, because the March contract has been trading for only a little more than a month.
4. "Open Interest" refers to the number of contracts still in effect at the end of the previous day's trading session. Each unit represents a buyer *and* a seller who still have a contract position. Notice how most of the trading is in the nearest-maturity contract. Open interest in the March'01 contract is minimal, and there has not even been any trading that day. (There are no open, high, and low data.) The settlement price for the March'01 contract has been set by the CME on the basis of bid–ask quotes.
5. The line below the price information gives an estimate of the volume traded that day and the previous day (Wednesday). Also shown are the total open interest (the total of the right column) across the three contracts, and the change in open interest from the prior trading day.

E. Currency Futures Quotations Reported Online

Exhibit 46 displays a currency quotation from Commodities, Charts & Quotes—Free (http://www.tfc-charts.w2d.com).

EXHIBIT 46

Commodity Futures Price Quotes For

(2) **CME Australian Dollar** (1)

(Price quotes for this commodity delayed at least 10 minutes as per exchange requirements)

Click here to refresh data

(3)

Month	(4) Session (5)				(6)	(7)		Prior Day			Options
Click for chart	Open	High	Low	Last	Time	Sett	Chg	Sett	Vol	O.Int	
Dec 99	6362	6410	6350	DN 6386	15:09	6388	+8	6378	82	22197	Call Put
Mar 00	6375	6412	6370	UP 6375	08:31	6398	-13	6388	1	112	Call Put
Jun 00	-	-	-	UC 6398	12:43	6408	-	6398	2	11	Call Put
Sep 00	-	-	-	UC 6408	09:28	6418	-	6408	-	1	Call Put

Explanatory Notes:

(1) The name of the currency, Australian Dollar.
(2) The name of the exchange, the Chicago Mercantile Exchange.
(3) The delivery date (closing date for the contract).
(4) The opening, high, and low prices for the trading day. The trading unit is 100,000 Australian Dollars. The price is $10 per point. Therefore, the settlement price for the trading unit is $63,620, and each Australian Dollar being delivered in December is worth $.6362.

(5) The last price at which the contract traded.
(6) The "official" daily closing price of a futures contract, set by the exchange for the purpose of settling margin accounts.
(7) The change in the price from the prior settlement price to the last price quoted.

F. Foreign Currency Futures Versus Forward Contracts

Foreign currency futures contracts differ from forward contracts in a number of important ways. Nevertheless, both futures and forward contracts are used for the same commercial and speculative purposes. Exhibit 47 provides a comparison of the major features and characteristics of the two instruments.

EXHIBIT 47
Basic Differences of Foreign Currency Futures and Forward Contracts

Characteristics	Foreign Currency Futures	Forward Contracts
Trading and location	In an organized exchange	By telecommunications networks
Parties involved	Unknown to each other	In direct contact with each other in setting contract specifications.
Size of contract	Standardized	Any size individually tailored
Maturity	Fixed	Delivery on any date
Quotes	In American terms	In European terms
Pricing	Open outcry process	By bid and offer quotes
Settlement	Made daily via exchange clearing house; rarely delivered	Normally delivered on the date agreed
Commissions	Brokerage fees for buy and sell (roundtrip)	Based on bid–ask spread
Collateral and margin	Initial margin required	No explicit margin specified
Regulation	Highly regulated	Self-regulating
Liquidity and volume	Liquid but low volume	Liquid but large volume
Credit risk	Low risk due to the exchange clearing house involved	Borne by each party

The following example illustrates how currency futures contracts are used for hedging purposes and gains and losses calculations, and how they compare with currency options.

EXAMPLE 50

TDQ Corporation must pay its Japanese supplier ¥125 million in three months. The firm is contemplating two alternatives:

1. Buying 20 yen call options (contract size is ¥6.25 millions) at a strike price of $0.00900 in order to protect against the risk of rising yen. The premium is $0.00015 per yen.
2. Buying 10 3-month yen futures contracts (contract size is ¥12.5 million) at a price of $0.00840 per yen. The *current* spot rate is $0.008823/¥. The firm's CFO believes that the most likely rate for the yen is $0.008900, but the yen could go as high as $0.009400 or as low as $0.008500.

Note: In all calculations, the current spot rate $0.008823/¥ is irrelevant.

1. For the call options, TDQ must pay a call premium of $\$0.00015 \times 125{,}000{,}000 = \$18{,}750$. If the yen settles at its minimum value, the firm will not exercise the option and it loses the entire call premium. But if the yen settles at its maximum value of $0.009400, the firm will

exercise at \$0.009000 and earn \$0.0004/¥ for a total gain of \$0.0004 × 125,000,000 = \$50,000. TDQ's net gain will be \$50,000 – \$18,750 = \$31,250.

2. By using a futures contract, TDQ will lock in a price of \$0.008940/¥ for a total cost of \$0.008940 × 125,000,000 = \$992,500. If the yen settles at its minimum value, the firm will lose \$0.008940 – \$0.008500 = \$0.000440/¥ (remember the firm is buying yen at \$0.008940, when the spot price is only \$0.008500), for a total loss on the futures contract of \$0.00044 × 125,000,000 = \$55,000. But if the yen settles at its maximum value of \$0.009400, the firm will earn \$0.009400 – \$0.008940 = \$0.000460/¥, for a total gain of \$0.000460 × 125,000,000 = \$57,500.

Exhibits 48 and 49 present profit and loss calculations on both alternatives and their corresponding graphs.

EXHIBIT 48
Profit or Loss on TDQ's Options and Futures Positions

Contract size:	125,000,000	Yens
Expiration date:	3.0	months
Exercise or strike price:	0.00900	\$/Yen
Premium or option price:	0.00015	\$/Yen

(1) CALL OPTION

Ending spot rate (\$/Yen)	0.00850	0.00894	0.00915	0.00940	0.00960
Payments					
Premium	(18,750)	(18,750)	(18,750)	(18,750)	(18,750)
Exercise cost	0	0	(1,125,000)	(1,125,000)	(1,125,000)
Receipts					
Spot sale of Yen	0	0	1,143,750	1,175,000	1,200,000
Net (\$)	(18,750)	(18,750)	0	31,250	56,250

(2) FUTURES	**Lock-in price =**		**0.008940\$/Yen**		
Receipts	1,062,500	1,117,500	1,143,750	1,175,000	1,200,000
Payments	(1,117,500)	(1,117,500)	(1,117,500)	(1,117,500)	(1,117,500)
Net (\$)	(55,000)	0	26,250	57,500	82,500

Exhibit 50 provides a comparison between a futures contract and an option contract.

See CURRENCY OPTION; FORWARD CONTRACT; FUTURES.

FOREIGN CURRENCY OPERATIONS

Also called *foreign-exchange market intervention*, foreign currency operations involve purchase or sale of the currencies of other countries by a central bank in order to influence foreign exchange rates or maintaining orderly foreign exchange markets.

EXHIBIT 49
TDQ's Profit (Loss) on Options and Futures Positions

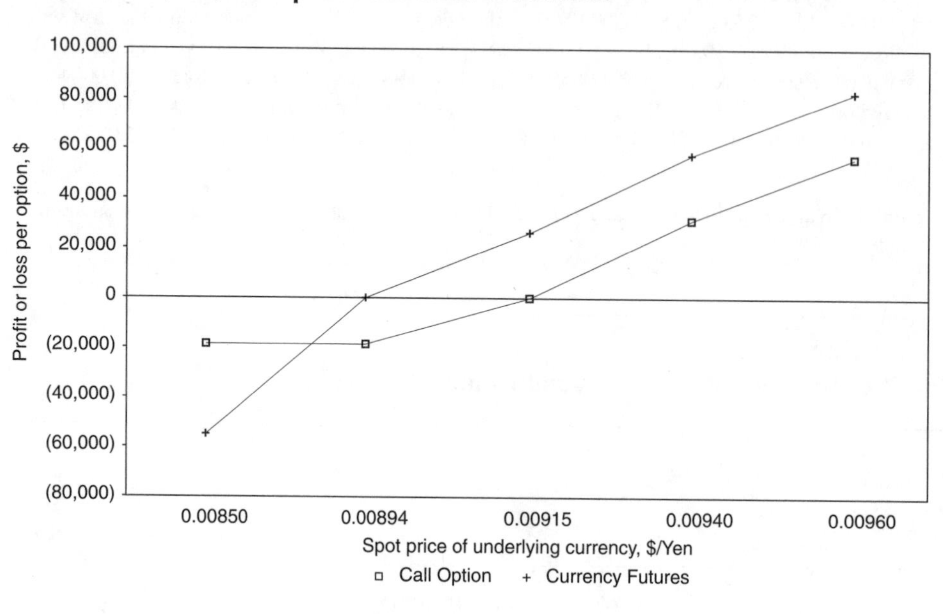

EXHIBIT 50
Currency Futures Versus Currrency Options

Futures	Options
A futures contract is most valuable when the quantity of foreign currency being hedged is *known*.	An options contract is most valuable when the quantity of foreign currency is unknown.
A futures contract provides a "two-sided" hedge against currency movements.	An options contract enables the hedging of a "one-sided" risk either with a call or with a put.
A buyer of a currency futures contract *must* take delivery.	A buyer of a currency options contract has the right (*not* the obligation) to complete the contract.

FOREIGN CURRENCY OPTION

A financial contract that gives the holder the right (but not the obligation) to exercise it to purchase/sell a given amount of foreign currency at a specified price during a fixed time period. Contract specifications are similar to those of *futures contracts* except that the option requires a premium payment to purchase (and it does not have to be exercised).
See also CURRENCY OPTION.

FOREIGN DIRECT INVESTMENT

Foreign direct investment (FDI) involves ownership of a company in a foreign country. In exchange for the ownership, the investing company usually transfers some of its financial, managerial, technical, trademark, and other resources to the foreign country. In the process, productive activities in different countries come under the ownership control of a single firm (MNC). It is the control aspect which distinguishes FDI from its near relations such as exporting, *portfolio investments*, and licensing. Such investment may be financed in many ways: the parent firm can transfer funds to the new affiliate and issue ownership shares to itself; the parent firm can borrow all of the required funds locally in the new country to pay for ownership; and many combinations of these and other alternatives can be adopted. The key point is that no international transfer of funds is necessary; it requires only transfer of ownership to the investing MNC. Types of foreign direct investments include *extractive, market-serving*, horizontal, vertical, agricultural, industrial, and service industries. *Horizontal FDI* is FDI in the same industry in which a firm operates at home, while *vertical FDI* is FDI in an industry that provides inputs for a firm's domestic operations. FDI is distinguished from *portfolio investments* that are made to earn investment income or capital gains. Exhibit 51 presents some basic reasons for FDI.

EXHIBIT 51
Strategic Reasons for Foreign Direct Investment

Demand Side	Supply Side
To serve a portfolio of markets and to explore new markets	To lower production and delivery costs
To enter an export market closed by restrictions such as a quota or tariff	To acquire a needed raw material
To establish a local presence	To do offshore assembly
To accommodate "buy national" regulations	To respond to rivals' threats
To gain visibility as a local firm, employing local workforce, paying local taxes, etc.	To build a "portfolio" of manufacturing resources

See also PORTFOLIO INVESTMENTS.

FOREIGN EXCHANGE

Foreign exchange (FOREX) is not simply currency printed by a foreign country's central bank. Rather, it includes such items as cash, checks (or drafts), wire transfers, telephone transfers, and even contracts to sell or buy currency in the future. Foreign exchange is really any financial transaction that fulfills payment from one currency to another. The most common form of foreign exchange in transactions between companies is the *draft* (denominated in a foreign currency). The most common form of foreign exchange in transactions between banks is the telephone transfer. A *foreign exchange market* is available for trading foreign exchanges.

FOREIGN EXCHANGE ARBITRAGE

Foreign exchange arbitrage involves simultaneous contracting in two or more foreign exchange markets to buy and sell foreign currency, profiting from exchange rate differences *without* suffering *currency risk*. Foreign exchange arbitrage may be *two-way (simple), three-way (triangular)*, or *intertemporal*; and it is generally undertaken by large commercial banks that can exchange large quantities of money to exploit small rate differentials.

See also INTERTEMPORAL ARBITRAGE; SIMPLE ARBITRAGE; TRIANGULAR ARBITRAGE.

FOREIGN EXCHANGE CONTRACT

See FUTURES; FORWARD CONTRACT.

FOREIGN EXCHANGE HEDGING

Foreign exchange hedging involves protecting against the possible impact of exchange rate changes on the firm's business by balancing foreign currency assets with foreign currency liabilities. Elimination of all foreign exchange risk is not necessarily the objective of a financial manager. Risk is a two-way street; gains are possible as well as losses. If gains from exchange-rate fluctuations appear more likely than losses, then it may make sense to bear the currency risk (that is, retain the exchange-rate exposure) so that gains may be realized. Another important consideration is that elimination of exchange risk entails a cost. In a cost-benefit analysis, elimination of all exchange risk may not be beneficial, while elimination of part of the risk may be. Exchange risk may be neutralized or hedged by a change in the asset and liability position in the foreign currency. An exposed asset position can be hedged or covered by creating a liability of the same amount and maturity denominated in the foreign currency. An exposed liability position can be covered by acquiring assets of the same amount and maturity in the foreign currency.

The objective, in the hedging strategy, is to have zero net asset position in the foreign currency. This eliminates exchange risk, because the loss (gain) in the liability (asset) is exactly offset by the gain (loss) in the value of the asset (liability) when the foreign currency appreciates (depreciates). Two popular forms of hedge are the *money-market hedge* and the *forward market hedge*. In both types of hedge the *amount* and the *duration* of the asset (liability) positions are matched. Many MNCs take long or short positions in the foreign exchange market, not to speculate, but to offset an existing foreign currency position in another market. Note: The forward market hedge is not adequate for some types of exposure. If the foreign currency asset or liability position occurs on a date for which forward quotes are not available, then the forward hedge cannot be accomplished. In these situations, MNCs may have to rely on other techniques such as *leading* and *lagging*.

See also BALANCE SHEET HEDGING; CURRENCY RISK MANAGEMENT; FORWARD MARKET HEDGE; HEDGE; LEADING AND LAGGING; MONEY-MARKET HEDGE.

FOREIGN EXCHANGE MARKET

A foreign exchange (FOREX) market is a market where foreign exchange transactions take place, that is, where different currencies are bought and sold. Or, more broadly, a foreign exchange market is a market for the exchange of financial instruments denominated in different currencies. The most common location for foreign exchange transactions is a commercial bank, which agrees to "make a market" for the purchase and sale of currencies other than the local one. This market is not located in any one place, most transactions being conducted by telephone, wire service, or cable. It is a global network of banks, investment banks, and brokerage houses that makeup an electronically linked infrastructure servicing international corporations, banks, and investment funds. FOREX trading follows the sun around the world—starting in Tokyo, the market activity moves through to London, the last banking center in Europe, before traveling on to New York and finally returning to Japan via Sydney. The interbank market has three sessions of trading. The first begins on Sunday at 7:00 p.m., NYT, which is the Asia session. The second is the European (London) session, which begins at approximately 3:00 a.m.; and the third and final session is the New York, which begins at approximately 8:00 a.m. and ends at 5:00 p.m. The majority of all trading occurs during the London session and the first half of the New York session. As a result, buyers and sellers are available 24 hours a day. Investors can respond to currency fluctuations caused by economic, social, and political events at the time they occur—day or night.

The functions of the foreign exchange market are basically threefold: (1) to transfer purchasing power, (2) to provide credit, and (3) to minimize exchange risk. The foreign exchange market is dominated by:

1. Large commercial banks (thus often called *bank market*);
2. Nonbank foreign exchange dealers, commercial customers (primarily MNCs) conducting commercial and investment transactions;
3. Foreign exchange brokers who buy and sell foreign currencies and make a profit on the difference between the exchange rates and interest rates among the various world financial centers; and
4. Central banks, which intervene in the market from time to time to smooth exchange rate fluctuations or to maintain target exchange rates (for example, the Federal Reserve Bank of New York in the U.S.).

In addition to the settlement of obligations incurred through investment, purchases, and other trading, the foreign exchange market involves speculation in exchange futures. New York, Tokyo, and London are the major centers for these transactions. The various linkages between banks and their customers in the foreign exchange market are displayed in Exhibit 52. Exhibit 53 is a list of a variety of participants in the foreign exchange market.

EXHIBIT 52
Structure of Foreign Exchange Markets

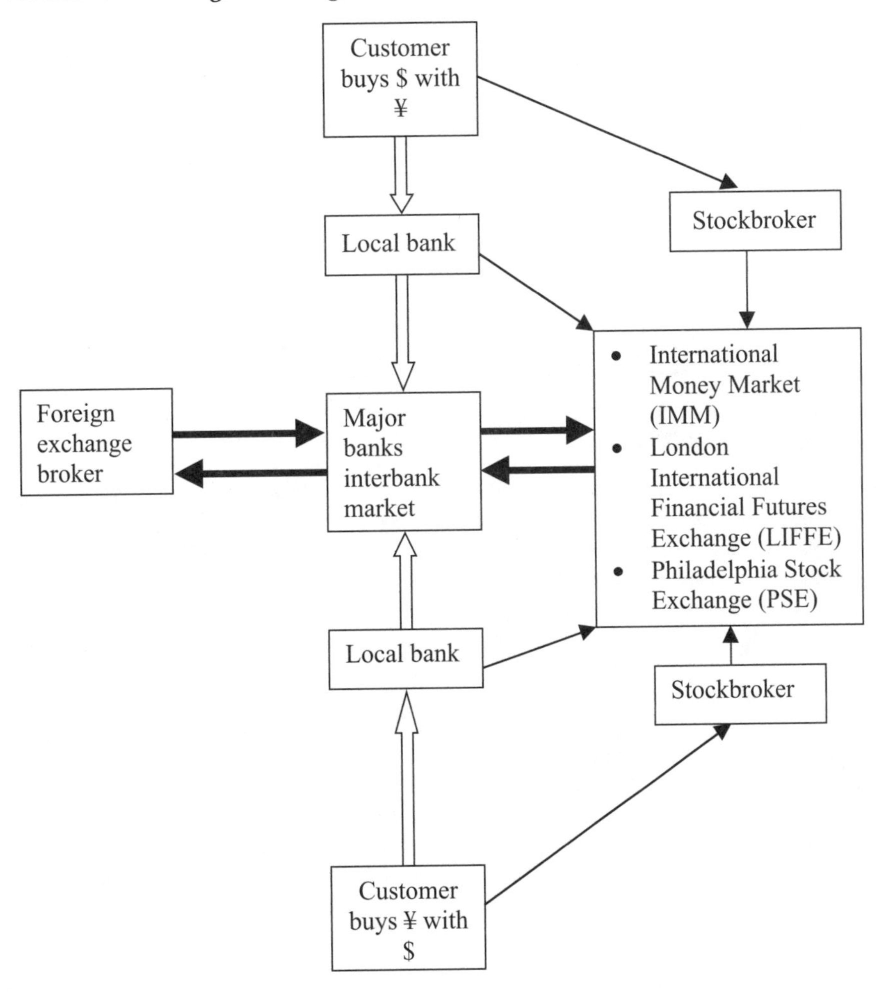

EXHIBIT 53
Participants in the Foreign Exchange Market

Suppliers of Foreign Currency (e.g., ¥)	Demanders for Foreign Currency (e.g., ¥)
U.S. exporters	U.S. direct investors in Japan
Japanese direct investors remitting profits	Japanese foreign investors in the U.S. remitting their profits
Japanese portfolio investors	U.S. portfolio investors
Bear speculators	Bull speculators
	Arbitrageurs
	Government interveners

Exhibit 54 shows the major economics forces moving today's FOREX markets, while Exhibit 55 chronicles the history of the FOREX market.

EXHIBIT 54
Economics Forces Moving Today's FOREX Markets

Economic Indicators	Leading Indicators
Balance of Payments	Employment Cost Index
Trade Deficits	Consumer Price Index (CPI)
Gross Domestic Product (GDP)	Produce Price Index (PPI)
Industrial Production	Retail Sales
Unemployment Rate	Prime Rate
Business Inventories	Discount Rate
	Federal Funds Rate
	Personal Income

EXHIBIT 55
History of the FOREX Market

1944—The major Western Industrialized nations agree to attempt to "stabilize" their currencies. The International Monetary Fund (IMF) is created and the U.S. dollar is "pegged" at $35.00 to the Troy ounce of gold. This was known as the *Gold Standard.*

1971—President Nixon abandons the Gold Standard, directly pegging major currencies to the U.S. dollar.

1973—Following the second major devaluation in the U.S. dollar, the fixed-rate mechanism is totally discarded by the U.S. Government and replaced by the Floating Rate. All major currencies seek to move independently of each other.

1999—Eleven European nations unite under one currency, the euro, and resolve to have a single monetary policy set by the European Central Bank (ECB).

Today—Leading global investment firms have made programs available for individual investors to capitalize on the opportunities offered. The *FOREX interbank market* was developed as a means of handling the huge transaction business in trading, lending, and consolidating deposits of foreign currencies. The vast size and volume of the FOREX market makes it impossible to manipulate the market for any length of time. As a result, FOREX is an action-based, decentralized international forum that allows various major currencies of the world to seek their true value.

Source: http://www.knewmoney.com.

FOREIGN EXCHANGE MARKET INTERVENTION

See FOREIGN CURRENCY OPERATIONS.

FOREIGN EXCHANGE RATE

A foreign exchange rate, or *exchange rate* for short, specifies the number of units of a given currency that can be purchased with one unit of another currency. Exchange rates appear in the business sections of newspapers each day and financial dailies such as *Investor's Business Daily* and *The Wall Street Journal*. Exchange rates are also available at many finance Web sites such as the *Bloomberg World Currency Values* Web site (http://www.bloomberg.com/markets/wcv.html). The actual exchange rate is determined by the interaction of supply and demand in the *foreign exchange market*. Such supply and demand conditions are determined by various economic factors (e.g., whether the country's basic balance of payments position is in surplus or deficit). Exchange rates can be quoted in *direct* or *indirect* terms and in *American terms* or *European terms*. Exchange rates can be fixed at a predetermined level (fixed exchange rate), or they can be flexible to reflect changes in demand (floating exchange rate). Foreign exchange is the instruments used for international payments. Exchange rates can be *spot (cash)* or *forward*. Foreign exchange rates are determined in various ways:

1. *Fixed exchange rates* are an international financial arrangement in which governments directly intervene in the foreign exchange market to prevent exchange rates from deviating more than a very small margin from some central or parity value. From the end of World War II until August 1971, the world was on this system.
2. *Flexible (floating exchange) rates* are an arrangement by which currency prices are allowed to seek their own levels without much government intervention (i.e., in respond to market demand and supply). Arrangements may vary from free float (i.e., absolutely no government intervention) to managed float (i.e., limited but sometimes aggressive government intervention in the foreign exchange market).
3. *Forward exchange rate* is the exchange rate in contract for receipt of and payment for foreign currency at a specified date, usually 30, 90, or 180 days in the future, at a specified current or "spot" (or cash) price. By buying and selling forward exchange, importers and exporters can protect themselves against the risks of fluctuations in the current exchange market.

FOREIGN EXCHANGE RATE FORECASTING

The forecasting of exchange rates is different under the freely floating and fixed exchange rate systems. Nevertheless, exchange rate forecasting in general is a very difficult and time-consuming task, because so many possible factors are involved. Therefore it might be worthwhile for firms to always hedge their future cash flows in order to eliminate the exchange rate risk and to have predictable cash flows and earnings. With the world becoming smaller and the businesses getting more and more multinational, companies are increasingly exposed to transaction risk, the risk that comes from a fluctuation in the exchange rate between the time a contract is signed and when the payment is received. With most of the exchange rates floating nowadays, they can vary easily as much as 5% within a week, and the exchange rate risk is therefore very considerable. There are four primary reasons why it is imperative to forecast the foreign exchange rates.

1. *Hedging Decision:* Multinational companies (MNCs) are constantly faced with the decision of whether to hedge payables and receivables in foreign currency. An exchange rate forecast can help MNCs determine if they should hedge their transactions. As an example, if forecasts determine that the Swiss franc is going to

appreciate in value relative to the dollar, a company expecting payment from a Swiss partner in the future may decide not to hedge the transaction. However, if the forecasts showed that the Swiss franc is going to depreciate relative to the dollar, the U.S. partner should hedge the transaction.

2. *Short-Term Financing Decision for MNC:* A large corporation has several sources of capital market and several currencies in which it can borrow. Ideally, the currency it would borrow would exhibit low interest rate and depreciate in value over the financial period. For example, a U.S. firm could borrow in Japanese yens; during the loan period, the yens would depreciate in value. At the end of the period, the company would have to use fewer dollars to buy the same amount of yens and would benefit from the deal.
3. *International Capital Budgeting Decision:* Accurate cash flows are imperative in order to make a good capital budgeting decision. In case of international projects, it is not only necessary to establish accurate cash flows but it is also necessary to convert them into an MNC's home country currency. This necessitates the use of foreign exchange forecast to convert the cash flows and there after, evaluate the decision.
4. *Subsidiary Earning Assessment for MNC:* When an MNC reports its earning, international subsidiary earnings are often translated and consolidated in the MNC's home country currency. For example, when Nokia makes a projection for its earnings, it needs to project its earnings in Japan, then it needs to translate these earnings from Japanese yens into U.S. dollars. A depreciation in yens would decrease a subsidiary's earnings and vice versa. Thus, it is necessary to generate an accurate forecast of yens to create a legitimate earnings assessment.

A. Forecasting Under a Freely Floating Exchange Rate System

Future spot rates are theoretically determined by the interplay of various economic factors:

- The expected inflation differential between countries (*purchasing power parity*)
- The difference between national interest rates (*Fisher* and *International Fisher Effects*)
- The forward premium or discount compared with the differences in short-term interest rates (interest rate parity)

Under the assumption that the foreign exchange market is reasonably efficient, the forward rate can be used as an unbiased predictor of the spot rate, which is called *market-based forecasting*. Under floating exchange rates the expected rates of change in the spot rate, differential rates of national inflation and interest, and the forward discount or premium are directly proportional to each other and mutually determined. With an efficient market all the variables adjust very quickly to changes in any one of them. Market efficiency assumes that (a) all relevant formation is quickly reflected in both the spot and forward markets, (b) transaction costs are low, and (c) instruments denominated in different currencies are perfect substitutes for one another.

As a result, forecasting success depends heavily on having prior information that one of the variables is going to change. In an efficient market the possession of such information doesn't bring a competitive edge, because all information is quickly reflected in both the spot and forward markets. Recent tests however are contradicting the efficient market theory, and, therefore, it might well be that the market is inefficient. In that case it pays for a firm to spend resources on forecasting exchange rates. Therefore, forecasters spend a lot of time with *fundamental analysis*. This approach involves examining the macroeconomic variables and policies that are likely to influence a currency's performance, basically spot and forward exchange rates, interest rates, and rates of inflation. Another forecasting approach is *technical analysis*, which focuses exclusively on past price and volume movements, while completely ignoring economic and political factors. The two primary methods of technical analysis are charting and trend analysis. The objective is to find particular trends from where you can make predictions for the future. Probably the best forecasting method is a mixture of the three methods, because there are more variables involved and this might produce a better result than relying on only a single forecast. Exhibit 56 lists variables used to forecast future exchange rate movements.

EXHIBIT 56
Variables Used to Forecast Future Exchange Rate Movements

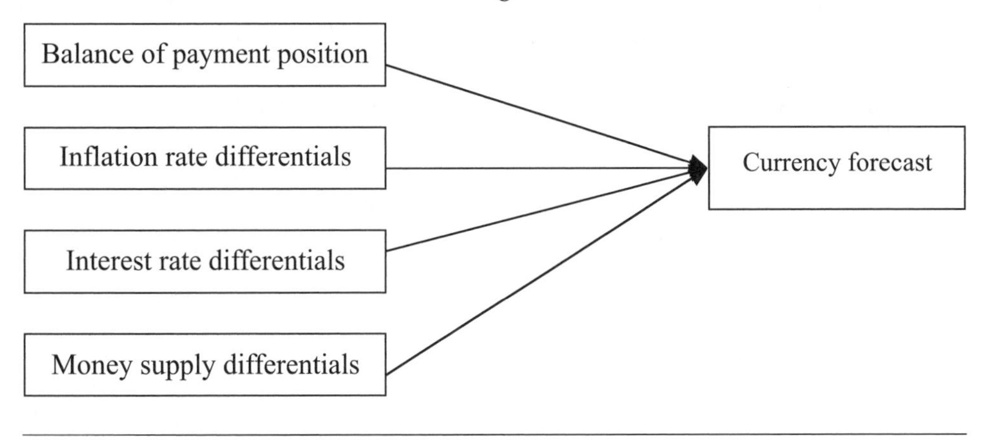

B. Forecasting under a Managed or Fixed Exchange Rate System

Under fixed exchange rates the movement toward equilibrium in spot and forward exchange rates, interest rates, and inflation rates, which occurs under freely floating rates, is artificially prevented from happening. The government is responsible for keeping the exchange rate at a certain level with all means, even if it means accepting huge losses. However, even the government must resign to the market pressure and allow the exchange rate to change. The objective of exchange rate forecasting is how to predict when this change will occur. The timing and amount of fixed exchange rates are primarily a political decision and, therefore, forecasters have to consider various economic factors:

- *Balance of payments deficit*: If more imports than exports and hence more monetary claims against the country exist, a devaluation becomes more likely.
- *Differential rates of inflation*: Goods in the country with lesser inflation become cheaper, which leads to a balance of payment surplus in this country, because there is more demand for the goods. Hence this currency would appreciate.
- *Growth in money supply*: If money supply grows faster than is warranted by real economic growth, inflation will incur and there will be pressure on the exchange rate as explained above.
- *Decline in international monetary reserves*: Balance of payments deficits will lead to a decline in international monetary reserves, which might lead to a run on the currency.
- *Excessive government spending*: If resources of a nation are overextended, it might lead to currency devaluation.

Exchange rate forecasting is, in reality, neglected in many firms. It is often argued that forecasting is useless because it does not provide an accurate estimate. It is certainly true that forecasting will not provide an accurate estimate, because the future is uncertain and every forecast can be only an approximation, which is based on uncertain variables. Despite the fact that forecasting will not provide an exact result, it is nevertheless not useless because it gives the company a general idea about the overall trend of the future, which ensures that they won't be caught sleeping when new trends happen. Besides, exchange rate forecasting gives companies an idea about when and how to hedge. Moreover, it is important to forecast exchange rates for the predictability of future cash flows. All in all, exchange rate forecasting is a very difficult and time-consuming task, because so many possible factors are involved. Therefore, it might be worthwhile for firms to always hedge their future cash flows in order to eliminate the exchange rate risk and to have predictable cash flows and earnings.

Exhibit 57 compares forecasts of Malaysia's ringgit drawn from each forecasting technique.

EXHIBIT 57
Forecasts of Malaysia's Ringgit Drawn from Each Forecasting Method

Method	Factors Considered	Situation	Possible Forecast
Fundamental	Inflation, interest rates, economic growth	Malaysia's interest rates are high while inflation rate is stable	The ringitt's value will go up as U.S. investors take advantage of the high interest rates by investing in Malaysian securities
Technical	Recent movements—chart and trend	The ringgit's value fell in the past few weeks below a specific threshold level	The ringitt's value will continue to decline now that it is beyond the threshold level
Market-based	Spot rate, forward rate	The ringitt's forward rate shows a significant discount, which is due to Malaysia's relatively high interest rates	Based on the predicted forward rate discount, the ringitt's value will fall
Mixed	Weighted average	Interest rates are high; forward rate discount displayed; the ringitt's value of recent weeks has headed lower	There is better than 50% chance that the ringitt's value will decline

C. Currency Forecasting Service

The corporate need to forecast currency values has prompted the emergence of several consulting firms, including Business International, Conti Currency, Predex, and Wharton Econometric Forecasting Associates (WEFA). In addition, some large investment banks, such as Goldman Sachs, and commercial banks, such as Citibank, Chemical Bank, and Chase Manhattan Bank, offer forecasting services. Many consulting services use at least two different types of analysis to generate separate forecasts, and then determine the weighted average of the forecasts.

Some forecasting services, such as Capital Techniques, FX Concepts, and Preview Economics, focus on technical forecasting, while other services, such as Corporate Treasury Consultants Ltd. and WEFA, focus on fundamental forecasting. Many services, such as Chemical Bank and Forexia Ltd., use both technical and fundamental forecasting. In some cases, the technical forecasting techniques are emphasized by the forecasting firms for short-term forecasts, while fundamental techniques are emphasized for long-term forecasts.

Recently, most forecasting services have been inaccurate regarding future currency values. Given the recent volatility in foreign exchange markets, it is quite difficult to forecast currency values. One way for a corporation to determine whether a forecasting service is valuable is to compare the accuracy of its forecasts to those of alternative publicly available and free forecasts. The forward rate serves as a benchmark for comparison here, because it is quoted in many newspapers and magazines. A recent study compared the forecasts of several currency forecasting services regarding nine different currencies to the forward rate. Only 5% of the forecasts (when considering all forecasting firms and all currencies forecasted) for one month ahead were more accurate than the forward rate, while only 14% of forecasts for three months ahead were more accurate. These results are frustrating to the MNCs that have paid $25,000 or more per year for expert opinions. Perhaps some corporate clients of these consulting firms believe the fee is worth it even when the forecasting performance is poor, if other services (such as cash management) are included in the package. It is also possible that a corporate treasurer, in recognition of the potential for error in forecasting exchange rates, may prefer to pay a consulting firm for its forecasts. Then the treasurer is not (in a sense) directly responsible for corporate problems that result from inaccurate currency forecasts. Not all MNCs hire consulting firms to do their forecasting. For example, Kodak, Inc. once used a service, but became dissatisfied with it and has now developed its own forecasting system.

See also FUNDAMENTAL FORECASTING; MARKET-BASED FORECASTING; MIXED FORECASTING; TECHNICAL FORECASTING.

FOREIGN EXCHANGE RATE RISK

See CURRENCY RISK.

FOREIGN EXCHANGE RISK

See CURRENCY RISK.

FOREIGN EXCHANGE RISK PREMIUM

The foreign exchange risk premium is the difference between the forward rate and the expected future spot rate.

FOREIGN MARKET BETA

A foreign market beta is a measure of foreign market risk which is derived from the *Capital Asset Pricing Model* (*CAPM*). The formula for the computation is:

$$\text{Foreign market beta} = \text{Correlation with the U.S. market} \times \frac{\text{Standard deviation of foreign market}}{\text{Standard deviation of U.S. market}}$$

EXAMPLE 51

If the correlation between the German and U.S. markets is 0.44, and the standard deviations of German and U.S. returns are 20.1% and 12.5%, respectively, then the German market beta is

$$0.76 = \left(0.44 \times \frac{0.201}{0.125}\right).$$

In other words, despite the greater riskiness of the German market relative to the U.S. market, the low correlation between the two markets leads to a German market beta (.76) which is lower that the U.S. market beta (1.00).

FOREIGN PORTFOLIO INVESTMENT

See PORTFOLIO INVESTMENTS.

FOREIGN SUBSIDIARIES AND AFFILIATES

A foreign subsidiary bank is a separately incorporated bank owned entirely or in part by a U.S. bank, a U.S. bank holding company, or an *Edge Act Corporation*. A foreign subsidiary provides local identity and the appearance of a local bank in the eyes of potential customers in the host country, which often enhances its ability to attract additional local deposits. Furthermore, management is typically composed of local nationals, giving the subsidiary bank better access to the local business community. Thus, foreign-owned subsidiaries are generally in a stronger position to provide domestic and international banking services to residents of the host country.

Closely related to foreign subsidiaries are foreign affiliate banks, which are locally incorporated banks owned in part, but not controlled, by an outside parent. The majority of the ownership and control may be local, or it may be other foreign banks.

FOREX

See FOREIGN EXCHANGE.

FORFAITING

Forfaiting is the simplest, most flexible and convenient form of medium-term fixed rate finance for export transactions. It is a form of export financing similar to *factoring*, which is the sale of export receivables by an exporter for cash. The word comes from the French *à forfait*, a term meaning "to forfeit or surrender a right." It is the name given to the purchase of trade receivables maturing at various future dates without recourse to the exporter. The basic idea is to get the exporter's goods shipped and received and for the exporter to receive payment for the goods. The exporter accepts one or a series of bills of exchange payable over a specified period of years. The exporter immediately turns around and sells these notes or bills of exchange to a forfait house. This sale is made without recourse to the exporter, who now has his goods delivered and the cash in hand. He is out of the picture. The only obligation that he still has is that the goods shipped were of the quality stated. The exporter is now free to go on to the next deal without the worry of extending credit and collecting. This keeps his cash flow moving in a positive direction. A specialized finance firm called a *forfaiter* buys trade receivables, notes, or bills of exchange and then repackages them for sale to investors. Forfaiting may also carry the guarantee of the foreign government. Under a typical arrangement, the exporter receives cash up front and does not have to worry about the financial ramifications of nonpayment, this risk being transferred to the forfaiter. The forfaiter carries all political, country, and commercial risk.

Forfaiting is the discounting—at a fixed rate without recourse—of medium-term export receivables denominated in fully convertible currencies (U.S. dollar, Swiss franc, Deutsche mark). This technique is typically used in the case of capital-goods exports with a five-year maturity and repayments in semiannual installments. The discount is set at a fixed rate—about 1.25% above the local cost of funds or above the *London interbank offer rate* (*LIBOR*). In a typical forfaiting transaction, an exporter approaches a forfaiter before completing a transaction's structure. Once the forfaiter commits to the deal and sets the discount rate, the exporter can incorporate the discount into the selling price. Forfaiters usually work with bills of exchange or promissory notes, which are unconditional and easily transferable debt instruments that can be sold on the secondary market. Forfaiting differs from factoring in four ways: (1) forfaiting is relatively quick to arrange, often taking no more than two weeks; (2) forfaiting may be for years, while factoring may not exceed 180 days; (3) forfaiting typically involves political and transfer uncertainties but factoring does not; and (4) factors usually want access to a large percentage of an exporter's business, while most forfaiters will work on a one-shot basis. A third party, usually a specialized financial institution, guarantees the financing. Many forfaiting houses are subsidiaries of major international banks, such as Credit Suisse. These houses also provide help with administrative and collection problems.

FORWARD CONTRACT

A forward contract is a written agreement between two parties for the purchase or sale of a stipulated amount of a commodity, currency, financial instrument, or other item at a price determined at the present time with delivery and settlement to be made at a future date, usually in 30, 90, or 180 days.

See also FUTURES; FOREIGN EXCHANGE.

FORWARD DIFFERENTIAL

See FORWARD PREMIUM OR DISCOUNT.

FORWARD-SPOT DIFFERENTIAL

See FORWARD PREMIUM OR DISCOUNT.

FORWARD EXCHANGE RATE

The forward exchange rate is the contracted exchange rate for receipt of and payment for foreign currency at a specified date in the future, at a stipulated current or spot price. In the forward market, you buy and sell currency for a future delivery date, usually 1, 3, or 6 months in advance. If you know you need to buy or sell currency in the future, you can hedge against a loss by selling in the forward market. By buying and selling forward exchange contracts, importers and exporters can protect themselves against the risks of fluctuations in the current exchange market. Suppose that the spot and the 90-day forward rates for the British pound are quoted as $1.5685 and $1.5551, respectively. You are required to pay £10,000 in 3 months to your English supplier. You can purchase £10,000 today by paying $15,551 (10,000 × $1.5551). These pounds will be delivered in 90 days. In the meantime you have protected yourself. No matter what the exchange rate of the pound or U.S. dollar is in 90 days, you are assured delivery at the quoted price. As can be seen in the example, the cost of purchasing pounds in the forward market ($15,551) is less than the price in the spot market ($15,685). This implies that the pound is selling at a *forward discount* relative to the dollar, so you can buy more pounds in the forward market. It could also mean that the U.S. dollar is selling at a *forward premium*. *Note*: It is extremely unlikely that the forward rate quoted today will be exactly the same as the future spot rate.

See also SPOT EXCHANGE RATE.

FORWARD FOREIGN EXCHANGE MARKET

Simply called *forward market*, the forward foreign exchange market is the market where foreign exchange dealers can enter into a *forward contract* to buy or sell any amount of a currency at any date in the future. Forward contracts are negotiated between the offering bank and the client as to size, maturity, and price. The rationale for forward contracts in currencies is analogous to that for futures contracts in commodities; the firm can lock in a price today to eliminate uncertainty about the value of a future receipt or payment of foreign currency. Such a transaction may be undertaken because the firm has made a previous contract, for example, to pay a certain sum in foreign currency to a supplier in three months for some needed inputs to its operations. This concept is called *hedging*. The principal users of the forward market are currency arbitrageurs, hedgers, importers and exporters, and speculators.

Arbitrageurs wish to earn risk-free profits; hedgers, importers, and exporters want to protect the home currency values of various foreign currency-denominated assets and liabilities; and speculators actively expose themselves to exchange risk to benefit from expected movements in exchange rates. It differs from the *futures market* in many significant ways.

See also FUTURES; HEDGE.

FORWARD MARKET

See FORWARD FOREIGN EXCHANGE MARKET.

FORWARD MARKET HEDGE

A forward market hedge is a hedge in which a net asset (liability) position is covered by a liability (asset) in the *forward market*.

EXAMPLE 52

XYZ, an American importer, enters into a contract with a British supplier to buy merchandise for £4,000. The amount is payable on delivery of the goods, 30 days from today. The company knows the exact amount of its pound liability in 30 days. However, it does not know the payable in dollars. Assume further that today's foreign exchange rate is \$1.50/£ and the 30-day forward exchange rate is \$1.49. In a forward market hedge, XYZ may take the following steps to cover its payable.

Step 1. Buy a forward contract today to purchase (buy the pounds forward) £4,000 in 30 days.
Step 2. On the 30th day pay the foreign exchange dealer \$5,960.00 (4,000 pounds × \$1.49/£) and collect £4,000. Pay this amount to the British supplier.

By using the forward contract, XYZ knows the exact worth of the future payment in dollars (\$5,960.00). The currency risk in pounds is totally eliminated by the net asset position in the forward pounds.

Note: (1) In the case of the net asset exposure, the steps open to XYZ are the exact opposite: Sell the pounds forward (buy a forward contract to sell the pounds), and on the future day receive and deliver the pounds to collect the agreed-upon dollar amount. (2) The use of the forward market as a hedge against *currency risk* is simple and direct. That is, it matches the liability or asset position against an offsetting position in the forward market.

See also MONEY-MARKET HEDGE.

FORWARD PREMIUM OR DISCOUNT

The forward rate is often quoted at a premium to or discount from the existing spot rate. The forward premium or discount is the difference between spot and forward rates, expressed as an *annual* percentage, also called *forward-spot differential, forward differential*, or *exchange agio*. When quotations are on an *indirect* basis, a formula for the percent-per-annum forward premium or discount is as follows:

$$\text{Forward premium } (+) \text{ or discount } (-) = \frac{\text{Spot} - \text{Forward}}{\text{Forward}} \times \frac{12}{n} \times 100$$

where n = the number of months in the contract.

EXAMPLE 53

Assume that the spot exchange rate = ¥110/$ and the one-month forward rate = ¥109.66/$. Since the spot rate is greater than the one-month forward rate (in *indirect quotes*), the yen is selling forward at a premium.

The 1-month (30-day) forward premium (discount) is:

$$\frac{¥110.19 - ¥109.66}{¥109.66} \times \frac{12}{1} \times 100 = +5.80\%$$

The 3-month (90-day) forward premium (discount) is:

$$[(¥110.19 - ¥108.55)/¥108.55] \times 12/3 \times 100 = +6.04\%$$

The 6-month (180-day) forward premium (discount) is:

$$[(¥110.19 - ¥106.83)/¥106.83] \times 12/6 \times 100 = +6.29\%$$

Note: A currency is said to be selling at a premium (discount) if the forward rate expressed in *indirect quotes* is less (greater) than the spot rate.

With *direct quotes:*

$$\text{Forward premium or discount} = \frac{\text{Forward} - \text{Spot}}{\text{Spot}} \times \frac{12}{1} \times 100$$

$$\text{Forward premium or discount} = \frac{\$0.009119 - \$0.009075}{\$0.009075} \times \frac{12}{1} \times 100 = +5.80\%$$

Note: A currency is said to be selling at a premium (discount) if the forward rate expressed in *direct quotes* is greater (less) than the spot rate.

Exhibit 58 shows forward rate quotations and annualized forward premiums (discounts). *Note:* In Exhibit 58, since a dollar would buy *fewer* yen in the forward than in the spot market, the forward yen is selling at a *premium.*

EXHIBIT 58
Forward Rate Quotations and Annualized Forward Premiums (Discounts)

Quotation		¥/$ (Indirect Quote)	$/¥ (Direct Quote)	% per Annum
Spot Rate		¥110.19	$0.009075	
Forward				
	1-month	109.66	0.009119	+5.80%
	3-month	108.55	0.009212	+6.04%
	6-month	106.83	0.009361	+6.29%

FORWARD RATE

See FORWARD EXCHANGE RATE.

FORWARD RATE QUOTATIONS

The quotations for forward rates can be made in two ways. They can be made in terms of the amount of local currency at which the quoter *will* buy and sell a unit of foreign currency. This is called the *outright rate* and it is used by traders in quoting to customers. The forward rates can also be quoted in terms of points of discount and premium from spot, called the *swap rate*, which is used in interbank quotations. The outright rate is the spot rate adjusted by the swap rate. To find the outright forward rates when the premiums or discounts on quotes of forward rates are given in terms of *points* (swap rate), the points are *added* to the spot price if the foreign currency is trading at a forward *premium*; the points are *subtracted* from the spot price if the foreign currency is trading at a forward *discount*. The resulting number is the outright forward rate. It is usually well known to traders whether the quotes in points represent a premium or a discount from the spot rate, and it is not customary to refer specifically to the quote as a premium or a discount. However, this can be readily determined in a mechanical fashion. If the first forward quote (the bid or buying figure) is smaller than the second forward quote (the offer or selling figure), then it is a premium—that is, the swap rates are added to the spot rate. Conversely, if the first quote is larger than the second, then it is a discount. (A 5/5 quote would require further specification as to whether it is a premium or discount.) This procedure assures that the buy price is lower than the sell price, and the trader profits from the spread between the two prices. For example, when asked for spot, 1-, 3-, and 6-month quotes on the French franc, a trader based in the United States might quote the following:

.2186/9 2/3 6/5 11/10

In outright terms these quotes would be expressed as indicated as follows:

Maturity	Bid	Offer
Spot	.2186	.2189
1-month	.2188	.2192
3-month	.2180	.2184
6-month	.2175	.2179

Notice that the 1-month forward franc is at a premium against the dollar, whereas the 3- and 6-month forwards are at discounts. *Note*: The literature usually ignores the existence of bid and ask prices, and, instead, uses only one rate, which can be treated as the *midrate* between bid and ask prices.

FORWARD RATES AS UNBIASED PREDICTORS OF FUTURE SPOT RATES

Because of a widespread belief that foreign exchange markets are "efficient," the forward currency rate should reflect the expected future spot rate on the date of settlement of the forward contract. This theory is often called the *expectations theory of exchange rates*.

EXAMPLE 54

If the 90-day forward rate is DM 1 = \$0.456, arbitrage should ensure that the market expects the spot value of DM in 90 days to be about \$0.456.

An "unbiased predictor" intuitively implies that the distribution of possible future actual spot rates is centered on the forward rate. This, however, does not mean the future spot rate will actually be equal to what the future rate predicts. It merely means that the forward rate will, on average, under- and over-estimate the actual future spot rates in equal frequency and degree. As a matter of fact, the forward rate may never actually equal the future spot rate.

The relationship between these two rates can be restated as follows:

The forward differential (premium or discount) equals the expected change in the spot exchange rate.

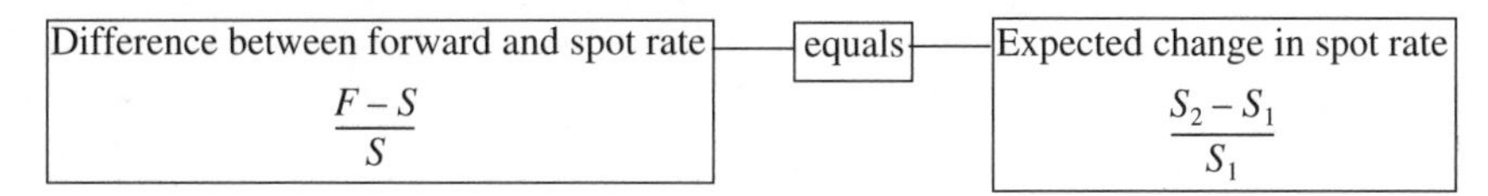

Algebraically,

With indirect quotes:

$$\frac{\text{Spot} - \text{forward}}{\text{Spot}} = \frac{\text{Beginning rate} - \text{ending rate}}{\text{Ending rate}}$$

With direct quotes:

$$\frac{\text{Forward} - \text{Spot}}{\text{Forward}} = \frac{\text{Ending rate} - \text{beginning rate}}{\text{Beginning rate}}$$

The relationship between the forward rate and the future spot rate is illustrated in Exhibit 59. See also APPRECIATION OF THE DOLLAR; PARITY CONDITIONS.

EXHIBIT 59
Relationship Between the Forward Rate and the Future Spot Rate

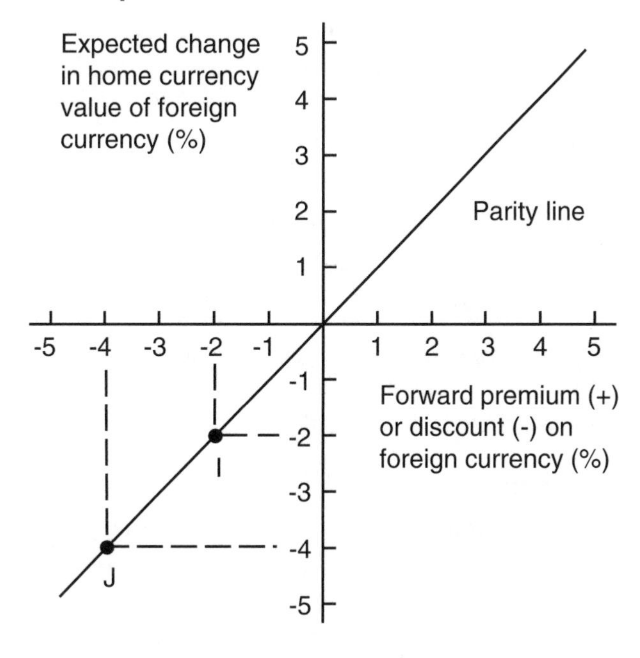

FORWARD TRANSACTION

Forward transactions are types of transactions that take place in the forward foreign exchange market (*forward market*). In the forward market, unlike in the *spot market* where currencies are traded for immediate delivery, trades are made for future dates, usually less than one year away. The forward market and the *futures market* perform similar functions, but with a difference. In the forward market, foreign exchange dealers can enter into a contract to buy or sell any amount of a currency at any date in the future. In contrast, futures contracts are for a given month (March, June, September, or December), with the third Wednesday of the month as delivery date.

FORWARD WITH OPTION EXIT

The forward with option exit (FOX) refers to *forward contracts* with an option to break out of the contract at a future date. With FOX, the forward exchange rate price includes an option premium for the right to break the forward contract. This type of forward is used by customers desiring to have insurance provided by a forward contract when the exchange rate moves against them and yet not lose the potential for profit available with favorable exchange rate movements.

FOX

See FORWARD WITH OPTION EXIT.

FRANC

Monetary unit of the following nations: Belgium, Benin, Burundi, Cameroons, Central Africa, Chad, Comoros, Congo, Dahomey, Djibouti, France, French Somalialand, Gabon, Guadeloupe, Ivory Coast, Liechtenstein, Luxembourg, Madagascar, Malagasy, Mali, Martinique, Monaco, New Caledonia, New Hebrides Islands, Niger, Oceania, Reunion Island, Rwanda, Senegal, Switzerland, Tahiti, Togo, and Upper Volta.

FRANC AREA

The group of former French colonies that use the French franc as a suitable currency and/or link their currency values to the French franc.

FREE ALONGSIDE (FAS)

After the seller delivers the goods alongside the ship that will transport the goods, within reach of the ship's loading apparatus, the buyer is responsible for the goods beyond this point.

FREE FLOAT

Also called a *clean float*, a free float is a system in which free-market currency rates are determined by the interaction of currency demands and supplies.

See also FLOATING EXCHANGE RATES; MANAGED FLOAT.

FREE ON BOARD (FOB)

The title to the goods passes to the buyer when the goods are loaded aboard ship (or airplane, or however the goods are being shipped). The seller is responsible for all costs until the goods are on board; the buyer then pays all further costs.

FRONTING LOAN

A fronting loan is a loan between a parent and its subsidiary channeled through a financial intermediary, usually a large international bank.

See also BACK-TO-BACK LOANS.

FUNCTIONAL CURRENCY

As defined in *FASB No. 52*, in the context of translating financial statements, functional currency is the currency of the primary economic environment in which the subsidiary operates. It is the currency in which the subsidiary realizes its cash flows and conducts its operations. To help management determine the functional currency of its subsidiary, SFAS 52 provides a list of six salient economic indicators regarding cash flows, sales price, sales market, expenses, financing, and intercompany transactions. Depending on the circumstances:

1. The functional currency can be the local currency. For example, a Japanese subsidiary manufactures and sells its own products in the local market. Its cash flows, revenues, and expenses are primarily in Japanese yen. Thus, its functional currency is the local currency (Japanese yen).
2. The functional currency can be the U.S. dollar. For foreign subsidiaries that are operated as an extension of the parent and integrated with it, the functional currency is that of the parent. For example, if the Japanese subsidiary is set up as a sales outlet for its U.S. parent (i.e., it takes orders, bills and collects the invoice price, and remits its cash flows primarily to the parent), then its functional currency would be the U.S. dollar.

The functional currency is also the U.S. dollar for foreign subsidiaries operating in highly inflationary economies (defined as having a cumulative inflation rate of more than 100% over a three-year period). The U.S. dollar is deemed the functional currency for translation purposes because it is more stable than the local currency.

FUNDAMENTAL ANALYSIS

1. An analysis based on economic theory drawn to construct econometric models for forecasting future exchange rates. The variables examined in these models include relative inflation and interest rates, national economic growth, money supply growth rates, and variables related to countries' *balance-of-payment* positions.

   See also FUNDAMANTAL FORECASTING; TECHNICAL FORECASTING.
2. The assessment of a company's financial statements, fundamental analysis is used primarily to select *what* to invest in, while *technical analysis* is used to help decide *when* to invest in it. Fundamental analysis concentrates on the future outlook of growth and earnings. The analyst studies such elements as earnings, sales, management, and assets. It looks at three things: the overall economy, the industry, and the company itself. Through the study of these elements, an analyst is trying to determine whether the stock is undervalued or overvalued compared with the current market price.

FUNDAMENTAL FORECASTING

Fundamental forecasting is based on fundamental relationships between economic variables and exchange rates. Given current values of these variables along with their historical impact on a currency's value, corporations can develop exchange rate projections. Based on exchange rate theories (*Purchasing Power Parity, Interest Rate Parity Theory*, and the *Fisher Effect*) involving a basic relationship between exchange rates, inflation rates, and interest rates, one can develop a simple *regression model* for forecasting Deutsche mark.

$$DM = a + b\,(INF) + c\,(INT)$$

where DM = the quarterly percentage change in the German mark, INF = quarterly percentage change in inflation differential (U.S. inflation rate *minus* German inflation rate), and INT = quarterly percentage change in interest rate differential (U.S. interest rate *minus* German interest rate). *Note*: This model is relatively simple with only two explanatory variables. In many cases, several other variables are added but the essential methodology remains the same.

The following example illustrates how exchange rate forecasting can be accomplished using the fundamental approach.

EXAMPLE 55

Exhibit 60 shows the basic input data (for an illustrative purpose only) for the ten quarters. Exhibit 61 shows a summary of the regression output, based on the use of *Microsoft Excel.*

EXHIBIT 60
Quarterly Percentage Change
(For 10 Quarters)

Period	DM/$	Inflation Differential	Interest Differential
1	–0.0058	–0.5231	–0.0112
2	–0.0161	–0.1074	–0.0455
3	–0.0857	2.6998	–0.0794
4	0.0012	–0.4984	0.0991
5	–0.0535	0.5742	–0.0902
6	–0.0465	–0.2431	–0.2112
7	–0.0227	–0.1565	–0.8033
8	0.1695	0.0874	3.8889
9	0.0055	–1.4329	–0.2955
10	–0.0398	3.0346	–0.0161

EXHIBIT 61
Regression Output for the Forecasting Model

SUMMARY OUTPUT

Regression Statistics	
Multiple R	0.9602
R Square	0.9219
Adjusted R Square	0.8996
Standard Error	0.0218
Observations	10.0000

ANOVA

	df	SS	MS	F	Significance F
Regression	2	0.03933	0.01966	41.32289	0.00013
Residual	7	0.00333	0.00048		
Total	9	0.04266			

	Coefficients	Standard Error	t Stat	P-value	Lower 95%	Upper 95%
Intercept	–0.01492	0.00725	–2.05762	0.07864	–0.03206	0.00223
INF Diff.	–0.01709	0.00510	–3.35276	0.01220	–0.02914	–0.00504
INT Diff.	0.04679	0.00557	8.39921	0.00007	0.03362	0.05997

One forecasting model that can be used to predict the DM/$ exchange rate for the next quarter is:

$$\begin{aligned} DM &= -0.0149 - 0.0171(INF) + 0.0468(INT) \\ R^2 &= 92.19\% \end{aligned}$$

Assuming that INT = –0.9234 and INF = 0.1148 for the next quarter:

$$\begin{aligned} DM &= -0.0149 - 0.0171(-0.9234) + 0.0468(0.1148) = 0.00623 \\ DM/\$ &= (1 + 0.00623) \times (1.6750) = 1.6854 \end{aligned}$$

Note: This model relies on relationships between macroeconomic variables. However, there are certain problems with this forecasting technique.

1. This technique will not be very effective with *fixed exchange rates*.
2. The precise timing of the impact of some factors on a currency's value is not known. It is possible that the impact of inflation on exchange rates will not completely occur until two, three, or four quarters later. The regression model would need to be adjusted accordingly.
3. Another limitation is related to those that exhibit an immediate impact on exchange rates. Their inclusion in a forecasting model would be useful only if forecasts could be obtained. Forecasts of these factors should be developed for a period that corresponds to the period in which a forecast for exchange rates is necessary. The accuracy of the exchange rate forecasts will be somewhat dependent on forecasting accuracy of these factors. Even if firms knew exactly how movements in these factors affected exchange rates, their exchange rate projections could be inaccurate if they could not predict the values of the factors. *Note*: These estimates, however, are frequently published in trade publications and bank reports.
4. This technique often ignores other variables that influence the foreign exchange rate.
5. There may be factors that deserve consideration in the fundamental forecasting process that cannot be easily quantified. For example, what if large Japanese exporting firms experienced an unanticipated labor strike, causing shortages? This would reduce the availability of Japanese goods for U.S. consumers and, therefore, reduce U.S. demand for Japanese yens. Such an event, which would place downward pressure on the Japanese yen value, is not normally incorporated into the forecasting model.
6. Coefficients derived from the *regression analysis* will not necessarily remain constant over time.

These limitations of fundamental forecasting are discussed to emphasize that even the most sophisticated forecasting techniques are not going to provide consistently accurate forecasts. MNCs that use forecasting techniques must allow for some margin of error and recognize the possibility of error when implementing corporate policies.

See also FOREIGN EXCHANGE RATE FORECASTING; REGRESSION ANALYSIS.

FUTURES

In the futures market, investors and MNCs trade in commodities and financial instruments. A future is a contract to purchase or sell a given amount of an item for a given price by a certain date (in the future—thus the name *futures market*). The seller of a futures contract agrees to deliver the item to the buyer of the contract, who agrees to purchase the item.

The contract specifies the amount, valuation, method, quality, month and means of delivery, and exchange to be traded in. The month of delivery is the expiration date—in other words, the date on which the commodity or financial instrument must be delivered. Commodity contracts are guarantees by a seller to deliver a commodity (e.g., cocoa or cotton). Financial contracts are a commitment by the seller to deliver a financial instrument (e.g., a Treasury bill) or a specific amount of foreign currency. Futures can by risky; to invest in them, you will need specialized knowledge and great caution. Exhibits 62 and 63 show some of the commodity and financial futures available. Quotations for futures can be obtained from the Commodity Charts & Quotes—Free Internet site (tfc-charts.w2d.com).

EXHIBIT 62
Commodities Futures

Grains & Oilseeds	Livestock & Meat	Food, Fiber, & Wood	Metals & Petroleum
Barley	Beef—Boneless	Butter	Copper
Canola	Broilers	Cheddar Cheese	Gold
Corn	Cattle—Feeder	Cocoa	Heating Oil
Flaxseed	Cattle—Live	Coffee	High-Grade Copper
Oats	Cattle—Stocker	Cotton #2	Light Sweet Crude
Peas—Feed	Hogs—Lean	Lumber	Mercury
Rice—Rough	Pork Bellies—Fresh	Milk Bfp	Natural Gas
Rye	Pork Bellies—Frozen	Milk—Non-Fat Dry	Palladium
Soybean Meal	Turkeys	Orange Juice	Palo Verde Electricity
Soybean Oil		Oriented Strand Board	Platinum
Soybeans		Potatoes	Propane
Wheat		Rice	Silver
Wheat—Duram		Shrimp—Black Tiger	Silver—1000 oz.
Wheat—Feed		Shrimp—White	Twin City Electricity
Wheat—Spring		Sugar	Unleaded Gasoline
Wheat—White		Sugar—World	
Wheat—Winter			

EXHIBIT 63
Financial Futures

Currencies	Interest Rates	Securities	Indexes
Australian Dollar	Eurodollars	Bank CDs	Dow Jones Industrials
Brazilian Real	Federal Funds—30 Days	GNMA Passthrough	Eurotop 100 Index
British Pound	Libor—1-Month	Stripped Treasuries	Goldman Sachs
Canadian Dollar	Treasury Bills		Major Market
Euro	Treasury Bonds—30-Year		Municipal Bond Index
French Franc	Treasury Notes—2-Year		NASDAQ 100
German Mark	Treasury Notes—5-Year		Nikkei 225
Japanese Yen	Treasury Notes—10-Year		NYSE Composite
Mexican Peso			PSE 100 Tech
Russian Ruble			Russell 1000
S. African Rand			Russell 2000
Swiss Franc			S&P 400 MidCap

Thai Baht	S&P 500
U.S. Dollar	S&P 500—Mini
	S&P Barra—Growth
	S&P Barra—Value
	Value Line
	Value Line—Mini

A long position is the acquisition of a contract in the hope that its price will rise. A short position is selling it in anticipation of a price drop. The position may be terminated through reversing the transaction. For instance, the long buyer can later take a short position of the same amount of the commodity or financial instrument.

Almost all futures are offset (canceled out) before delivery. It is rare for delivery to settle the futures contract.

Trading in futures is conducted by hedgers and speculators. Hedgers protect themselves with futures contracts in the commodity they produce or in the financial instrument they hold. For instance, if a producer of wheat anticipates a decline in wheat prices, he can sell a futures contract to guarantee a higher current price. Then, when future delivery is made, he will receive the higher price. Speculators use futures contracts to obtain capital gain on price rises of the commodity, currency, or financial instrument. Commodity futures trading is accomplished by open outcry auction.

A futures contract can be traded in the futures market. Trading is done through specialized brokers, and certain commodity firms deal only in futures. The fees for futures contracts are based on the amount of the contract and the price of the item. The commissions vary according to the amount and nature of the contract. The trading in futures is basically the same as dealing in stocks, except that the investor must establish a commodity trading account. The margin buying and kinds of orders are the same, however. The investor can purchase or sell contracts with desired terms.

Futures trading can help an investor cope with inflation. However, future contracts are a specialized, high-risk area because of the numerous variables involved, one of which is the international economic situation. Futures contract prices can be quite volatile.

A. Commodities Futures

In a commodity contract, the seller promises to deliver a given commodity by a certain date at a predetermined price. The contract specifies the item, the price, the expiration date, and the standardized unit to be traded (e.g., 50,000 pounds). Commodity contracts may run up to one year. Investors must continually evaluate the effect of market activity on the value of the contract.

Let's say that you buy a futures contract for the delivery of 1,000 units of a commodity five months from now at $4.00 per unit. The seller of the contract does not have to have physical possession of the item, and you, as the contract buyer, need not take custody of the commodity at the "deliver" date. Typically, commodity contracts are reversed, or terminated, prior to their consummation. For instance, as the initial buyer of 1,000 bushels of corn, you may enter into a similar contract to sell the same quantity, thus in effect closing out your position. *Note*: Besides investing in futures contracts directly, an investor can invest directly in a commodity or indirectly through a mutual fund. A third method is to buy into a limited partnership involved in commodity investments. The mutual fund and partnership strategies are more conservative, since risk is spread and management experience provided.

Investors may engage in commodity trading in the hope of high return rates and inflation hedges. In inflation, commodities move favorably since they are tied into economic trends. But high risk and uncertainty exist because commodity prices vacillate and because there is a great deal of low-margin investing. Investors must have plenty of cash available in the event of margin calls and to cover their losses. To reduce risk, commodities investors should hold a diversified portfolio, and they should determine the integrity and reliability of the salesperson.

The buyer of a commodity always has the option for terminating the contract or letting it run to gain possible higher profits. On the other hand, he or she may use the earnings to put up margins on another futures contract. This is referred to as an inverse pyramid in a futures contract.

Commodity futures exchanges enable buyers and sellers to negotiate cash (spot) prices. Cash is paid for immediately upon receiving physical possession of a commodity. Prices in the cash market rely to some degree on prices in the futures market. There may be higher prices for the commodity over time, incorporating holding costs and anticipated inflation. Commodity and financial futures are traded in the Chicago Board of Trade (CBT), which is the largest exchange. Other exchanges exist, some specializing in given commodities. Examples of commodity exchanges are the New York Cotton Exchange, Chicago Mercantile Exchange, and Amex Commodities Exchange. Since there is a chance of significant gains and losses in commodities, exchanges have restrictions on the highest daily price movements for a commodity. Regulation of the commodities exchanges is by the Federal Commodity Futures Trading Commission.

B. Return on a Futures Contract

The return on a futures contract comes from capital gain (selling price minus purchase price), as no current income is involved. High capital gain is possible due to price volatility of the commodity and the effect of leverage from the low margin requirement. However, if things go sour, the entire investment in the form of margin could be lost quickly. The return on investment when dealing in commodities (whether a long or short position) equals:

$$\frac{\text{Selling price} - \text{purchase price}}{\text{Margin deposit}}$$

EXAMPLE 56

Assume that you purchase a contract on a commodity for \$60,000, putting up an initial deposit of \$5,000. You later sell the contract for \$64,000. The return is:

$$\text{Return} = \frac{\$64{,}000 - \$60{,}000}{\$5{,}000} = 80\%$$

Margin requirements for commodity contracts are relatively low, usually ranging from 5% to 10% of the contract's value. (For stocks, you will remember, the margin requirement is 50% of the cost of the security.) In commodities trading, no money is really lent, and so no interest is paid.

An "initial margin" is required as a deposit on the futures contract. The purpose of the deposit is to cover a market value decline on the contract. The amount of the deposit depends on the nature of the contract and the commodity exchange involved. Investors also have to put up a maintenance deposit, which is lower than the initial deposit and provides the minimum margin that must always be maintained in the account. It is usually about 80% of the initial margin.

EXAMPLE 57

On July 1, you enter into a contract to buy 37,500 pounds of coffee at \$5 a pound to be delivered by October. The value of the total contract is \$187,500. Assume the initial margin requirement is 10%, or \$18,750.

The margin maintenance requirement is 70%, or $13,125. If there is a contract loss of $1,500, you must put up the $1,500 to cover the margin position; otherwise, the contract will be terminated with the resulting loss.

EXAMPLE 58

As a second example, assume you make an initial deposit of $10,000 on a contract and a maintenance deposit of $7,500. If the market value of the contract does not decrease by more than $2,500, you'll have no problem. However, if the market value of the contract declines by $4,500, the margin on deposit will go to $5,500, and you will have to deposit another $2,000 in order to keep the sum at the maintenance deposit level. If you don't come up with the additional $2,000, the contract will be canceled.

Commodity trading may be in the form of hedging, speculating, or spreading. Investors use hedging to protect their position in a commodity. For example, a citrus grower (the seller) will hedge to get a higher price for his products while a processor (or buyer) of the item will hedge to obtain a lower price. By hedging, an investor minimizes the risk of loss but loses the prospect of sizable profit.

EXAMPLE 59

Let's say that a commodity is currently selling at $120 a pound, but the potential buyer (assume a manufacturer) expects the price to rise in the future. To guard against higher prices, the buyer acquires a futures contract selling at $135 a pound. Six months later, the price of the commodity moves to $180. The futures contract price will similarly increase to say, $210. The buyer's profit is $75 a pound. If 5,000 pounds are involved, the total profit is $375,000. At the same time, the cost on the market rose by only $60 a pound, or $300,000. In effect, the manufacturer has hedged his position, coming out with a profit of $75,000, and has kept the rising costs of the commodity under control.

Some people invest in commodities for speculative purposes.

EXAMPLE 60

Suppose that you purchase an October futures contract for 37,500 pounds of coffee at $5 a pound. If the price rises to $5.40, you'll gain $.40 a pound for a total gain of $15,000. The percent gain, considering the initial margin requirement, say 10%, is 80% ($.40/$.50). If the transactions occurred over a two-month period, your annual gain would be 480% (80% × 12 months/2 months). This resulted from a mere 8% ($.40/$5.00) gain in the price of a pound of coffee.

Spreading attempts to take advantage of wide swings in price and at the same time puts a cap on loss exposure. Spreading is similar to stock option trading. The investor enters into at least two contracts to obtain some profit while limiting loss potential. He or she purchases one contract and sells the other in the hope of achieving a minimal but reasonable profit. If the worst happens, the spread helps to minimize the investor's loss.

EXAMPLE 61

Suppose you acquire Contract 1 for 10,000 pounds of commodity Z at $500 a pound. At the same time, you sell short Contract 2 for 10,000 pounds of the same commodity at $535 a pound. Subsequently, you sell Contract 1 for $520 a pound and buy Contract 2 for $543 a pound. Contract 1 yields a profit of $20 a pound while Contract 2 takes a loss of $8 a pound. On net, however, you earn a profit of $12 a pound, so your total gain is $120,000.

C. Financial Futures

The basic types of financial futures are (1) interest rate futures, (2) foreign currency futures, and (3) stock-index futures. Financial futures trading is similar in many ways to commodity trading and now constitutes about two-thirds of all contracts. Because of the instability in interest and exchange rates, financial futures can be used to hedge. They can also be used as speculative investments because of the potential for significant price variability. In addition financial futures have a *lower* margin requirement than commodities do. The margin on a U.S. Treasury bill, for example, may be a low as 2%.

Financial futures are traded in the New York Futures Exchange, AMEX Commodities Exchange, International Monetary Market (part of Chicago Mercantile Exchange), and Chicago Board of Trade. Primarily, financial futures are for fixed income debt securities to hedge or speculate on interest rate changes and foreign currency.

An interest rate futures contract provides the holder with the right to a given amount of the related debt security at a later date (usually no more than three years). They may be in Treasury bills and notes, certificates of deposit, commercial paper, and "Ginnie Mae" (GNMA) certificates, among others. Interest rate futures are stated as a percentage of the par value of the applicable debt security. The value of interest rate futures contracts is directly tied into interest rates. For example, as interest rates decrease, the value of the contract increases. As the price or quote of the contract goes up, the purchaser of the contract has the gain while the seller loses. A change of one basis point in interest rates causes a price change. A basis point is 1/100 of 1%.

Those who trade in interest rate futures do not usually take possession of the financial instrument. In essence, the contract is used either to hedge or to speculate on future interest rates and security prices. For example, a banker might use interest rate futures to hedge his or her position.

As an example of hedging, assume a company will issue bonds in ninety days, and the underwriters are now working on the terms and conditions. Interest rates are expected to rise in the next three months. Thus, investors can hedge by selling short their Treasury bills. A rise in interest rates will result in a lower price to repurchase the interest rate future with the resulting profit. This will net against the increased interest cost of the debt issuance.

Speculators find financial futures attractive because of their potentially large return on a small investment. With large contracts (say a $1,000,000 Treasury bill), even a small change in the price of the contract can provide significant gain. However, significant risk also exists with interest futures. They may involve volatile securities with great gain or loss potential. If you are a speculator hoping for increasing interest rates, you will want to sell an interest rate future, because it will soon decline in value.

A currency futures contract gives you a right to a specified amount of foreign currency at a future date. The contracts are standardized, and secondary markets do exist. Currency futures are expressed in dollars or cents per unit of the related foreign currency. They typically have a delivery period of no more than one year. Currency futures can be used for either hedging or speculation. The purpose of hedging in a currency is to lock into the best money exchange possible. Here's an example of hedging an exposed position: A manager enters into an agreement to get francs in four months. If the franc decreases compared to the dollar, the manager obtains less value. To hedge his exposure, the manager can sell a futures contract in francs by going short. If the franc declines in value, the futures contract will make a profit, thus offsetting the manager's loss when he receives the francs.

EXAMPLE 62

Assume a standardized contract of 100,000 pounds. In February you buy a currency futures contract for delivery in June. The contract price is $1 which equals 2 pounds. The total value of the contract is $50,000, and the margin requirement is $6,000. The pound strengthens until it equals 1.8 pounds to $1. Hence, the value of your contract increases to $55,556 ($50,000 × 2/1.8), giving you a return of 92.6% ($5,556/$$6,000). If the pound had weakened, you would have taken a loss on the contract.

A stock-index futures contract is tied into a broad stock market index. Introduced in 1982, futures contracts at the present time can apply to the S & P 500 Stock Index, New York Stock Exchange Composite Stock Index, and Value Line Composite Stock Index. However, smaller investors can avail themselves of the S & P 100 futures contract that involves a smaller margin deposit. Stock-index futures allow you to participate in the general change in the entire stock market. You can buy and sell the "market as a whole" rather than a specific security. If you anticipate a bull market but are unsure which particular stock will rise, you should buy (long position) a stock-index future. Because of the risks involved, you should trade in stock-index futures only for the purpose of hedging or speculation. Exhibit 64 displays specifications for some stock-index futures contracts.

EXHIBIT 64
Stock Index Futures Contracts Specifications

Index and Exchange	Trading Hours	Index	Contract Size and Value	Contract Months
S&P 500 Index				
Index an Optins Market (IOM) of the Chicago Mercantile Exchange (CME)	10:00 am to 4:15 pm (NYT)*	Value of 500 selected stocks Traded on NYSE, AMEX, and OTC, weighted to reflect market value of issues	$500 x the S&P 500 Index	March, June, September, December
NYSE Composite Index				
New York Futures Exchange (NYFE) of the New York Stock Exchange	10:00 am to 4:15 pm (NYT)*	Total value of NYSE Market: 1550 listed common stock, weighted to reflected market value of issues	$500 x the NYSE Composite Index	March, June, September, December
Value Line Index				
Kansas City Board of Trade (KCBT)	10:00 am to 4:15 pm (NYT)*	Equally-weighted average 1700 NYSE, AMEX, OTC, and regional stock prices expressed in index form	$500 x the Value Line Index	March, June, September, December
Major Market Index				
Chicago Board of Trade (CBT)	10:00 am to 4:15 pm (NYT)*	Price-weighted average of 20 blue-chip companies	$250 x MMI Index	March, June, September, December

**New York Time*

D. How Do You Transact in Futures?

You may invest directly in a commodity or indirectly through a mutual fund. A third way is to buy a limited partnership involved with commodity investments. The mutual fund and partnership approaches are more conservative, because risk is spread and they have professional management.

Futures may be directly invested as follows:

- Commodity pools—Professional traders manage a pool. A filing is made with the Commodity Futures Trading Commission (CFTC).
- Full service brokers—They may recommend something when attractive.
- Discount brokers—You must decide on your own when and if.
- Managed futures—You deposit funds in an individual managed account and choose a commodity trading advisor (CTA) to trade it.

To obtain information on managed futures, refer to:

- ATA Research Inc. provides information on trading advisors and manages individuals' accounts via private pools and funds.
- Barclay Trading Group publishes quarterly reports on trading advisers.
- *CMA Reports* monitors the performance of trading advisers and private pools.
- *Management Account Reports* are monthly newsletters, tracking the funds and furnishing information on their fees and track records.
- *Trading Advisor* follows more than 100 trading advisers.

There are several drawbacks to managed futures, including:

- High cost of a futures program, ranging from 15 to 20% of the funds invested
- Substantial risk and inconsistent performance of fund advisors

Note: Despite its recent popularity, management futures is still a risky choice and should not be done apart from a well-diversified portfolio.

E. Printed Chart Service and Software for Futures

There are many printed chart services such as Future Charts [Commodity Trend Service, (800) 331-1069 or (407) 694-0960]. Also, there are many computer software packages and other resources for futures analysis and charting service, including:

Strategist: Iotinomics Corp., (800) 255-3374 or (801) 466-2111
Futures Pro: Essex Trading Co., (800) 726-2140 or (708) 416-3530
Futures Markets Analyzer: Investment Tools, Inc., (702) 851-1157
Commodities and Futures Software Package, Foreign Exchange Software Package: Programmed Press, (516) 599-6527
Understanding Opportunities and Risks in Futures Trading: This 45-page booklet, prepared by National Futures Association, 200 West Madison St., Suite 1600, Chicago, IL 60606, provides a plain language explanation on opportunities and risks associated with futures investing. It can be obtained by writing to the Consumer Information Center, Pueblo, CO 81009.
Some Useful Web Sites: There are many Internet sites available to educate the investor on the rewards and risks associated with investing options, futures, and financial derivatives (see the Appendix for a list these Web sites).

Futures is a contract to deliver a specified amount of an item by some given future date. Futures differs from forward contracts in many ways (see Exhibit 65).

EXHIBIT 65
Futures versus Forward Contracts

Futures Contracts	Forward Contracts
Standardized contracts in terms of size and delivery dates	Customized contracts in terms of size and delivery dates
Standardized contract between a customer and a clearinghouse	Private contracts between two parties
Contract may be freely traded on the market	Impossible to reverse a contract
All contracts are marked-to-market; profits and losses are realized immediately	Profit or loss on a position is realized only on the delivery date
Margins must be maintained to reflect price movements	Margins are set once, on the day of the initial transaction

The advantages and disadvantages of using futures rather than forwards are presented in Exhibit 66.

EXHIBIT 66
Pros and Cons of Futures

Advantages	Disadvantages
Because of the institutional arrangements, the default risk is low	Futures exist only for a few high-turnover (large volume) exchange rates
Because of standardization, transaction costs or commissions are low	Because of standardization, a hedger may have to settle for an imperfect but inexpensive hedge in terms of the size or the amount
Due to the liquid nature of futures, futures position can be closed out early	Futures is available only for short maturities
	Futures contracts involve cash flow and interest rate risk

Futures and forwards appear to cater to two different clientele. As a general rule, forward markets are used primarily by corporate hedgers, whereas futures tend be preferred by speculators.

See also FOREIGN CURRENCY FUTURES; FORWARD CONTRACT.

FUTURES CONTRACTS

See FUTURES.

FUTURES FOREIGN EXCHANGE CONTRACT

See FUTURES.

FUTURES OPTIONS

Options on futures contracts are widely available in many currencies including Deutsche marks, British pounds, Japanese yen, Swiss francs, and Canadian dollars. Trading involves purchases and sales of puts and calls on a contract calling for delivery of a standard IMM futures contract in the currency rather than the currency itself. When such a contract is exercised, the holder receives a short or long position in the underlying currency futures contract that is *marked-to-market*, providing the holder with a cash gain. (If there were a loss on the futures contract, the option wouldn't be exercised and the spot market would be used instead.) Specifically,

1. If a call futures option contract is exercised, the holder receives a *long* position in the underlying futures contract plus an amount of cash equal to the current futures price minus the *strike price*.
2. If a put futures option is exercised, the holder receives a *short* position in the futures contract plus an amount of cash equal to the *strike price* minus the *current futures* price.

The seller (writer) of these options has the opposite position to the holder following exercise: a cash outflow plus a short futures position on a call and a long futures position on a put option. *Note*: The advantage of a futures *option* contract over a futures contract is that with a futures contract, the holder must deliver one currency against the other or reverse the contract, irrespective of whether this move is profitable. With the futures option contract, the holder is protected against an adverse movement in exchange rate but may allow the option to expire unexercised if it would be more profitable to use the spot market.

EXAMPLE 63

You are holding a pound call futures option contract for June delivery (representing £62,500) at a strike price of $1.4050. The current price of a pound futures contract due in June is $1.4148. You will receive a long position in the June futures contract established at a price of $1.4050 and the option writer has a short position in the same futures contract. These positions are immediately marked-to-market, triggering a cash payment to you from the option writer of 62,500 ($1.4148 – $1.4050) = $612.50. If you desire, you can immediately close out your long futures position at no cost, leaving you with the $612.50 payoff.

EXAMPLE 64

Suppose that you are holding one Swiss franc March put futures option contract (representing SFr 125,000) at a strike price of $0.6950. The current price of a Swiss franc futures contract due in March is $0.7132. Then you will receive a short position in the March futures contract established at a price of $0.6950 and the option writer has a long position in the same futures contract. These positions are immediately marked-to-market and you will receive a cash payment from the option writer of 125,000 ($0.7132 – 0.6950) = $2,275. If you wish, you can immediately close out the short futures position at no cost, leaving you with the $2,275 payoff.

A. Reading Futures Options

Exhibit 67 shows the Chicago Mercantile Exchange (IMM) options on a futures contract.

EXHIBIT 67
Futures Options

DEUTSCHE MARK (CME)—125,000 marks; cents per mark

	Calls—Settle[3]			Puts—Settle[4]		
Strike Price[1]	**Feb[2]**	**Mar**	**Apr**	**Feb**	**Mar**	**Apr**
5650	1.20	1.37	1.44	0.06	0.24	0.63
5700	0.75	1.04	1.15	0.11	0.40	0.83
5750	0.41	0.75	0.90	0.27	0.61	1.08
5800	0.20	0.52	0.69	0.56	0.88	....
5850	0.09	0.34	0.52	0.95	1.19	....
5900	0.02	0.22	0.39	1.38	1.57	....

5. Est. vol. 12,585; Wed. vol. 7,875 calls; 9,754 puts

6. Open interest Wed. 111,163 calls; 74,498 puts

Explanations:

1. Most active strike prices

2. Expiration months

3. Closing prices for call options

4. Closing prices for put options

5. Volume of options transacted in the previous two trading sessions, each unit representing both the buyer *and* the seller

6. The number of options in still open positions at the end of the previous day's trading session

To interpret the numbers in this column, consider the call options. These are rights to buy the June DM futures contract at specified prices—the strike prices. For example, the call option with a strike price of 5800 means that you can purchase an option to buy a June DM futures contract, up to the June settlement date, for $0.5800 per mark. This option will cost $0.0134 per Deutsche mark, or $1,675, plus brokerage commission, for a DM 125,000 contract. The price is high because the option is in-the-money. In contrast, the June futures option with a strike price of 6000, which is out-of-the-money, costs only $0.0044 per mark, or $550 for one contract. These option prices indicate that the market expects the dollar price of the Deutsche mark to exceed $0.5800 but not to go up much above $0.6000 by June. *Note*: A *futures call option* allows you to buy the relevant futures contract, which is settled at maturity. In contrast, a *call options* contract is an option to buy foreign exchange spot, which is settled when the call option is exercised; the buyer receives foreign currency immediately.

See CURRENCY OPTION; FOREIGN CURRENCY FUTURES; MARKED-TO-MARKET.

G

GAMMA

Gamma is a measure of the sensitivity of an option's *delta* to small changes in the price of the underlying security (or foreign exchange).
See also CURRENCY OPTION PRICING SENSITIVITY; DELTA.

GATT

See GENERAL AGREEMENT ON TARIFFS AND TRADE.

GENERAL AGREEMENT ON TARIFFS AND TRADE (GATT)

An agreement signed at the Geneva Conference in 1947 which became effective on January 1, 1948. It set a framework of policies and guidelines for international trade, including a negotiation of lower international trade barriers, and settling trade disputes. GATT also acts as international arbitrator with respect to trade agreement abrogation. More specifically, GATT has four basic long-run objectives: (1) reduction of tariffs by negotiation, (2) elimination of import quotas (with some exceptions), (3) nondiscrimination in trade through adherence to unconditional most-favored-nation treatment, and (4) resolution of differences through arbitration and consultation.

GENERIC

See PLAIN-VANILLA.

GEOMETRIC AVERAGE RETURN

See ARITHEMATIC AVERAGE RETURN VS. COMPOUND (GEOMETRIC) AVERAGE RETURN; RETURN RELATIVE.

GILT (OR GILT-EDGED)

A U.K. government bond.

GLOBAL BOND INDEXES

Today, investors do not have to settle for only the U.S. bond market to get fixed-income results. The *J.P. Morgan Government Bond Index* is considered the most widely-used benchmark for measuring performance and quantifying risk across international fixed income bond markets. The index and its underlying subindexes measure the total, principal, and interest returns in each of 13 countries and can be reported in 19 different currencies. The index limits inclusion to markets and issues that are available to international investors, to provide a more realistic measure of market performance.

Other global bond indexes include:

- *J.P. Morgan Emerging Markets Bond Index Plus:* Total return index of U.S. dollar and other external currency denominated Brady bonds, loans, Eurobonds, and local market debt instruments traded in emerging markets.
- *Salomon Smith Barney World Government Bond Index:* Tracks debt issues traded in 14 world government bond markets. Issues included in the Index have fixed-rate coupons and maturities of at least one year.

While finding bond yields is relatively easy, locating bond index results can be trickier. *The Wall Street Journal* and *Barron's* have extensive bond index coverage. And business news cable TV channels such as *CNBC* and *CNNfn* also track these indexes. Internet users can check Web sites such as www.bloomberg.com or www.bondsonline.com for bond index results.

GLOBAL FUND

A global fund is a mutual fund that invests globally—in both U.S. and foreign securities. Unlike an *international fund*, it invests anywhere in the world, including the United States. However, most global funds keep the majority of their assets in foreign markets.

GLOBAL REGISTERED SHARES

Global registered shares are equity shares that are registered and traded in many foreign equity markets. They contrast with *American Depository Receipts (ADRs)*, which are foreign company shares placed in trust with a U.S. bank, which in turn issues depository receipts to U.S. investors. For quotations on global shares, log on to www.adr.com by J.P. Morgan.

See also AMERICAN DEPOSITORY RECEIPTS; AMERICAN SHARES.

GOLD EXCHANGE STANDARD

The Gold Exchange Standard emerged as a result of negotiations at the *Bretton Woods Conference* in 1944. Under the Agreement, all countries were to fix the value of their currencies in terms of gold but were not required to exchange their currencies for gold. Only the U.S. dollar remained convertible into gold at $35 per ounce. All participating countries agreed to maintain the value of their currencies within 1% of par by buying or selling foreign exchange or gold as needed. But if a currency became weak, a devaluation of up to 10% would be allowed. Larger devaluations required the *IMF*'s approval.

GOLD STANDARD

The Gold Standard is the setting by most major countries of fixed values for their currencies in relation to gold. A country's currency is expressed on its equivalent value to gold such as one U.S. dollar equating to 23.22 grains of fine gold. It displays how much of the units of a currency are exchangeable into a certain amount of gold. If a significant outflow of gold occurs, a deficit in the balance of payments may arise. A gold standard may aid stability in exchange rates. It is a system in which currencies are defined in terms of their gold content and payment imbalances between countries are settled in gold.

GUILDER

Monetary unit of the Netherlands, Antilles, and Surinam.

GULF RIYAL

Monetary unit of Dubai and Qatar.

H

HARD CURRENCY

Often referred to as *convertible currency*, hard currency is the currency of a country that is widely accepted in the world and may be exchanged for that of another nation without restriction. Hard currency nations typically have sizeable surpluses in their balance of payments and foreign exchange reserves. The U.S. dollar and British pound are good examples.
See also SOFT CURRENCY.

HARD LANDING

See SOFT LANDING VS HARD LANDING.

HEDGE

Hedge is the process of protecting oneself against unfavorable changes in prices. One may enter into an offsetting purchase or sale agreement for the express purpose of balancing out any unfavorable changes in an already consummated agreement due to price fluctuations. Hedge transactions are commonly used to protect positions in (1) foreign currency, (2) commodities, and (3) securities. For example, MNCs engage in *forward contracts* to protect home currency value of various foreign-currency-denominated assets and liabilities on their balance sheet that are not to be realized over the life of the contracts. Also, the importer must consider the basis for the expected future spot rate and why that value diverges from the forward rate, the willingness to bear risk, and whether it has any offsetting currency assets.

HEDGE FUND

A hedge fund is a mutual fund that seeks to make money betting on a particular bond market, currency movements, or directional movements based on certain events such as mergers and acquisitions. The initial concept of the hedge fund, developed by Alfred Winslow Jones in the 1950s, was that securities of two different companies in similar businesses would have different characteristics in up and down markets. By buying the one likely to do the best in a rising market and selling short the one likely to do the worst in a falling market, the investor would be hedged and should make money no matter what direction the market took. Hedge funds offer plays *against* markets, using options, short-selling, futures, and other derivative products

HEDGER

Hedgers, mostly MNCs, are individuals or businesses engaged in *hedging* activities. They engage in *forward contracts* to protect the home-currency value of foreign-currency-denominated assets and liabilities on their balance sheets that are not to be realized over the life of the contracts.

HEDGE RATIO

A ratio comparing the amount you are hedging with the size of the position being hedged against. For example, a 25% hedge ratio means you have 1/4 of a portfolio with a neutral return.

HEDGING

Hedging involves entering into a contract at the present time to buy or sell a security (such as foreign exchange) at a specified price on a given future date.
See also HEDGE.

HERSATT RISK

Hersatt risk is *settlement risk*, named after a German bank that went bankrupt after losing a huge sum of money on foreign currencies.
See SETTLEMENT RISK.

HOLDING PERIOD RETURN

See TOTAL RETURN.

HOT MONEY

Used to describe money that moves internationally from one currency to another, either for speculation or because of *interest rate differentials*, and swings away immediately when the interest difference evaporates. An MNC is likely to withdraw funds from a foreign country having currency problems.

HYBRID FOREIGN CURRENCY OPTIONS

Hybrid foreign currency options involves the purchase of a put option and the simultaneous sale of a call—or vice versa—so that the overall cost is less than the cost of a straight option.

I

IBRD

See INTERNATIONAL BANK FOR RECONSTRUCTION AND DEVELOPMENT.

IMF

See INTERNATIONAL MONETARY FUND.

IMM

See INTERNATIONAL MONETARY MARKET; INTERNATIONAL MONEY MANAGEMENT.

IMPORT AND EXPORT PRICE INDEXES

The import and export price indexes measure price changes in agricultural, mineral, and manufactured products for goods bought from and sold to foreigners. They represent increases and decreases in prices of internationally traded goods due tô changes in the value of the dollar and changes in the markets for the items. Import and export price indexes are provided monthly by the Bureau of Labor Statistics in the U.S. Department of Labor. The data are published in a press release and in the BLS monthly journal, *Monthly Labor Review.* The import and export price indexes cover most foreign traded goods. The broad product categories of the indexes are food, beverages and tobacco, crude materials, fuels, intermediate manufactured products, machinery and transportation equipment, and miscellaneous manufactured products. The monthly figures cover approximately 10,000 products. Additional product detail is provided quarterly. Military equipment, works of art, commercial aircraft, and ships are excluded. Prices represent the actual transaction value including premiums and discounts from list prices and changes in credit terms and packaging. Prices usually are based on the time the item is delivered, not the time the order is placed. The indexes reflect movements for the same or similar items exclusive of enhancement or reduction in the quality or quantity of the item. The import and export price indexes are not seasonally adjusted.

INDIRECT QUOTE

The price of a unit of home currency, expressed in terms of a foreign currency. For example, in the Unites States, a quotation of 110 yens per dollar is an indirect quote for the Japanese yen. Indirect and *direct* quotations are reciprocals.

$$\text{Indirect quote} = \frac{1}{\textit{Direct quote}} = \frac{1}{\$0.00909} = 110 \text{ yens}$$

An indirect quote is the general method used in the over-the-counter market. Exceptions to this rule include British pounds, Irish punts, Australian dollars, and New Zealand dollars, which are quoted via *direct quote* for historical reasons. (e.g., 1 pound sterling = $1.68). In their foreign exchange activities, U.S. banks follow the European method of *direct quote*. See also AMERICAN TERMS; DIRECT QUOTE; EUROPEAN TERMS.

INFLATION

Inflation is the general rise in prices of consumer goods and services. The federal government measures inflation with four key indices: Consumer Price Index (CPI), Producer Price Index (PPI), Gross Domestic Product (GDP) Deflator, and Employment Cost Index (ECI). Price indexes are designed to measure the rate of inflation of the economy. Various price indexes are used to measure living costs, price level changes, and inflation.

- *Consumer Price Index:* The Consumer Price Index, the most well-known inflation gauge, is used as the cost-of-living index, to which labor contracts and social security are tied. The CPI measures the cost of buying a fixed bundle of goods (some 400 consumer goods and services), representative of the purchase of a typical working-class urban family. The fixed basket is divided into the following categories: food and beverages, housing, apparel, transportation, medical care, entertainment, and other. Generally referred to as a *cost-of-living index*, it is published by the Bureau of Labor Statistics of the U.S. Department of Labor. The CPI is widely used for escalation clauses. The base year for the CPI index was 1982–84 at which time it was assigned 100.
- *Producer Price Index:* Like the CPI, the PPI is a measure of the cost of a given basket of goods priced in wholesale markets, including raw materials, semifinished goods, and finished goods at the early stage of the distribution system. The PPI is published monthly by the Bureau of Labor Statistics of the Department of Commerce. The PPI signals changes in the general price level, or the CPI, some time before they actually materialize. (Because the PPI does not include services, caution should be exercised when the principal cause of inflation is service prices.) For this reason, the PPI and especially some of its subindexes, such as the index of sensitive materials, serve as one of the leading indicators that are closely watched by policy makers.
- *GDP Deflator:* This index of inflation is used to separate price changes in GDP calculations from real changes in economic activity. The Deflator is a weighted average of the price indexes used to deflate GDP so true economic growth can be separated from inflationary growth. Thus, it reflects price changes for goods and services bought by consumers, businesses, and governments. Because it covers a broader group of goods and services than the CPI and PPI, the GDP Deflator is a very widely used price index that is frequently used to measure inflation. The GDP deflator, unlike the CPI and PPI, is available only quarterly—not monthly. It is also published by the U.S. Department of Commerce.
- *Employment Cost Index:* This index is the most comprehensive and refined measure of underlying trends in employee compensation as a cost of production. It measures the cost of labor and includes changes in wages, salaries, and employer costs for employee benefits. ECI tracks wages and bonuses, sick and vacation pay plus benefits such as insurance, pension and Social Security, and unemployment taxes from a survey of 18,300 occupations at 4,500 sample establishments in private industry and 4,200 occupations within about 800 state and local governments.
- *CRB Bridge Spot Price Index* and the *CRB Bridge Futures Price Index:* These are two widely watched benchmarks for commodity prices by Bridge/CRB, formerly Commodity Research Bureau. The CRB Spot Price Index is based on prices of 23 different commodities, representing livestock and products, fats and oils, metals, and textiles and fibers, and it serves as an indicator of inflation. Higher commodity prices, for example, can signal inflation, which in turn can lead to higher interest rates and yields. The CRB Bridge Futures Price Index is

the composite index of futures prices that tracks the volatile behavior of commodity prices. As the best known commodity index, the CRB Futures Index, produced by Bridge Information Systems, was designed to monitor broad changes in the commodity markets. The CRB Futures Index consists of 21 commodities. In addition to the CRB Futures Index, nine subindexes are maintained for baskets of commodities representing currencies, energy, interest rates, imported commodities, industrial commodities, grains, oil, seeds, livestock and meats, and precious metals. All indexes have a base level of 100 as of 1967, except the currencies, energy, and interest rates indexes, which were set at 100 in 1977.

- *The Economist Commodities Index:* This is the gauge of commodity spot prices and their movements. The index is a geometric weighted-average based on the significance in international trade of spot prices of major commodities. The index is design to measure inflation pressure in the world's industrial powers. It includes only commodities that freely trade in open markets, excluding items such as iron and rice. Also, this index does not track precious metal or oil prices. The commodities tracked are weighted by their export volume to developed economies. The index information may be obtained from Reuters News Services or in *The Economist* magazine. The index may be used as a reflection of worldwide commodities prices, enabling the investor to determine the attractiveness of specific commodities. The MNC may enter into futures contracts. The indicators also may serve as barometers of global inflation and global interest rates.

Price indexes get major coverage, appearing in daily newspapers and business dailies, on business TV cable networks such *CNNfn* and *CNBC* and on Internet financial news services. Government Internet Web sites www.stats.bls.gov and www.census.gov/econ/ also provide this data.

A Word of Caution: Inflation results in an increase in all prices, but relative price changes indicate that not all prices move together. Some prices increase more rapidly than others, and some go up while others go down. Inflation is like an elevator carrying a load of tennis balls, which represent the prices of individual goods. As the inflation continues, the balls are carried higher by the elevator, which means that all prices are increasing. But as the inflation continues and the elevator rises, the balls, or individual prices, are bouncing up and down. So while the elevator lifts all the balls inside, the balls do not bounce up and down together. The balls bouncing up have their prices rising relative to the balls going down.

See also PURCHASING POWER PARITY; PURCHASING POWER RISK.

INFLATION RISK

See PURCHASING POWER RISK.

INITIAL MARGIN

The minimum amount of money (or equity) that must be provided by a margin investor at the time of purchase. It is used to prevent overtrading and excessive speculation. Margin requirements for stock have been 50% for some time. Margin requirements for foreign currency futures are determined and periodically revised by the *International Money Market (IMM)*, a division of the Chicago Mercantile Exchange, in line with changing currency volatilities using a computerized risk management program called SPAN (Standard Portfolio Analysis of Risk). See also MAINTENANCE MARGIN; MARGIN TRADING.

INTERBANK MARKET

The interbank market is the market for exchange of financial instruments—especially, foreign exchange—between commercial banks. Although the foreign exchange dealings of most managers involve a company buying from or selling to a bank, the vast majority of large-scale foreign exchange transactions are interbank. These transactions tend to determine exchange rates—with which occasional market participants such as companies must deal. Local and regional commercial banks *may* offer clients a foreign exchange service, which such banks provide on request by dealing through a larger bank, typically in a large city (such as New York, San Francisco, Chicago, Miami, and perhaps half a dozen more U.S. cities). Another surface of large-scale foreign exchange dealing in the United States is the *brokers' market*.

See also BROKERS' MARKET; FOREIGN EXCHANGE MARKET.

INTERBANK OFFERED RATE

Interbank Offered Rate (IBOR) is the rate of interest at which banks lend to other major banks. Terms are established for the length of loan and individual foreign currencies. A number of financial centers offer an IBOR, including: Abu Dhabi (DIBOR), Bahrain (BIBOR), Brussels (BRIBOR), Hong Kong (HKIBOR), London (LIBOR), Luxembourg (LUXIBOR), Madrid (MIBOR), Paris (PIBOR), Saudi Arabia (SAIBOR), Singapore (SIBOR), and Zurich (ZIBOR).

See also LONDON INTERBANK OFFERED RATE.

INTEREST AGIO

See INTEREST RATE DIFFERENTIAL.

INTEREST ARBITRAGE

Also called *intertemporal arbitrage*, interest arbitrage is an exchange arbitrage across maturities. It involves buying foreign exchange in the spot market, investing in a foreign currency asset, and converting back to the initial currency through a *forward contract*. It is similar to two- or three-way (*triangular*) arbitrage, in that it requires starting and ending with the same currency and incurring no exchange rate risk. In this case, profits are made by exploiting *interest rate differentials* as well as exchange rate differentials. In other words, this works when *the interest parity theory* is not valid. Also, an interest arbitrageur must use funds for the time period between contract maturities, while the two- and three-way arbitrageurs need funds only on the delivery date.

See also COVERED INTEREST ARBITRAGE.

INTEREST PARITY

See INTEREST RATE PARITY.

INTEREST RATE DIFFERENTIAL

Also called *interest agio*, the interest rate differential is the difference in interest rates between two nations. *The International Fisher effect* proposes that the spot exchange rate should change by the same amount as the interest rate differential between two countries.

INTEREST RATE PARITY

Interest rate parity (IRP) holds that investors should expect to earn the same return in all countries after adjusting for risk. It recognizes that when you invest in a country other than your home country, you are affected by two forces—returns on the investment itself and changes in the exchange rate. It follows that your overall return will be higher than the investment's stated return if the currency your investment is denominated in appreciates relative to your home currency. By the same token, your overall return will be lower if the overseas currency you are holding declines in value. The IRP is expressed as follows:

$$\frac{F}{S} = \frac{(1 + r_h)}{(1 + r_f)}$$

where F = forward rate, S = spot rate, and r_h and r_f = home (domestic) and foreign interest rates, respectively. Subtracting 1 from both sides yields:

$$\frac{F - S}{S} = \frac{(r_h - r_f)}{(1 + r_f)}$$

This implies that the differences between national interest rates for securities of similar risk and maturity should be equal to (but opposite in sign) the forward exchange rate differential between two currencies, except transaction costs. Specifically, the premium or discount (in *direct quotes*) should be:

$$P \text{ (or } D) = \frac{F - S}{S} = \frac{(r_h - r_f)}{(1 + r_f)} = -\frac{(r_f - r_h)}{(1 + r_f)}$$

Difference in interest rates		Difference between forward and spot rate
$-\frac{(r_f - r_h)}{(1 + r_f)}$	equals	$\frac{F - S}{S}$

When interest rates are relatively low, this equation can be approximated by

$$P \text{ (or } D) = \frac{F - S}{S} = -(r_f - r_h)$$

If this relationship, which is defined as interest rate parity, does not hold, then currency traders will buy and sell currencies—that is, engage in *arbitrage*—until it does hold. *Note:* The theory is applicable only to securities with maturities of one year or less.

EXAMPLE 65

To illustrate interest rate parity, suppose that you, as an investor, can buy default-free 90-day German bonds that promise a 4% nominal return and are denominated in German marks. The 90-day rate, r_f, equals 1% per 90 days (4% × 90/360). Assume also that the spot exchange rate is S = \$0.4982/DM, which means that you can exchange 0.4982 dollar for one mark, or DM 2.0074 per dollar. Finally, assume that the 90-day forward exchange rate, F, is \$0.5013, which means that you can exchange one mark for 0.5013 dollar, or receive DM 1.9949 per dollar exchanged, 90 days from now. You can receive a 4% annualized return denominated in marks, but if you want a dollar return, those marks must be converted to dollars. The dollar return of the investment depends, therefore, on what happens to exchange rates over the next three months.

However, you can lock in the dollar return by selling the foreign currency in the forward market. For example, you could

- Convert $1,000 to 2,007.40 marks in the spot market.
- Invest the 2,007.40 marks in 90-day German bonds that have a 4% annualized return or a 1% quarterly return, hence paying (2,007.40) (1.01) = 2,027.47 marks in 90 days.
- Agree today to exchange these 2,027.47 marks 90 days from now at the 90-day forward exchange rate of 1.9949 marks per dollar, for a total of $1,016.33.

This investment, therefore, has an expected 90-day return of [($1,016.33/$1,000) – 1] = 1.63%, which translates into a nominal return of 4(1.63%) = 6.52%. In this case, 4% of the expected 6.52% return is coming from the bond itself, and 2.52% arises because the market believes the mark will strengthen relative to the dollar.

Note: (1) By locking in the forward rate today, the investor has eliminated any exchange rate risk. And, since the German bond is assumed to be default-free, the investor is assured of earning a 6.52% dollar return. (2) The IRP implies that an investment in the United States with the same risk as a German bond should have a return of 6.52%. Solving for r_h in the preceding equation would yield the predicted interest rate in the United States of 6.52%.

$$\frac{\$0.5013 - 0.4982}{0.4982} \times \frac{12 \text{ months}}{3 \text{ months}} = 0.0250 = 2.50\%$$

$$P(\text{or } D) = \frac{F-S}{S} = 2.49\% = -(4\% - i_h)$$

Solving for r_h yields 6.50% (due to rounding errors).

(3) The IRP shows why a particular currency might be at a forward premium or discount. Notice that a currency (in *direct quotes*) is at a forward premium ($F > S$) whenever domestic interest rates are higher than foreign interest rates ($r_h > r_f$). Discounts prevail if domestic interest rates are lower than foreign interest rates. If these conditions do not hold, then arbitrage will soon power interest rates back to parity.

EXAMPLE 66

The 180-day U.S. T-bill interest rate in the U.S. is about 8%; the rate on 180-day United Kingdom instruments is 16%, annualized; and the 180-day forward premium for the pound is –5.86%, expressed as an annual rate. Given these data, does a riskless profit (or arbitrage) opportunity exist? The IRP says:

$$P\text{ (or } D) = \frac{F-S}{S} = \frac{(r_h - r_f)}{(1+r_f)} = -\frac{(r_f - r_h)}{(1+r_f)}$$

Step 1: Compute the forward premium P (or D).

$$P\text{ (or } D) = \frac{F-S}{S}$$

This is given in this example as $P = -5.86\%$.

Step 2: Compute the interest rate differential using

$$-\frac{(r_f - r_h)}{(1+r_f)}$$

$$-\frac{(0.16 - 0.09)}{(1+0.16)} = -0.603 = -6.03\%$$

The premiums obtained by the use of the two equations are not identical. The IRP is violated. Profit opportunity exists. An arbitrage profit can be made by doing the following:

1. Borrow U.S. dollars at 8%.
2. Exchange the U.S. dollars for British pounds, at the spot rate.
3. Invest the pounds at 16% for 180 days in the British money market.
4. Sell the pounds at the 180-day forward rate.
5. Accept U.S. dollars in 180 days by delivering the pounds invested in the U.K. market.
6. Repay loan of U.S. dollars, pocketing the profit.

This set of transactions is called *covered-risk arbitrage* since risk is *covered* by the forward market.

EXAMPLE 67

On May 3, 20x1, a 30-day forward contract in Japanese yens (in *direct quotes*) was selling at a 4.82% premium using the equation

$$P\,(\text{or } D) = \frac{F - S}{S}$$

to annualize yields would give:

$$\frac{\$.012003 - .011955}{.011955} \times \frac{12 \text{ months}}{1 \text{ month}} = 0.482 = 4.82\%$$

The 30-day U.S. T-bill rate is 8% annualized. Then, the 30-day Japanese rate,

using the equation:

$$P\,(\text{or } D) = -\frac{(r_f - r_h)}{(1 + r_f)}$$

would be:

$$.0482 = -\frac{r_f - .08}{1 + r_f}$$

$$-0.0318 = -1.0482\, r_f$$

$$r_f = 0.0303 = 3.03\%$$

The 30-day Japanese rate should be 3.03%.

EXAMPLE 68

Country	Contract	U.S. Dollar Equivalent
Japan (Yen)	Spot	$.009524
	90-day forward	?

The interest rate in Japan on 90-day government securities is 4%; it is 8% in the U.S. If the interest rate parity theory holds, then the implied 90-day forward exchange rate in yen per dollar would be calculated as follows:

Step 1: Compute the premium (discount) on the forward rate using:

$$P(\text{or } D) = -\frac{(r_f - r_h)}{(1 + r_f)} = -\frac{.04 - .08}{1 + .04} = .0385$$

Step 2: Compute the forward rate using the equation:

$$P(\text{or } D) = \frac{F - S}{S} \times \frac{12 \text{ months}}{n} \times 100$$

$$.0385 = \frac{F - .009524}{.009524} \times \frac{12}{3} \times 100$$

Solving this equation for F yields \$.009616.

EXAMPLE 69

Almond Shoe, Inc. sells to a wholesaler in Germany. The purchase price of a shipment is 50,000 DMs with terms of 90 days. Upon payment, Almond will convert the DM to dollars.

Country	Contract	U.S. Dollar Equivalent	Currency per U.S. \$
Germany (Mark)	Spot	.730	1.37
	90-day future	.735	1.36

Note that the DM is at a premium because the 90-day forward rate of DM per dollar is less than the current spot rate. The DM is expected to strengthen (*fewer* DM to buy a dollar). The differential in interest rates is –2.73%, as shown below.

$$\frac{(7.35 - 7.30)}{7.30} \times \frac{12}{3} \times 100 = -(r_f - r_d) = 2.73\%$$

This means that if the interest parity theory holds, interest rates in the U.S. should be 2.73% higher than those in Germany.

Exhibit 68 illustrates the conditions necessary for equilibrium. The vertical axis shows the difference in interest rates in favor of the foreign currency, and the horizontal axis shows the forward premium or discount on that currency. The interest rate parity line shows the equilibrium state, but transaction costs cause the line to be a band rather than a thin line. Point X shows one possible equilibrium position, where a 3% lower rate of interest on yen securities would be offset by a 3% premium on the forward yen. The disequilibrium situation is illustrated by point U. It is located off the interest rate parity line because the lower interest on the yen is 3% (annual basis), whereas the premium on the forward yen is slightly over 4.5% (annual basis). The situation depicted by point U is unstable because all investors have an incentive to execute a covered interest arbitrage whose gain is virtually risk free.

EXHIBIT 68
Interest Rate Parity and Equilibrium

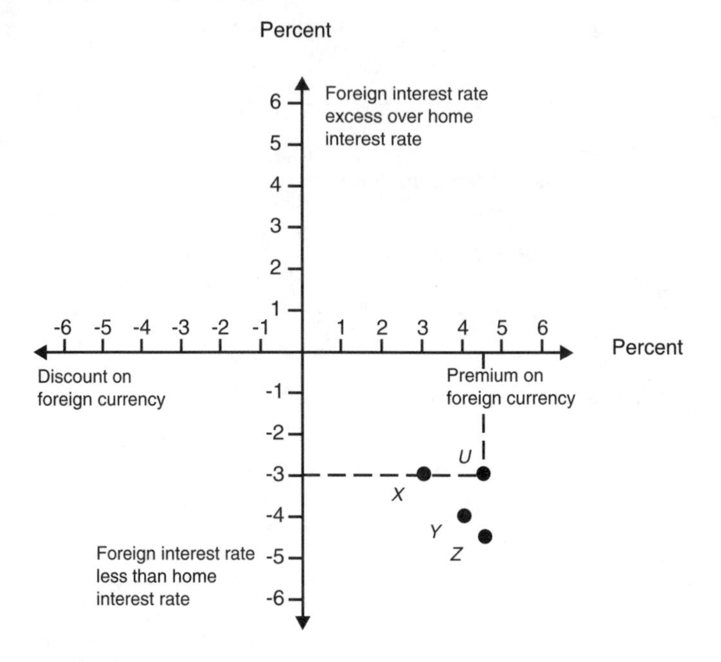

INTEREST RATE RISK

The variability in the value of a security as the interest rates and conditions of the money and capital markets change, interest rate risk relates to fixed-income securities such as bonds. For example, if interest rates rise (fall), bond prices fall (rise). Price changes induced by interest-rate changes are greater for long-term than for short-term bonds. There are different types of interest rate risk: (1) *basis risk* in which the interest rate base is mismatched and (2) *gap risk* in which the timing of maturities is mismatched.

INTEREST RATE SWAPS

An interest rate swap is an exchange by two parties of interest rates on borrowings made in the two markets—fixed for floating and floating for fixed. Consequently, each party obtains the type of liability it prefers and at a more attractive rate. The advantages to interest rate swaps are the following: (1) The swap contract is simple and straightforward. (2) Swaps provide alternative sources of financing. (3) Swaps give the corporation the flexibility to convert floating-rate debt to fixed-rate, and vice versa. (4) There are potential interest rate savings. (5) Swaps may be based on outstanding debt and may thus avoid increasing liabilities. (6) Swaps are private transactions. (7) Rating agencies, such as Moody's and S&P, take a neutral to positive position on corporate swaps. (8) Tax treatment on swaps is uncomplicated, as there are no withholding taxes levied on interest payments to overseas swap partners, and the interest expense of the fixed-rate payer is treated as though it were on a fixed-rate obligation. One major drawback to swaps is the risk that one swap partner may fail to make the agreed payments to the other swap partner. It is a growing trend that MNCs will use swaps to match assets to liabilities and to protect investments in capital assets, such as plant and equipment, from floating-rate interest fluctuations. Financial institutions also see swaps as a way to match receivables (loans made) to liabilities (investors deposits).

See also SWAPS.

INTEREST RATES

Interest rates take many different forms. They are as follows:

A. Discount, Fed Funds, and Prime Rates

These are three key interest rates closely tied to the banking system. The discount rate is the rate the Federal Reserve Board charges on loans to banks that belong to the Fed system. The fed or federal funds rate is the rate that bankers charge one another for very short-term loans, although the Fed heavily manages this rate as well. The prime rate is the much discussed benchmark rate that bankers charge customers. The three rates, at times, work in tandem. The Fed Funds rate is the major tool that the nation's central bank, the Federal Reserve, has to manage interest rates. Changing the target for the Fed Funds rate is done when the Fed wants to use monetary policy to alter economic patterns. The discount rate was once the Fed's key tool. Now it takes a backseat, as a largely ceremonial nudge to markets made often after Fed Funds changes are implemented. The prime rate is a heavily tracked rate although it is not as widely used as a corporate loan benchmark as it has been in the past. The prime is set by bankers to vary loan rates to smaller businesses and on consumers' home equity loans and credit cards.

B. 30-Year Treasury Bonds

The most widely watched interest rate in the world, the security known as the *T-bond* is seen as the daily barometer of how the bond market is performing. The 30-year Treasury bond is a fixed-rate direct obligation of the U.S. government. There are no call provisions on Treasury bonds. Traders watch the price of the U.S. Treasury's most recently issued 30-year bond, often called the *bellwether*. The price is decided by a series of dealers who own the exclusive right to make markets in the bonds in U.S. markets. (The bond trades around the clock in foreign markets.) Bond yields are derived from the current trading price and its fixed coupon rate.

Because of its long-term nature, the T-bond is extra sensitive to inflation that could ravage the buying power of its fixed-rate payouts. Thus, the T-bond market also is watched as an indicator of where inflation may be headed. Also, T-bond rates somewhat impact fixed-rate mortgages. (These loans are more often tied to 10-year Treasury rate.) Still, the T-bond yield is also seen as a barometer for the housing industry, a key leading indicator for the economy.

C. Three-Month Treasury Bills

The Treasury bill rate is a widely watched rate for secure cash investments. In turbulent times the rate can be volatile and can be viewed as a signal of the economy's health. T-bills, both three-month and six-month issues, are auctioned every Monday by the U.S. Treasury through the Federal Reserve. The T-bill rate shows what can be expected to be earned on no-risk investments. Historically, T-bills have returned little more than the inflation rate. Many conservative investors buy T-bills directly from the government. T-bill rates approximate rates on money-market mutual funds or statement savings accounts, also popular savings tools for the small investor.

INTERNAL FINANCIAL TRANSFER SYSTEM

The internal financial transfer system of the MNC covers various mechanisms for transferring funds internally. It includes transfer prices on goods and services traded internally, dividend payments, *leading and lagging* intercompany payments, payments for fees and royalty charges, intercompany loans, and equity investments.

INTERNAL RATE OF RETURN

Internal rate of return (IRR) is defined as the rate of interest that equates initial capital outlay (I) with the present value (PV) of future cash inflows. Or at IRR, $I = PV$.

Decision rule: Accept the project if the IRR exceeds the cost of capital. Otherwise, reject it.

EXAMPLE 70

Consider the following foreign investment project:

Initial investment (I)	\$12,950,000
Estimated life	10 years
Annual cash inflows (CF)	\$3,000,000
Cost of capital	12%

We set the following equality ($I = PV$):

$$\$12{,}950{,}000 = \$3{,}000{,}000 \times T4(k, 10 \text{ years})$$

where $T4$ is a present value of an annuity factor.

$$T4(k, 10 \text{ years}) = \frac{\$12{,}950{,}000}{\$3{,}000{,}000} = 4.317$$

which stands somewhere between 18% and 20% in the 10-year line of Exhibit 4 in the Appendix. The interpolation follows:

PV of an Annuity of \$1 Factor
$T4(k,10$ years)

18%	4.494	4.494
IRR	4.317	
20%		4.192
Difference	0.177	0.302

Therefore,

$$\text{IRR} = 18\% + \frac{0.177}{0.302}(20\% - 18\%) = 18\% + 0.586(2\%) = 18\% + 1.17\% = 19.17\%$$

Since the IRR of the investment is greater than the cost of capital (12%), accept the project. The advantage of using the IRR method is that it considers the time value of money. The shortcomings of this method are that (1) it fails to recognize the varying sizes of investment in competing project, and (2) it is time-consuming to compute, especially when the cash inflows are not even. However, spreadsheet software and financial calculators can be used in making IRR calculations. For example, MSExcel has a function IRR(*values*, *guess*). Excel considers negative numbers such as the initial investment as cash outflows and positive numbers as cash inflows. Many financial calculators have similar features. As in the example, suppose you want to calculate the IRR of a \$12,950,000 investment (the value –12950000 entered in year 0 that is followed by 10 year cash inflows of \$3,000,000. Using a guess of 12% (0.12), which is in effect the cost of capital, your formula would be @IRR(values, 0.12) and Excel would return 19.15%, as shown below.

Year 0	1	2	3	4	5	6	7	8	9	10
–12,950,000	3,000,000	3,000,000	3,000,000	3,000,000	3,000,000	3,000,000	3,000,000	3,000,000	3,000,000	3,000,000
IRR =	19.15%									

INTERNATIONAL ACCOUNTING STANDARDS COMMITTEE

The International Accounting Standards Committee (IASC), founded in 1973, aims at the development of international accounting standards. It also works toward the improvement and harmonization of accounting standards and procedures relating to the presentation and comparability of financial statements (or at least through enhanced disclosure, if differences are present). To date, it has developed a conceptual framework and issued a total of 32 International Accounting Standards (IAS) covering a wide range of accounting issues. It is currently working on a project concerned with the core standards in consultation with other international groups, especially the International Organization of Securities Commissions (IOSCO), to develop worldwide standards for all corporations to facilitate multilisting of foreign corporations on various stock exchanges. At the inception, its members consisted of the accountancy bodies of Australia, Canada, France, Japan, Mexico, the Netherlands, the United Kingdom, Ireland, the United States, and Germany. Since its founding, membership has grown to around 116 accountancy bodies from approximately 85 countries.

INTERNATIONAL BANK FOR RECONSTRUCTION AND DEVELOPMENT (IBRD)

The International Bank for Reconstruction and Development, also called the *World Bank* (http://www.worldbank.org), was established in December 1945 to help countries reconstruct their economies after World War II. IBRD assists developing member countries by lending to government agencies and by guaranteeing private loans for such projects as agricultural modernization or infrastructural development. It attempts to promote economic and social progress through the creation of modern economic and social infrastructures. It makes loans to countries or firms for such purposes as roads, irrigation projects, and electric generating plants. Bank headquarters are in Washington, D.C.

See also WORLD BANK.

INTERNATIONAL BANKING FACILITY (IBF)

An international Banking Facility (IBF), authorized in December 1981, is a separate banking operation within a domestic U.S. bank, created to allow that bank to accept *Eurocurrency deposits* from foreign residents without the need for domestic reserve requirements, interest rate regulations, or deposit insurance premiums applicable to normal U.S. banking. IBFs simply require a different set of books to receive deposit from, and make loans to, nonresidents of the U.S. or other IBFs. IBFs are not institutions in the organizational sense, but accounting entities that represent a separate set of asset and liability accounts of their establishing offices. They are actually a set of asset and liability accounts segregated on the books of the establishing institutions. IBFs are allowed to conduct international banking operations that, for the most part, are exempt from U.S. regulation. Deposits, which can be accepted only from non-U.S. residents or other IBFs and must be at least $100,000, are exempt from reserve requirements and interest rate ceilings. The deposits obtained cannot be used domestically; they must be used for making foreign loans. In fact, to ensure that U.S.-based firms and individuals comply with this requirement, borrowers must sign a statement agreeing to this stipulation, when taking out the loan.

INTERNATIONAL CAPITAL ASSET PRICING MODEL

The international capital asset pricing model (ICAPM) is an international version of the *Capital Asset Pricing Model (CAPM)*. It differs from a domestic CAPM in two respects. First, the relevant market risk is world market risk, not domestic market risk. Second, additional risk premium is linked to an asset's sensitivity to currency movements. The ICAPM can be used to estimate the required return on foreign projects, taking into account the world market risk.

INTERNATIONAL CAPITAL BUDGETING

See ANALYSIS OF FOREIGN INVESTMENTS.

INTERNATIONAL CASH MANAGEMENT

See INTERNATIONAL MONEY MANAGEMENT.

INTERNATIONAL DEVELOPMENT ASSOCIATION

The International Development Association (IDA), a part of the *World Bank* Group, was created in 1959 (and began operations in November 1990) to lend money to developing countries at no interest and for a long repayment period. IDA provides development assistance through soft loans to meet the needs of many developing countries that cannot afford development loans at ordinary rates of interest and in the time span of conventional loans. The Association's headquarters are in Washington, D.C.
See also WORLD BANK.

INTERNATIONAL DIVERSIFICATION

International diversification is an attempt to reduce the multinational company's risk by operating facilities in more than one country, thus lowering the *country risk*. It is also an effort to reduce risk by investing in more than one nation. By diversifying across nations whose *business cycles* do not move in tandem, investors can typically reduce the variability of their returns. Adding international investments to a portfolio of U.S. securities diversifies and reduces your risk. This reduction of risk will be enhanced because international investments are much less influenced by the U.S. economy, and the correlation to U.S. investments is much less. Foreign markets sometimes follow different cycles from the U.S. market and from each other. Although foreign stocks can be riskier than domestic issues, supplementing a domestic portfolio with a foreign component can actually reduce your portfolio's overall volatility. The reason is that by being diversified across many different economies which are at different points in the economic cycle, downturns in some markets may be offset by superior performance in others.

There is considerable evidence that global diversification reduces systematic risk (*beta*) because of the relatively low correlation between returns on U.S. and foreign securities. Exhibit 69 illustrates this, comparing the risk reduction through diversification within the United States to that obtainable through global diversification. A fully diversified U.S. portfolio is only 27% as risky as a typical individual stock, while a globally diversified portfolio appears to be about 12% as risky as a typical individual stock. This represents about 44% less than the U.S. figure.

Exhibit 70 demonstrates the effect over the past ten years. Notice how adding a small percentage of foreign stocks to a domestic portfolio actually decreased its overall risk while increasing the overall return. The lowest level of volatility came from a portfolio with about 30% foreign stocks and 70% U.S. stocks. And, in fact, a portfolio with 60% foreign holdings and only 40% U.S. holdings actually approximated the risk of a 100% domestic portfolio, yet the average annual return was over two percentage points greater.

The benefits of international diversification can be estimated by considering the portfolio risk and portfolio return in which a fraction, w, is invested in domestic assets (such as stocks, bonds, investment projects) and the remaining fraction, $1 - w$, is invested in foreign assets:

EXHIBIT 69
Risk Reduction from International Diversification

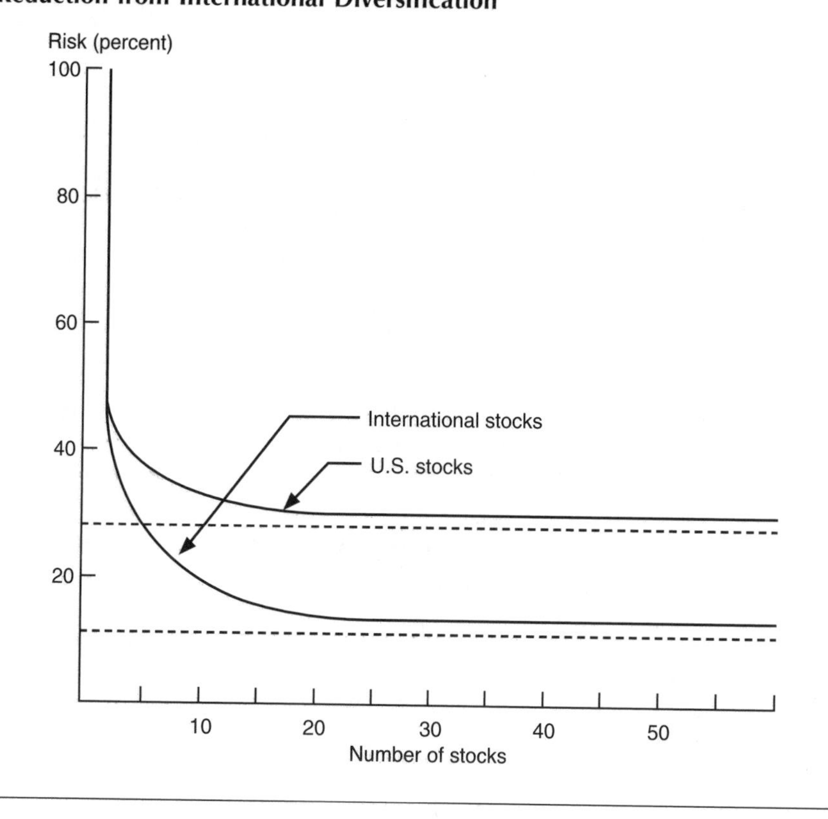

EXHIBIT 70
How Foreign Stocks Have Benefitted a Domestic Portfolio

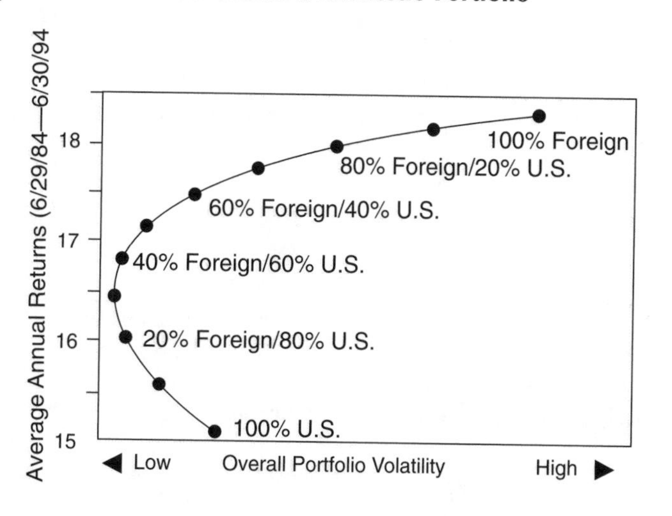

The expected portfolio return is calculated as follows:

$$r_p = wr_d + (1 - w)r_f$$

where r_d = return on domestic assets and r_f = return on foreign assets.

The expected portfolio standard deviation is calculated as follows:

$$\sigma_p = \sqrt{w^2\sigma_d^2 + (1-w)^2\sigma_f^2 + 2\rho_{d,f}^{2}w(1-w)\sigma_d\sigma_f}$$

where σ_d and σ_f = standard deviation on domestic and foreign assets, respectively, and ρ_{df} = correlation coefficient between domestic and foreign assets.

The risk of an internationally diversified portfolio is less than the risk of a fully diversified domestic portfolio.

EXAMPLE 71

Suppose that three projects are being considered by U.S. Minerals Corporation: Nickel projects in Australia and South Africa and a zinc mine project in Brazil. The firm wishes to invest in two plants, but it is unsure of which two are preferred. The relevant data are given below.

	Component Projects		
	Nickel Projects		Zinc Mine
	Australia	South Africa	Brazil
Mean return	0.20	0.25	0.20
Standard deviation	0.10	0.25	0.12
Correlation coefficient			

Correlation coefficient (Australia–South Africa): 0.8

Correlation coefficient (South Africa–Brazil): 0.2

Correlation coefficient (Australia–Brazil): 0.2

Possible portfolios and their portfolio returns and risks are the following:

A. Australian and South African Nickel Operations:

Mean return = 0.5(0.20) + 0.5(0.25) = 0.225 = 22.5%

Standard deviation

$$= \sqrt{(0.5)^2(0.10)^2 + (0.5)^2(0.25)^2 + 2(0.8)(0.5)(0.5)(0.10)(0.25)}$$

$$= \sqrt{0.028125} = 0.168 = 16.8\%$$

B. Australian Nickel Operation and Brazil Zinc Mine:

Mean return = 0.5(0.20) + 0.5(0.20) = 0.20 = 20%

Standard deviation

$$= \sqrt{(0.5)^2(0.10)^2 + (0.5)^2(0.25)^2 + 2(0.2)(0.5)(0.5)(0.10)(0.12)}$$

$$= \sqrt{0.0073} = 0.085 = 8.5\%$$

C. South African Nickel Operation and Brazil Zinc Mine:

Mean return = 0.5(0.25) + 0.5(0.20) = 0.225 = 22.5%

Standard deviation $= \sqrt{(0.5)^2(0.10)^2 + (0.5)^2(0.25)^2 + 2(0.2)(0.5)(0.5)(0.25)(0.12)}$

$$= \sqrt{0.02223} = 0.149 = 14.9\%$$

To summarize:

Portfolio	Mean Return	Standard Deviation
B. Australian Nickel Operation and Brazil Zinc Mine	20.0%	8.5%
C. South African Nickel Operation and Brazil Zinc Mine	22.5%	14.9%
A. Australian and South African Nickel Operations	22.5%	16.8%

The efficient portfolios, in increasing order of returns, are portfolios B, C , and A. Portfolio A can be eliminated as being inferior to portfolio C—both portfolios yield a mean return of 22.5%, but portfolio A has a higher risk than portfolio C. Management has to select between portfolios B and C, based on their risk–return trade-off.

See also PORTFOLIO THEORY.

INTERNATIONAL EXCHANGE RATE PARITY CONDITIONS

See PARITY CONDITIONS.

INTERNATIONAL FINANCIAL CENTERS

International banking is heavily concentrated on cities in which international money center banks are located, such as New York, London, and Tokyo. Four major types of financial transactions transpire in an international financial center that is in effect an important domestic financial center. Exhibit 71 displays major transactions that occur in this arena.

EXHIBIT 71
Major Types of Financial Transactions in an International Financial Market Arena

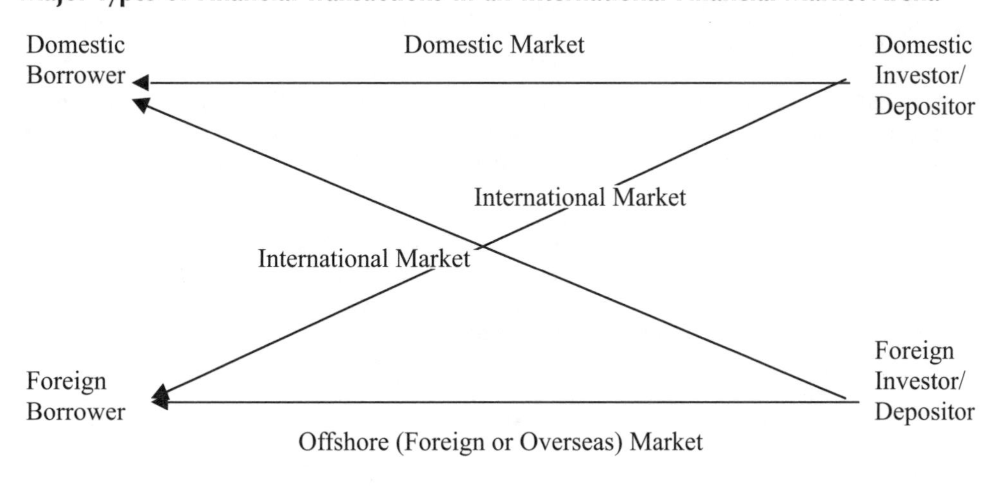

INTERNATIONAL FINANCING

1. Also called *foreign financing, overseas financing, or offshore financing*, raising capital in the *Eurocurrency* or *Eurobond* markets.
2. A strategy used by MNCs for financing *foreign direct investment*, international banking activities, and foreign business operations.

INTERNATIONAL FISHER EFFECT

Often, called *Fisher-open*, the theory states that the spot exchange rate should change by the same amount as the interest differential between two countries. The International Fisher effect is derived by combining the *purchasing power parity (PPP)* and the *Fisher effect.*

$$\frac{S_2}{S_1} = \frac{(1 + r_h)}{(1 + r_f)} \qquad \text{(Equation 1)}$$

where r_h and r_f = the respective national interest rates and S = the spot exchange rate (using *direct* quotes) at the beginning of the period (S_1) and the end of the period (S_2).

According to Equation 1, the expected return from investing at home, $(1 + r_\eta)$, should equal the expected home currency (HC) return form investing abroad, $(1 + r_f)\ S_2/S_1$.

EXAMPLE 72

In March, the one-year interest rate is 4% on Swiss francs and 13% on U.S. dollars.

(a) If the current exchange rate is *SFr* 1 = \$0.63, the expected future exchange rate in one year would be \$0.6845:

$$S_2 = S_1\ (1 + r_h)/(1 + r_f) = 0.613 \times 1.13/1.04 = \$0.6845$$

(b) Assume that the Swiss interest rate stays at 4% (because there has been no change in expectations of Swiss inflation). If a change in expectations regarding future U.S. inflation causes the expected future spot rate to rise to \$0.70, according to the international Fisher effect, the U.S. interest rate would rise to 15.56%:

$$S_2/S_1 = (1 + r_h)/(1 + r_f)$$
$$0.70/0.63 = (1 + r_h)/1.04$$
$$r_h = 15.56\%$$

A simplified version states that, for any two countries, the spot exchange rate should change in an equal amount but in the opposite direction to the difference in the nominal interest rates between the two countries. It can be stated more formally:

$$\frac{S_2 - S_1}{S_1} = r_h - r_f \qquad \text{(Equation 2)}$$

Subtracting 1 from both sides of Equation 1 yields:

$$\frac{S_2 - S_1}{S_1} = \frac{(r_h - r_f)}{(1 + r_f)}$$

Equation 2 follows if r_f is relatively small.

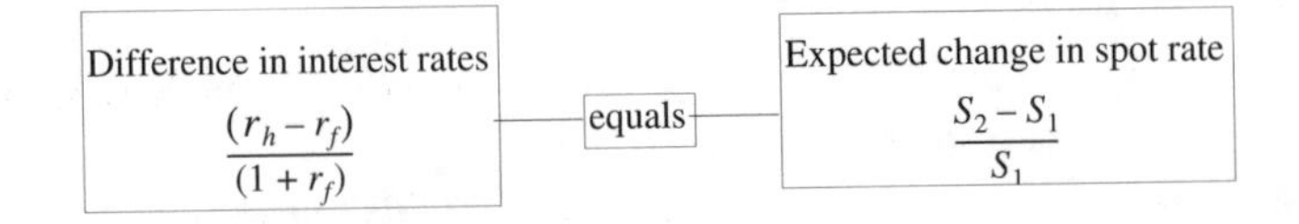

The rationale behind this theory is that investors must be rewarded or penalized to offset the change in exchange rates. Thus, the currency with the lower interest rate is expected to appreciate relative to the currency with the higher interest rate.

EXAMPLE 73

If a U.S. dollar-based investor buys a one-year yen deposit earning 4% interest, compared with 10% interest in dollars, the investor must be expecting the yen to appreciate vis-à-vis the dollar by about 6% (10% – 4% = 6%) during the year. Otherwise, the dollar-based investor would be better off staying in dollars.

A graph of Equation 2 in Example 72 is presented in Exhibit 72. The vertical axis shows the expected change in the home currency value of the foreign currency, and the horizontal axis shows the interest differential between the two countries for the same time period. The parity line shows all points for which $r_h - r_f = (S_2 - S_1)/S_1$. Point A is a position of equilibrium because it lies on the parity line, with the 4% interest differential in favor of the home country just offset by the anticipated 4% appreciation in the home currency value of the foreign currency. Point B, however, illustrates a case of disequilibrium. If the foreign currency is expected to appreciate by 3% in terms of the home currency but the interest differential in favor of the foreign country is only 2%, then funds flow from the home to the foreign country to take advantage of the higher exchange-adjusted returns there. This capital flow will continue until exchange-adjusted returns are equal in the two nations.

EXHIBIT 72
International Fisher Effect

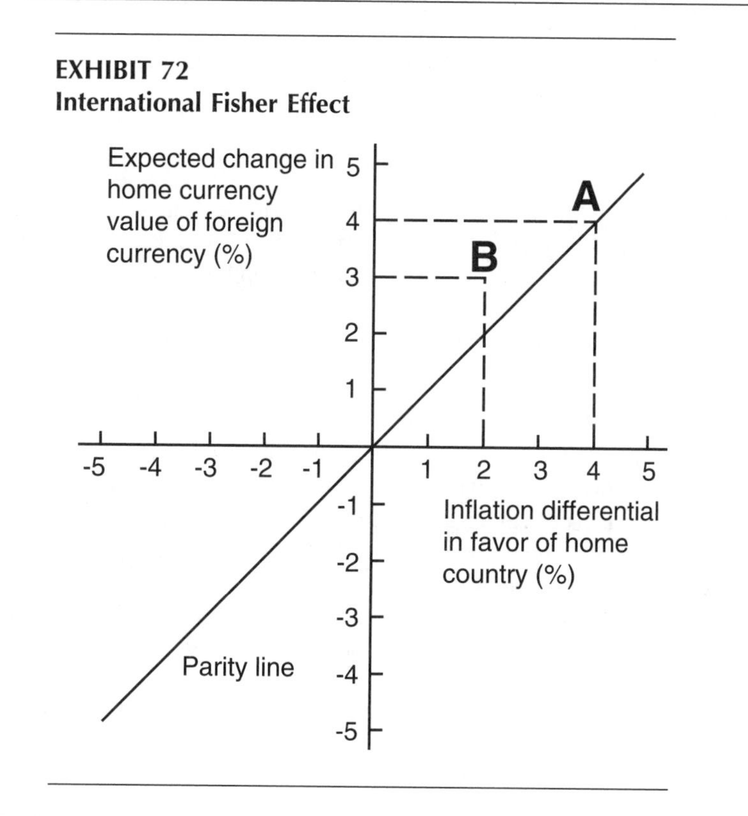

INTERNATIONAL FUND

Also called a *foreign fund*, an international fund is a mutual fund that invests only in foreign stocks. Because these funds focus only on foreign markets, they allow investors to control what portion of their personal portfolio they want to allocate to non-U.S. stocks. There exists currency risk associated with international fund investing. *Note:* General Electric Financial Network (www.gefn.com), for example, has a tool "How do exchange rates affect my foreign fund?" (www.calcbuilder.com/cgi-bin/calcs/MUT12.cgi/gefa).

INTERNATIONAL LENDING

International lending involves some risks: (1) *Commercial risk* (business risk) as in domestic lending, and (2) the added risk comes from cultural differences and lack of information (especially due to differing accounting standards and disclosure practices)—*Country risk* including *political risk* and *currency risk.* Further, the central role played by the enforcement problem and the absence of collateral make international lending fundamentally different from domestic lending.

See also COMMERCIAL RISK; CURRENCY RISK; POLITICAL RISK.

INTERNATIONAL MONETARY FUND (IMF)

International Monetary Fund (IMF) (www.imf.org) is an international financial institution that was created in 1946 after the 1944 *Bretton Woods Conference.* It aims at promoting international monetary harmony, monitoring the exchange rate and monetary policies of member nations, and providing credit for member countries which experience temporary balance of payments deficits. Each member has a quota, expressed in *Special Drawing Rights,* which reflects both the relative size of the member's economy and that member's voting power in the Fund. Quotas also determine members' access to the financial resources of, and their shares in the allocation of Special Drawing Rights by, the Fund. The IMF, funded through members' quotas, may supplement resources through borrowing.

INTERNATIONAL MONETARY MARKET

International Monetary Market (IMM) is a division of the Chicago Mercantile Exchange where currency futures contracts, patterned after grain and commodity contracts, are traded. Futures contracts are currently traded in the British pound, Canadian dollar, German mark, Swiss franc, French franc, Japanese yen, Australian dollar, and U.S. dollar. Most recently, the IMM has introduced a cross-current futures contract (e.g., DM/¥).

INTERNATIONAL MONETARY SYSTEM

1. The financial market for transactions between countries that belong to the International Monetary Fund (IMF), or between one of these countries and the IMF itself. A market among the central banks of these countries, functioning as a kind of central banking system for the national governments of its 137 members. Each member country deposits funds at the IMF, and in return each may borrow funds in the currency of any other member nation. This system is *not* open to private sector participants, so it is not directly useful to company managers. However, agreements made between member countries and the IMF often lead to major changes in government policies toward companies and banks (such as exchange rate changes and controls and trade controls), so an understanding of the international monetary system may be quite important to managers. Regulation in this system comes through rules passed by the IMF's members. The major financial instruments used in the international monetary system are national currencies, gold, and a currency issued by the IMF itself, called the *SDR* (*special drawing right*).
2. The sum of all of the devices by which nations organize their international economic relations.

3. The set of policies, arrangements, mechanisms, legal aspects, customs, and institutions dealing with money (investments, obligations, and payments) that determine the rate at which one currency is exchanged for another.

INTERNATIONAL MONEY MANAGEMENT

Also called *international working capital management* or narrowly *international cash management*, international money management (IMM) is concerned with financial policies used by MNCs aiming at optimizing profitability from currency and interest rate fluctuation while controlling risk exposure. It can be considered as comprising a series of interrelated subsystems that perform the following functions: (1) positing of funds—choice of location and currency of denomination for all liquid funds, (2) pooling funds internationally, (3) keeping costs of intercompany funds transferred at a minimum, (4) increasing the speed with which funds are transferred internationally between corporate units, and (5) improving returns on liquid funds.

INTERNATIONAL MONEY MARKET

The international money market is the *Eurocurrency market* and its linkages with other segments of national markets for credit. One unique feature of the international money market is the diversity of its participants, the wide range of borrowers and lenders that compete with one another on the same basis. It is simultaneously an *interbank market*, a market where governments raise funds, and a lending and deposit market for corporations. The market is extremely homogeneous in its treatment of borrowers and lenders. While in national markets there is invariably credit rationing during periods of tight credit, often mandated by government, in the Euromarkets the funds are always available for those willing and able to pay the price. Equally important, the market's size assures that the marginal cost of funds is less. Another advantage to borrowers is that funds raised in the international money market have no restrictions attached as where they can be deployed. And also, the Euromarkets provide corporate borrowers with flexibility as to terms, conditions, covenants, and even currencies. The international money market parallels the *foreign exchange market*. It is located in the same centers as its foreign exchange counterparts. The market operates only in those currencies for which forward exchange market exists and that are easily convertible and available in sufficient quantity.

INTERNATIONAL RETURNS

When investors buy and sell assets in other countries, they must consider exchange rate risk. This risk can convert a gain from an investment into a loss or a loss from an investment into a gain. An investment denominated in an appreciating currency relative to the investor's domestic currency will experience a gain from the currency movement, while an investment denominated in a depreciating currency relative to the investor's domestic currency will experience a decrease in the return because of the currency movement. To calculate the return from an investment in a foreign country, we use the following formula:

$$\text{Total return (TR) in domestic terms} = \text{Return relative (RR)} \times \frac{\text{Ending value of foreign currency}}{\text{Beginning value of foreign currency}} - 1.0$$

The foreign currency is stated *direct* terms; that is, the amount of domestic currency necessary to purchase one unit of the foreign currency.

EXAMPLE 74

Consider a U.S. investor who invests in UniMex at 175.86 pesos when the value of the peso stated in dollars is $0.29. One year later UniMex is at 195.24 pesos, and the stock did not pay a dividend. The peso is now at $0.27, which means that the dollar *appreciated* against the peso. Return relative for UniMex = 195.24/175.86 = 1.11

Total return to the U.S. investor after currency adjustment is

$$
\begin{aligned}
\text{TR denominated in \$} &= 1.1 \times \frac{\$0.27}{\$0.29} - 1.0 \\
&= [1.11 \times 0.931] - 1.0 \\
&= 1.0334 - 1.0 \\
&= .0334 \text{ or } 3.34\%
\end{aligned}
$$

In this example, the U.S. investor earned an 11% total return denominated in Mexican currency, but only 3.34% denominated in dollars because the peso declined in value against the U.S. dollar. With the strengthening of the dollar, the pesos from the investment in UniMex buy fewer U.S. dollars when the investment is converted back from pesos, pushing down the 11% return a Mexican investor would earn to only 3.34% for a U.S. investor.

INTERNATIONAL SOURCES OF FINANCING

An MNC may finance its activities abroad, especially in countries in which it is operating. A successful company in domestic markets is more likely to be able to attract financing for international expansion. The most important international sources of funds are the *Eurocurrency market* and the *Eurobond market*. Also, MNCs often have access to national capital markets in which their subsidiaries are located. Exhibit 73 represents an overview of international financial markets.

EXHIBIT 73
International Financial Markets

Market	Instruments	Participants	Regulator
International monetary system	Special drawing rights; gold; foreign exchange	Central banks; International Monetary Fund	International Monetary Fund
Foreign exchange markets	Bank deposits; currency; futures and forward contracts	Commercial and central banks; firms; individuals	Central banks in each country
National money markets (short term)	Bank deposits and loans; short-term government securities; commercial paper	Banks; firms; individuals; government agencies	Central bank; other government agencies
National capital markets	Bonds; long-term bank deposits and loans; stocks; long-term government securities	Banks; firms; individuals; government agencies	Central bank; other government agencies

Eurocurrency markets (short term)	Bank deposits; bank loans; short-term and rolled-over credit lines; revolving commitment	Commercial banks; firms; government agencies	Substantially unregulated
Euro-commercial paper markets (short term)	Commercial paper issues and programs; note-issuing facility; revolving underwritten facilities	Commercial banks; firms; government agencies	Substantially unregulated
Eurobond market (medium and long term)	Fixed coupon bonds; floating-rate notes; higher-bound bonds; lower-bound bonds	Banks; firms; individuals; government agencies	Substantially unregulated
Euroloan market (medium and long term)	Fixed-rate loans; revolving loans; revolving loans with cap; revolving loans with floor	Banks; firms; individuals; government agencies	Substantially unregulated

The Eurocurrency market is a largely short-term (usually less than one year of maturity) market for bank deposits and loans denominated in any currency except the currency of the country where the market is located. For example, in London, the Eurocurrency market is a market for bank deposits and loans denominated in dollars, yen, franc, marks, and any other currency except British pounds. The main instruments used in this market are CDs and time deposits, and bank loans. *Note:* The term *market* in this context is not a physical market place, but a set of bank deposits and loans. The Eurobond market is a long-term market for bonds denominated in any currency except the currency of the country where the market is located. Eurobonds may be of different types such as straight, convertible, and with warrants. While most Eurobonds are fixed rate, variable rate bonds also exist. Maturities vary but 10 to 12 years are typical. Although Eurobonds are issued in many currencies, you wish to select a stable, fully convertible, and actively traded currency.

In some cases, if a Eurobond is denominated in a weak currency the holder has the option of requesting payment in another currency. Sometimes, large MNCs establish wholly owned offshore finance subsidiaries. These subsidiaries issue Eurobond debt and the proceeds are given to the parent or to overseas operating subsidiaries. Debt service goes back to bondholders through the finance subsidiaries. If the parent issued the Eurobond directly, the U.S. would require a withholding tax on interest. There may also be an estate tax when the bondholder dies. These tax problems do not arise when a bond is issued by a finance subsidiary incorporated in a tax haven. Hence, the subsidiary may borrow at less cost than the parent. In summary, the Euromarkets offer borrowers and investors in one country the opportunity to deal with borrowers and investors from many other countries, buying and selling bank deposits, bonds, and loans denominated in many currencies. Exhibit 74 provides a list of credit sources available to a foreign affiliate of an MNC.

EXHIBIT 74
International Sources of Credit

Borrowing	Domestic Inside the Firm	Domestic Market	Foreign Inside the Firm	Foreign Market	Euromarket
Direct, short-term	Intrafirm loans, transfer pricing, royalties, fees, service charges	Commercial paper	International intrafirm loans, international transfer pricing, dividends, royalties, fees		Euro-commercial paper
Intermediated short-term		Short-term bank loans, discounted receivables	Internal back-to-back loans	Short-term bank loans, discounted receivables	Euro short-term loans
Direct, long-term	Intrafirm loans, invested in affiliates	Stock issue, bond issue	International intrafirm long-term loans, foreign direct investment	Stock issue, bond issue	Eurobonds
Intermediated long-term		Long-term bank loans	Internal back-to-back loans	Long-term bank loans	Euro long-term loans

INTERNATIONAL STANDARD (ISO) CODE

Internationally agreed standard codes for foreign currencies are created by the International Standards Organization (ISO; www.xe.net/currency/iso_4217.htm). The following lists the commonly used symbols for several international currencies and their international standard (ISO) code.

Country	Currency	Symbol	ISO Code
Austrlia	Dollar	A$	AUD
Austria	Schilling	Sch	ATS
Belgium	Franc	BFr	BEF
Canada	Dollar	C$	CAD
Denmark	Krone	DKr	DKK
Finland	Markka	FM	FIM
France	Franc	FF	FRF
Germany	Deutsche mark	DM	DEM
Greece	Drachma	Dr	GRD
India	Rupee	Rs	INR
Iran	Rial	RI	IRR
Italy	Lira	Lit	ITL
Japan	Yen	¥	JPY

Kuwait	Dinar	KD	KWD
Mexico	Peso	Ps	MXP
Netherlands	Guilder	FL	NLG
Norway	Krone	NKr	NOK
Saudi Arabia	Riyal	SR	SAR
Singapore	Dollar	S$	SGD
South Africa	Rand	R	ZAR
Spain	Peseta	Pta	ESP
Sweden	Kronar	SKr	SEK
Switzerland	Franc	SF	CHF
United Kingdom	Pound	£	GBP
United States	Dollar	$	USD

INTERNATIONAL TRANSFER PRICING

A transfer price is defined as the price charged by a selling department, division, or subsidiary of a multinational national company (MNC) for a product or service supplied to a buying department, division, or subsidiary of the same MNC (in different countries). A major goal of transfer pricing is to enable divisions that exchange goods or services to act as independent businesses. It also encompasses the determination of interest rates for loans, charges for rentals, fees for services, and the methods of payments. International transfer pricing is an important issue for several reasons. First, raw materials not available or in short supply for an MNC unit in one country can be imported for sale or further processing by another unit of the MNC located in a different country. Second, some stages of an MNC's production process can be conducted more efficiently in countries other than where the MNC has its headquarters. Third, many MNCs operate sales offices in some countries but do no manufacturing there. To sell their products, the sales offices or subsidiaries must import products from manufacturing affiliates in other countries. Fourth, many services for MNC units are rendered by MNC headquarters or other affiliates of an MNC. Finally, there are many international financial flows between units of an MNC. Some are payments related to goods or services provided by other units; some are loans or loan repayment; some are dividends; and some are designed to lessen taxes or financial risks. Since the transfer price for a product has an important effect on performance of individual foreign subsidiary managers, their motivation, divisional profitability, and global profits, top management of MNCs should devote special attention to designing international transfer pricing policies.

A. Factors Influencing Transfer Price Determination

MNCs typically have a variety of objectives. Maximizing global after-tax profits is a major goal. Other goals often include increasing market share, maintaining employment stability and harmony, and being considered the "best" firm in the industry. However, not all of these goals are mutually compatible or collectively achievable. In addition, all MNCs face governmental and other constraints which influence their ability to achieve their objectives in the manner they would prefer. In determining international inter-corporate transfers and their prices, an MNC must consider both its objectives and the constraints it faces.

An MNC can also achieve further tax savings by manipulating its transfer prices to and from its subsidiaries. In effect, it can transfer taxable income out of a high-tax country into a lower-tax country. This tax scheme can be particularly profitable for MNCs based in a country that taxes only income earned in that country but does not tax income earned outside the country. But even if a country taxes the global income of its corporations, often income earned abroad is not taxable by the country of the corporate parent until it is remitted to the parent. If penetrating a foreign market is a company's goal, the company can underprice goods sold to foreign affiliates, and the affiliates can then sell them at prices which their local competitors cannot match. And if antidumping laws exist on final products, a company can underprice components and semi-finished products to its affiliates. The affiliates can then assemble or finish the final products at prices that would have been classified as dumping prices had they been imported directly into the country rather than produced inside. Transfer prices can be used in a similar manner to reduce the impact of tariffs.

Although no company can do much to change a tariff, the effect of tariffs can be lessened if the selling company underprices the goods it exports to the buying company. The underpricing of inter-corporate transfers can also be used to get more products into a country that is rationing its currency or otherwise limiting the value of goods that can be imported. A subsidiary can import twice as many products if they can be bought at half price. Artificially high transfer prices can be used to circumvent or lessen significantly the impact of national controls. A government prohibition on dividend remittances to foreign owners can restrict the ability of a firm to transfer income out of a country. However, overpricing the goods shipped to a subsidiary in such a country makes it possible for funds to be taken out. High transfer prices can also be considerable when a parent wishes to lower the profitability of its subsidiary. This may be caused by demands by the subsidiary's workers for higher wages or participation in company profits, political pressures to expropriate high-profit foreign-owned operations, or the possibility that new competitors will be lured into the industry by high profits. High transfer prices may be desired when increases from existing price controls in the subsidiary's country are based on production costs. Transfer pricing can also be used to minimize losses from foreign currency fluctuations, or shift losses to particular affiliates. By dictating the specific currency used for payment, the parent determines whether the buying or the selling unit has the exchange risk. Altering the terms and timing of payments and the volume of shipments causes transfer pricing to affect the net foreign exchange exposure of the firm.

International transfer pricing has grown in importance with international business expansion. It remains a powerful tool with which multinational companies can achieve a wide variety of corporate objectives. At the same time, international transfer pricing can cause relations to deteriorate between multinationals and governments because some of the objectives achievable through transfer price manipulation are at odds with government objectives. Complex manipulated transfer pricing systems can also make the evaluation of subsidiary performance difficult and can take up substantial amounts of costly, high-level management time. In spite of these problems, the advantages of transfer price manipulation remain considerable. These advantages keep international transfer pricing high on the list of important decision areas for multinational firms. Usually, multinational companies should be more considerate than domestic companies to set transfer prices, as they have to cope with different sets of laws, different competitive markets, and different cultures. Thus, it is not surprising that determining price for international sales is very difficult, especially for internal transactions among the segments.

When planning an internal sales price (transfer pricing) strategy, a corporation should be concerned about subsidiaries' contributions and competitive positions as well as the whole corporation's profitability, because subsidiaries' contributions do not always increase overall company profit. High income means more tax. Perhaps, for instance, the parent company wants to show losses and pass income to its segments in low tax rate areas also, transfer pricing should benefit both sides, seller and buyer. Otherwise, inappropriate transfer pricing may cause company conflicts and may even lower profits. For example, if the transfer price of the parent company is too high, the subsidiaries may buy from outside parties even though buying from the parent company may be better for the organization as a whole. On the contrary, if the subsidiaries want to buy at very low prices, the parent company may not make deals with them at such a low price because it could get more money elsewhere. Thus, how to set appropriate transfer prices is not easy. Multinational companies should know transfer price methods very well. They should focus on transfer price considerations, such as tax rates, competition, custom duties, currencies, and government legislation.

B. Transfer Price Structure

The four types of transfer prices used for management accounting purposes are: cost-based transfer price, market-based transfer price, negotiated transfer price, dictated transfer price.

- *Cost-based transfer price* is based on full or variable cost. It is simple for companies to apply, and it is a useful method to strengthen compatibility. The major disadvantages of cost-based transfer price are that it lacks incentives to control costs by selling divisions, and it is unable to provide information for companies to evaluate performance by the Return on Investment (ROI) formula. In order to increase the efficiency of cost control, companies should use standard costs rather than actual costs. Also, because many tax agencies require international firms to present transfer prices fairly, the use of the cost-based method may be deemed an unfair transfer price.
- *Market-based transfer price* is the one charged for products or services based on market value. It is the best approach to solve the transfer pricing problem. It connects costs with profits for managers to make the best decisions and provides an excellent basis for evaluating management performance. However, setting market-based transfer prices should meet the following two conditions: (1) the competitive market condition must exist, and (2) divisions should be autonomous from each other for decision making.
- *Negotiated transfer price* is set by the managers of the buying and selling divisions with an agreement. A major advantage of this transfer pricing is that both sides are satisfied. But, it has some disadvantages. The division in which the manager is a good negotiator may get more profits than those in which the manager is a poor negotiator. Also, managers may spend a lot of time and costs in the negotiations. Usually, companies use negotiated transfer prices in those situations where no intermediate market prices are available.
- *Dictated transfer price* is determined by top managers. They set the price in order to optimize profit for the organization as a whole. The disadvantage is that the dictated transfer price may conflict with the decisions that division managers make.

C. Transfer Price Considerations

In order to be successful in business, companies must consider any policy very carefully, and transfer pricing is no exception. Income taxes and the various degrees of competition are very important considerations. Custom duties, exchange controls, inflation, and currency exchange rates are usually considered. Moreover, multinational companies should think over the whole companies' profits when they set transfer prices. On the other hand, transfer pricing strategies by parent companies should not injure subsidiaries' interests. For example, an American firm has a subsidiary in a country with high tax rates. In order to minimize the subsidiary's tax liability and draw more money out of the host country, the parent company sets high transfer prices on products shipped to the subsidiary and sets low transfer prices on those imported from the subsidiary. However, this procedure may cause the subsidiary to have high duties, or it may increase its product cost and reduce its competitive position. Therefore, multinational companies should weigh the importance of each factor when planning transfer pricing strategies.

In the 1990s, many firms tend to hold the overall profit concept as the basic idea for transfer pricing strategies. Also, a lot of companies think of other considerations, such as differentials in income tax rates and income tax legislation among countries and the competitive position of foreign subsidiaries.

D. Tax Purposes

The basic idea of transfer pricing for tax purposes is to maneuver profit out of high tax rate countries into lower ones. The foreign subsidiaries can sell at or below cost to other family members in lower tax rate areas, thereby showing a loss in its local market, while contributing to the profit of buying members.

However, inappropriate transfer prices may cause companies to be exposed to tax penalties. For example, a parent company in the U.S. thinks that the U.S. tax rates are lower than its segment in a foreign country. The parent firm does not want to comply with the tax law of that foreign country. The company sets a high transfer price for its subsidiary so that profit of the subsidiary can be shifted to the parent company. But if the transfer price does not comply with the foreign country's tax law, the profit shifted may be lost due to a tax penalty.

In addition, revenue flights become significant as the countries grow to compete for international tax income. This evidence stimulated national treasuries to take actions to strengthen the power of controlling transfer pricing practices. During the past ten years, the United States and its major trading partners have revised or introduced new transfer pricing regulations.

E. IRS Transfer Pricing Regulations

Section 482 of the Internal Revenue Code of 1986 authorizes the IRS to allocate gross income, deductions, credits, or allowances among controlled taxpayers if such allocation is necessary to prevent evasion of taxes.

Another provision in this section defines intangibles to include (1) patents, (2) copyrights, (3) know-how, (4) trademarks and brand names, (5) franchises, and (6) customer lists. Guided by this rule, the IRS can collect royalties commensurate with the economic values of intangibles and can prevent many U.S. parent companies from transferring intangibles to related foreign subsidiaries at less than their value.

At the end of 1990, the IRS issued the proposed regulations. Under the proposed regulations, the U.S. subsidiaries owned by a foreign multinational company were required to submit the detailed records that reflect the profit or loss of each material industry segment. Noncompliance with the regulations may cause financial penalties.

In addition, other countries, such as Canada and Japan enforced new transfer pricing regulations. In June 1990, the European community countries reached agreements for the harmonization of direct taxes in Europe. They made a draft on transfer pricing arbitration to resolve transfer pricing disputes between member countries.

F. The Tax Implementation Problems Faced by MNCs

All of the changes above bring high pressure on managers and high cost to firms to maintain appropriate income allocation. First, traditional management transfer pricing methods are based on marginal revenue and marginal cost. These techniques do not satisfy the documentation and verification rules for tax purposes. Thus, managers must find appropriate transfer pricing methods to comply with tax complication requirements. Second, tax rules require that the transfer price methods should meet comparability and unrelated party standards. Following these tax codes will increase a global company's information costs. For example, multinational companies should submit various data and documents for different tax compliance requirements. They may even hire tax consultants to prepare all the necessary documents. Third, a manager must carefully analyze all the potential economic considerations of transfer pricing; otherwise, failure in following tax compliance requirements may cause heavy penalties.

G. Transfer Pricing Methods for Tax Purposes

When the transfer price does not satisfy tax requirements, the firm can reset its transfer pricing systems. However, this approach requires companies to apply multiple transfer pricing methods fluently. Usually, there are six transfer pricing methods for tax purposes. Exhibit 75 summarizes these six transfer pricing methods and an Other category.

EXHIBIT 75
Transfer Pricing Methods for Tax Purposes: Tangible Property

Method Description	Comparable Uncontrolled Price	Resale Price	Cost Plus	Comparable Profits	Profits Split	Other
Comparable factors	Comparable sales between unrelated parties	Price to unrelated party less related gross profit; nonmanufacturing	Production costs plus gross profit on unrelated sales	Priced to yield gross profits comparable to those for other firms	Split of combined operating profits of controlled parties	Gross profit reasonable for facts and circumstances
Comparability and Reliability Standards	Similarity of property; underlying circumstance	Comparable gross profit relative to comparable unrelated transfer	Gross profit from same type of goods in unrelated resale	Gross profit within range of profits for broadly similar product line	Allocation of combined profits of controlled parties	As appropriate
Measures of Comparability	Functional diversity; product category; terms in financing and sales; discounts; and the like	Functional diversity; product category; terms in financing and sales; intangibles; and the like	Functional diversity; accounting principles; direct vs. indirect costing; and the like	Business segment; functional diversity; different product categories acceptable if in the same industry	Profits split by unrelated parties or splits from transfers to unrelated parties	Fair allocation of profits relative to unrelated party sales
Same Geographic Market	Required	Required	Required	Required	Required, but some flexibility	Required, but some flexibility
Comments	Deemed the best method for all firms; minor accounting adjustments allowed to qualify as "substantially the same"	The best method for distribution operations; only used where little or no value added and no significant processing	Internal gross profit ratio is acceptable if there are both purchases from and sales to unrelated parties	Not if seller has unique technologies or intangibles because resale price is fixed; adjust the transfer price from seller	Controlled transaction allocations compared to profits split in uncontrolled transactions	Least reliable; uncertainty and costs of being wrong are severe

G.1. Comparable Uncontrolled Price

This price is based on comparable prices through transactions with unrelated parties. The company that focuses on a market-based organization uses this method. In a market-based structure, the company's segments are autonomic and independent from each other. The managers can decide to make transactions with unrelated parties if the prices offered by other members in the company are not reasonable.

To illustrate the use a comparable uncontrolled method, a parent company sells fiber to its foreign segment and to other parties in its domestic market. On the other hand, its foreign segment buys fiber from the parent company as well as from other manufacturers in the local market. Thus, under a comparable uncontrolled method, the parent company can set a transfer price according to both selling and buying comparable prices resulting from transactions with unrelated parties.

However, comparable uncontrolled prices are only acceptable for those global companies who make internal transactions among their segments, and they do not compete with each other in their backyards.

G.2. Resale Price Method

This method is the best for intermediate distributions, such as wholesalers and retailers. It also applies to market-based organizations. Usually these companies add little or no value to goods and do not have a significant manufacturing process.

The formula for this method is:

$$\text{Transfer Price} = \text{Resale Price to Unrelated Party} - \text{Gross Profit Ratio} \times \text{Resale Price}$$

Computing the gross profit ratio is based on information on the profit ratios in the same product categories used by unrelated parties. However, the information on profit ratios set by competitors is not readily obtainable and may be costly for global companies.

G.3. The Cost-Plus Method

This method is adaptable to manufacturing companies. Under this method the amount of company product cost is adjusted for gross profit ratios. The ratio can be internal gross profit ratio if both sides in the company purchase from and sell to unrelated parties and have comparable price standards, or the ratio can be based on comparable company's profit ratios for the same broad product category.

The formula of the cost-plus method is:

$$\begin{aligned}\text{Transfer Price} = {} & \text{Production Cost} + \text{Gross Profit Ratio} \\ & \times \text{Sales Price to Unrelated Parties}\end{aligned}$$

G.4. Comparable Profit Method

This is a profit markup method. The gross profit part of the transfer price should be compared to others within a range of profits for broadly similar product lines. The profit ratio should be based on some internal profit indicator, such as rate of return. However, if the product or process involved is unique in the market, setting transfer prices under this method is unacceptable.

G.5. Profits-Split Method

Under this method, MNCs allocate the combined profits of subsidiaries that are involved in internal transactions. Parent companies compute the combined profits after these goods to customers are sold outside of the group. Also, the profit for each member involved in intercompany transactions is comparable to unit profits where unrelated parties participate in similar activities with comparable products. The profits-split method requires companies to obtain reliable detailed data for comparable products. Usually, it is not difficult for the company to get aggregate profit data for the whole product line, but there is not enough detailed data for analysis and comparison. Thus, appropriate profits-split pricing relies on whether the information on profits is reliable.

G.6. Other Methods

One of the five transfer pricing methods cannot be adopted all the time. For example, an MNC trades products only among its members. Each member does not purchase from or sell to unrelated parties because those products are unique and no company outside uses them. Under this situation, when the parent company sets transfer prices, there are no reliability and comparability standards to match because comparable products in the markets do not exist. Therefore, the company cannot use any of the transfer pricing method mentioned above. When none of the five specific methods can be applied reasonably, the company may choose another method. The method should be reasonable under the facts and circumstances, and should fairly allocate profits relative to unrelated party sales. However, there are no objective guidelines under this approach. The company may face challenges by tax agencies that could result in high costs for noncompliance. To minimize the risk of penalties, companies should have the documents to prove why a method was chosen.

H. Competitive Position

For an MNC, its overall international competitive position is its major consideration in determining prices. It may adjust internal transfer prices to increase its segment competition so as to increase the whole company's profitability. Companies who use cost-based pricing methods especially present this approach. Under cost-based pricing, companies calculate cost on different bases, such as full cost, variable cost, and marginal cost. These cost bases allow the parent company to change its internal transfer price accordingly. For example, when a subsidiary faces serious competition in its local market, the parent company can charge the subsidiary at a reduced price based on full cost or only variable cost. In this way, the subsidiary can decrease its sale prices quickly to maintain its market share. On the other hand, in order to increase the subsidiary's competition, the parent company may raise transfer prices which it pays to its subsidiary so that the parent company can help its division make high profit. In this way, the parent company can improve its division's ability to get loans from local banks. As a result, the subsidiary can be supported by significant financing to defeat its local competitors. Therefore, through various transfer pricing strategies, multinational corporations create good conditions for their subsidiaries to improve divisional contribution to the profitability and competitive position of the overall corporation.

I. The Influence of Governments

As many governments become more sensitive to their loss of tax revenues or to the negative effects of competition suffered by purely local producers, they revise or even create international laws to control international transfer pricing practices. Therefore, multinational companies should focus on "fairness" of the transfer price and on regulations specifying market prices. Also, multinational companies should improve their long-term relationship with foreign governments.

J. The Influence of Currency Changes

Typically in many foreign countries, particularly in Latin America, the value of the local currency depreciates quite rapidly relative to the United States dollar. For those American corporations having subsidiaries in those areas, the parent companies may charge them high transfer prices for exported products or may pay them a low transfer price for imported products. By so doing, the parent companies can draw divisional profit out of the host countries and minimize the exchange risk.

K. Custom Duties

In the 1990s, MNCs tended to consider the rate of custom duties and customs legislation as an important factor when they planned international transfer pricing strategies. If there were a high custom rate in a divisional country, the parent company may have offered a low transfer price to its subsidiary for parts shipping to the division. Thus, the parent company reduced the custom duty payments and helped the subsidiary compete in foreign markets by keeping costs low.

INTERNATIONAL UNDERWRITING SYNDICATE

An international underwriting syndicate is a group of *investment bankers* engaged in public offerings of debt issues such as Eurobonds. The offering procedure for Eurobonds is much like that of a domestic issue. The offering is preceded by a prospectus and is then marketed by an international underwriting syndicate.

See also INVESTMENT BANKER.

INTERNATIONAL WORKING CAPITAL MANAGEMENT

See INTERNATIONAL MONEY MANAGEMENT.

INTERNATIONAL YIELD CURVES

International yield curves are graphical presentations of the *term structure of interest rates*—yield-to-maturity—for each currency, typically the yield curves for government bonds. International interest rate differentials are caused by a variety of factors, such as differences in national monetary and fiscal policies, and inflationary and foreign exchange expectations.

See also TERM STRUCTURE OF INTEREST RATES.

INTERTEMPORAL ARBITRAGE

Also called *interest arbitrage*, intertemporal arbitrage is exchange arbitrage across maturities, similar to *simple (two-way) arbitrage* and *triangular (three-way) arbitrage*, in that it requires starting and ending with the same currency and incurring no currency risk. Profits are made, however, by exploiting interest rate differentials, as well as exchange rate differentials. Further, the intertemporal *arbitrageur* (or interest arbitrageur) must employ funds for the time period between contract maturities, while the two- and three-way arbitrageurs need funds only on the delivery date. See also SIMPLE ARBITRAGE; TRIANGULATION.

INTI

Peru's currency.

INTRINSIC VALUE

1. An intrinsic value is the basic theoretical value of a *call* or *put option*. It is the amount by which the option is *in-the-money*; that is, the spot price *minus* the strike (or exercise) price.

See also CURRENCY OPTION; OPTION.

2. The present value of expected future cash flows of an asset.

See also VALUATION.

INVESTMENT BANKER

An investment banker is a professional who specializes in marketing primary offerings of securities. Investment bankers buy new securities (equity or debt issues) from issuers and resell them publicly, that is they underwrite the risk of distributing the securities at a satisfactory price. They can also perform other financial functions, such as (1) advising clients about the types of securities to be sold, the number of shares or units of distribution, and the timing of the sale; (2) negotiating mergers and acquisitions; and (3) selling secondary offerings.

Most investment bankers function as broker-dealers and offer a growing variety of financial products and services to their wholesale and retail clients. Investment bankers typically form an *international underwriting syndicate* in order to handle a large volume of Eurobond issues. See also INTERNATIONAL UNDERWRITING SYNDICATE.

INVESTMENT COMPANY

See INVESTMENT TRUST.

INVESTMENT TRUST

An investment company that invests in other companies after which it sells its own shares to the public. If it is a closed-end company, it sells its shares only. If it is an open-end company or a mutual fund, it repeatedly buys and sells its shares.

IRREVOCABLE LETTER OF CREDIT

An irrevocable letter of credit is a noncancelable *letter of credit (L/C)* in which the designated payment is guaranteed by the *issuing bank* if all terms and conditions are met by the drawee. An irrevocable letter of credit, once issued, cannot be amended or canceled without the agreement of all named parties. As such, it must have a fixed expiration date. It is as good as the issuing bank.

ISSUING BANK

An issuing bank is a bank which opens a straight or a negotiable *letter of credit*. This bank takes up the obligation to pay the beneficiary or a *correspondent bank* if the documents presented are compliant with the terms of the letters of credit.

J

JAPANESE TERM

A Japanese yen rate of one U.S. dollar, generally known as *European terms*. A Japanese yen quote of ¥105.65/$ is called *Japanese terms.*

J-CURVE EFFECT

Devaluation is conventionally believed to be a tool of improving a nation's trade deficit. It is believed that it takes time for an exchange rate change to affect *trade balance*. International trade transactions are prearranged and cannot be immediately adjusted. For this reason, the J-curve effect suggests that a decline in the value of a currency does not always improve a trade deficit. A currency devaluation will initially worsen the trade deficit before showing signs of improvement. The further decline in the trade balance before a reversal creates a trend that can look like the letter J (see Exhibit 76). The J-shaped curve traces the initial decline in the trade balance followed by an increase.

EXHIBIT 76
The J Curve

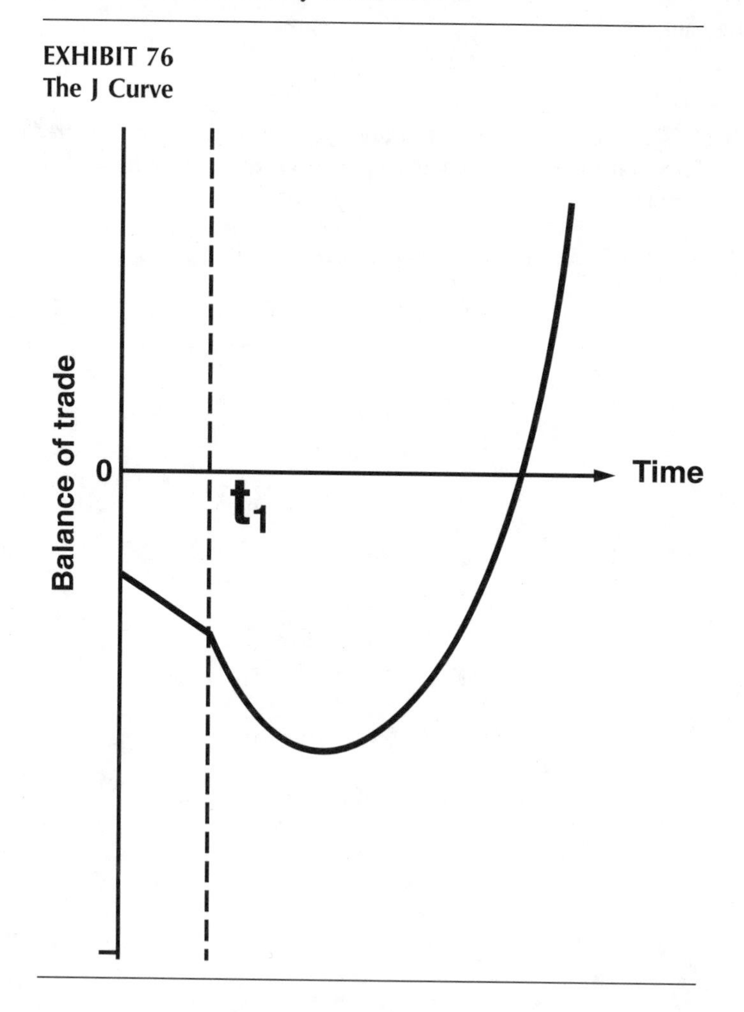

K

KANGAROO BONDS

Australian dollar-denominated bonds issued within Australia by a foreign firm.

KEEFE'S DOMESTIC AND FOREIGN BANK RATINGS

Keefe BankWatch ratings are based upon a quantitative analysis of all segments of the organization including, where applicable, holding company, member banks or associations, and other subsidiaries. Keefe BankWatch assigns only one rating to each company, based on consolidated financials. While the ratings are intended to be equally applicable to all operating entities of the organization, there may, in certain cases, be more liquidity and/or credit risk associated with doing business with one segment of the company versus another (i.e., holding company vs. bank). It should be further understood that Keefe BankWatch ratings are not merely an assessment of the likelihood of receiving payment of principal and interest on a timely basis. The ratings incorporate Keefe's opinion as to the vulnerability of the company to adverse developments which may impact the market's perception of the company, thereby affecting the marketability of its securities.

Keefe BankWatch ratings do not constitute recommendations to buy or sell securities of any of these companies. Further, Keefe BankWatch does not suggest specific investment criteria for individual clients. In those instances where disclosure, in its opinion, is either incomplete and/or untimely, a qualified rating (QR) is assigned to the institution. These ratings are derived exclusively from a quantitative analysis of publicly available information. Qualitative judgments have not been incorporated.

Generally, banks with assets of less than $500 million are assigned a numerical "score" based exclusively on a statistical model developed by BankWatch. These scores, which are compiled from regulatory reports, represent a performance evaluation of each company relative to a nationwide composite of similar sized banks. The score indicates the bank's percentile ranking (e.g., a score of 75 suggests that the company has outperformed 75% of its peer group). If a bank of this size is associated with a holding company that is not rated by BankWatch, an asterisk (*) may appear before the holding company name. This indicates that the analysis (score) on that company has been done on a bank only basis and is not reflective of the consolidated company's financial performance.

A Company possesses an exceptionally strong balance sheet and earnings record, translating into an excellent reputation and unquestioned access to its natural money markets. If weakness or vulnerability exist in any aspect of the company's business, it is entirely mitigated by the strengths of the organization.

A/B Company is financially very solid with a favorable track record and no readily apparent weaknesses. Its overall risk profile, while low, is not quite as favorable as for companies in the highest rating category.

B A strong company with a solid financial record and well received by its natural money markets. Some minor weaknesses may exist, but any deviation from the company's historical performance levels should be both limited and short-lived. The likelihood of a problem developing is small, yet slightly greater than for a higher-rated company.

B/C Company is clearly viewed as good credit. While some shortcomings are apparent, they are not serious and/or are quite manageable in the short term.

C Company is inherently a sound credit with no serious deficiencies, but financials reveal at least one fundamental area of concern that prevents a higher rating. The company may recently have experienced a period of difficulty, but those pressures should not be long term in nature. The company's ability to absorb a surprise, however, is less than that for organizations with better operating records.

C/D While still considered an acceptable credit, the company has some meaningful deficiencies. Its ability to deal with further deterioration is less than that for better-rated companies.

D The company's financials suggest obvious weaknesseses, most likely created by asset quality considerations and/or a poorly structured balance sheet. A meaningful level of uncertainty and vulnerability exists going forward. The ability to address further unexpected problems must be questioned.

D/E The company has areas of major weaknesses which may include funding/liquidity difficulties. A high degree of uncertainty exists as to the company's ability to absorb incremental problems.

E Very serious problems exist for the company, creating doubt as to its continued viability without some form of outside assistance—regulatory or otherwise.

A. Qualified Rating Characteristics

QR-A Exceptionally strong company as evidenced by its recent financial statements and its historical record.

QR-B Statistically a very sound institution. Financials reveal no abnormal lending or funding practices, while profitability, capital adequacy, and asset quality indicators consistently rank above peer group standards.

QR-C Statistical credentials should be viewed as average relative to peer group norms. A sound credit with one or more concerns preventing a higher rating.

QR-D The company's financial performance has typically fallen below average parameters established by its peers. Existing earnings weakness, exposure to margin contraction, asset quality concerns, and/or aggressive management of loan growth raise serious questions.

QR-E Several problems of a serious nature exist. Key financial indicators and/or abnormal growth patterns suggest significant uncertainty over the near term. The institution may well be under special regulatory supervision, and its continued viability with outside assistance may be at issue.

B. Country Rating

An assessment of the overall political and economic stability of a country in which a bank is domiciled.

I. An industrialized country with a long history of political stability complemented by an overall sound financial condition. The country must have demonstrated the ability to access capital markets throughout the world on favorable terms.

II. An industrialized country which has had a history of political and economic stability but is experiencing some current political unrest or significant economic difficulties. It enjoys continued ability to access capital markets worldwide but at increasingly higher margins. In the short run, the risk of default is minimal.

III. An industrialized or developing country with a wealth of resources, which may have difficulty servicing its external debt as a result of political and/or economic problems. Although it has access to capial markets worldwide, this cannot be assured in the future.

IV. A developing country which is currently facing extreme difficulty in raising external capital at all maturity levels.

V. A country which has defaulted on its external debt payments or which is in a position where a default is highly probable.

KNOCKOUT OPTION

Also known as *barrier option*, a knockout option is an option that is canceled (i.e., knocked out) if the exchange rate crosses, even momentarily, a predefined level called the *outstrike*. If the exchange rate breaks this barrier, the holder cannot exercise this option, even though it ends up in-the-money. This type of option is obviously less expensive than the standard option because of this risk of early termination.

KORUNA

Monetary unit of Czechoslovakia.

KRONA

Monetary unit of Iceland and Sweden.

KRONE

Monetary unit of Denmark and Norway.

KWACHA

Monetary unit of Angola.

KYAT

Monetary unit of Burma.

L

LABEL OF CONSUMER CONFIDENCE

This label was displayed in shop windows and in advertising during the transition period for the *euro* between January 1, 1999 and January 1, 2001, to inform customers that the prices of the products or services are displayed in both euro and national currency units.
See also EURO.

LAG

As a means of hedging transaction exposure of MNCs, lagging involves delaying the collection of receivables in foreign currency if that currency is expected to appreciate, and delaying conversion when payables are to be made in another currency in the belief the other currency will cost less when needed.

See also LEADING AND LAGGING.

LAMBDA

See CURRENCY OPTION PRICING SENSITIVITY.

LAW OF ONE PRICE

See PURCHASING POWER PARITY.

LEADING AND LAGGING

Leading and lagging are important risk-reduction techniques for a wide variety of working capital problems of MNCs. In many situations, forward and money-market hedges are not available to eliminate currency risk. Under such circumstances, MNCs can reduce their foreign exchange exposure by leading and lagging payables and receivables, that is, paying early or late. Leading is accelerating, rather than delaying (lagging). This can be accomplished by accelerating payments from soft-currency to hard-currency countries and by delaying inflows from hard-currency to soft-currency countries. Leading and lagging can be used most easily between affiliates of the same company, but the technique can also be used with independent firms. Most governments impose certain limits on leads and lags, since it has the effect of putting pressures on a weak currency. For example, some governments set 180 days as a limit for receiving payments for exports or making payment for imports. When the MNC makes use of leads and lags, it must adjust the performance measures of its subsidiaries and managers that are cooperating in the endeavor. The leads and lags can distort the profitability of individual divisions.

See also BALANCE SHEET HEDGING; CURRENCY RISK MANAGEMENT; FOREIGN EXCHANGE HEDGING; FORWARD MARKET HEDGE; MONEY-MARKET HEDGE.

LETTERS OF CREDIT

A letter of credit (L/C) is a credit letter normally issued by the buyer's bank in which the bank promises to pay money up to a stated amount for a specified period for merchandise when delivered. It substitutes the bank's credit for the importer's and eliminates the exporter's risk. It is used in international trade. The letter of credit (L/C) can be revocable or irrevocable. A *revocable L/C* is a means of arranging payment, but it does not guarantee payment. It can be altered or revoked, without notice, at any time up to the time a draft is presented to the *issuing bank*. An *irrevocable L/C*, on the other hand, cannot be revoked without the special permission of all parties concerned, including the exporter. A letter of credit can also be confirmed or unconfirmed. A *confirmed L/C* is an L/C issued by one bank and confirmed by another, obligating both banks to honor any drafts drawn in compliance. For example, "we hereby confirm this credit and undertake to pay drafts drawn in accordance with the terms and conditions of the letter of credit." An *unconfirmed L/C* is the obligation of only the issuing bank. The three main types of letters of credit, in order of safety for the exporter, are (1) the irrevocable, confirmed L/C; (2) the irrevocable, unconfirmed L/C; and (3) the revocable L/C. A summary of the terms and arrangements concerning these three types of L/Cs is shown in Exhibit 77.

EXHIBIT 77
Arrangements in Different Letters of Credit

	Irrevocable Confirmed L/C	Irrevocable Unconfirmed L/C	Revocable L/C
Who applies for	Importer	Importer	Importer
Who is obligated to pay	Issuing bank and confirming bank	Issuing bank	None
Who applies for amendment	Importer	Importer	Importer
Who approves amendment	Issuing bank, exporter, and confirming bank	Issuing bank and exporter	Issuing bank
Who reimburses paying bank	Issuing bank	Issuing bank	Issuing bank
Who reimburses issuing bank	Importer	Importer	Importer

There are also other types of letters of credit:

- *Letter of credit (Cumulative)*—A revolving letter of credit which permits any amount not used during any of the specified periods to be carried over and added to the amounts available in subsequent periods.
- *Letter of credit (Non-Cumulative)*—A revolving letter of credit which prohibits the amount not used during the specific period to be available in the subsequent periods.
- *Letter of credit (Deferred Payment)*—A letter of credit issued for the purchase and financing of merchandise, similar to acceptance letter of credit, except that it requires presentation of sight drafts which are payable on installment basis usually for periods of 1 year or more. Under this type of credit, the seller is financing the buyer until the stipulated time his drafts can be presented to the bank for payment. There is a significant difference in the bank's commitment, depending on whether the negotiating bank advised or confirmed the letter of credit.
- *Letter of credit (McLean)*—A letter of credit which requires the beneficiary to present only a draft or a receipt for specified funds before he receives payment.
- *Letter of credit (Negotiable)*—A letter of credit issued in such form that it allows any bank to negotiate the documents. Negotiable credits incorporate the opening bank's engagement, stating that the drafts will be duly honored on presentation, provided they comply with all terms of the credit. A negotiable letter of credit says specifically that the "Drafts must be negotiated or presented to the drawee not later

than...." In contrast, the straight letter of credit does not mention the word *negotiated.*

- *Letter of credit (Revolving)*—A credit which includes a provision for reinstating its face value after being drawn within a stated period of time. This kind of credit facilitates the financing of ongoing regular purchases.
- *Letter of credit (Standby)*—One issued for the express purpose of effecting payment in the event of default. The issuing bank is prepared to pay but does not expect to as long as the underlying transaction is properly fulfilled.
- *Letter of credit (Traveler's)*—A letter of credit which is issued by a bank to a customer preparing for an extended trip. The customer pays for the letter of credit at the time of issuance, and a bank issues the letter for a specified period of time in the amount purchased. The bank furnishes a list of correspondent banks where drafts against the letter of credit will be honored. The bank also identifies the customer by exhibiting a specimen signature of the purchaser in the folder enclosing the list of correspondent banks. Each bank, which honors a draft, endorses on the letter of credit the date when a payment was made, the bank's name, and the amount drawn against the letter of credit and charges the issuing bank's account.

The steps involved in a letter of credit transaction are summarized in Exhibit 78.

EXHIBIT 78
Steps in a Letter of Credit Transaction

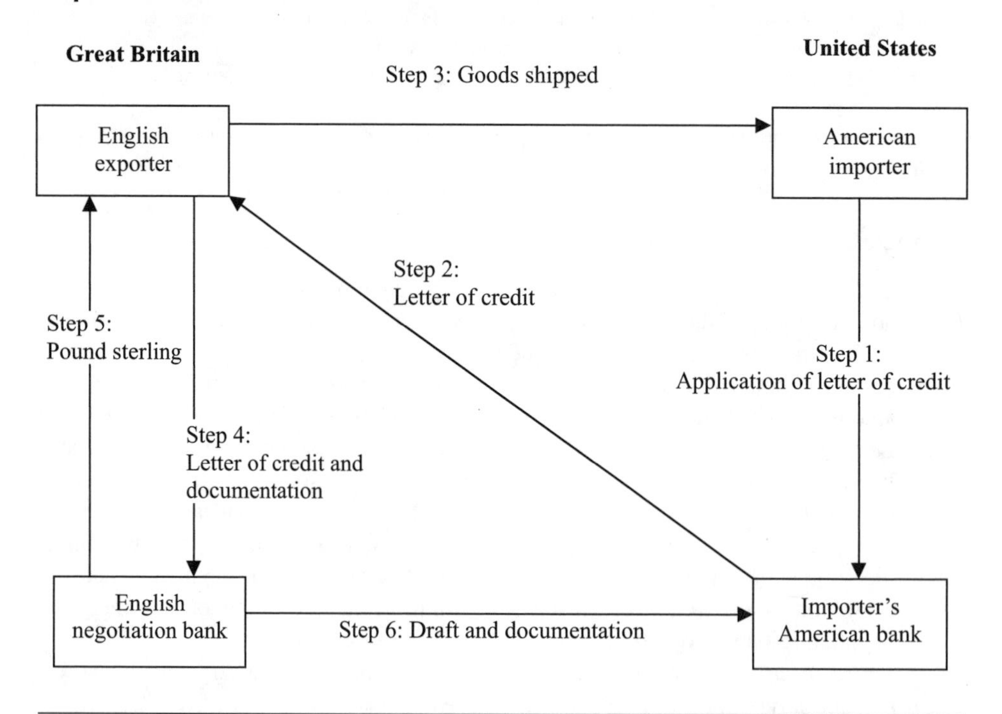

See also CONFIRMED LETTER OF CREDIT; DOCUMENTARY LETTER OF CREDIT; DRAFT; IRREVOCABLE LETTER OF CREDIT; NONDOCUMENTARY LETTER OF CREDIT; REVOCABLE LETTER OF CREDIT; REVOLVING LETTER OF CREDIT; TRADE CREDIT INSTRUMENTS; TRANSFERABLE LETTER OF CREDIT; UNCONFIRMED LETTER OF CREDIT.

LIABILITY MANAGEMENT OF BANKS

Traditionally, banks have taken the liability side of the balance sheet pretty much as outside of their control in the short run. They have taken liabilities as given and have been concerned with asset management. Liability management involves actions taken by banks and other depositary institutions to actively obtain funds at their own initiative by issuing negotiable certificates of deposit (CDs), by borrowing federal funds, and by using other procedures. It means altering the bank's liability structure (mix of demand deposits, time deposits, etc.) by changing the interest paid on nontransaction liabilities (such as CDs). The technique of liability management is certainly an important discretionary source of bank funds. However, the excessive use of the technique would make them vulnerable in future liquidity crises.

See also ASSET MANAGEMENT OF BANKS.

LIBOR

See LONDON INTERBANK OFFERED RATE.

LINK FINANCING

See BACK-TO-BACK LOANS.

LIRA

Monetary unit of Italy, Turkey, San Marino, and Vatican City.

LOCATIONAL ARBITRAGE

Locational arbitrage may occur if foreign exchange quotations differ among banks. The action of locational arbitrage should, however, force the currency quotations of banks to realign, and locational arbitrage will no longer be necessary.

LOMBARD RATE

The Lombard rate, one of the formal interest rates in Germany, is used to regulate the money market. Other countries use the term Lombard to describe rates which function somewhat like the Lombard rate. The Swiss, for example, have their own Lombard rate. In France, the Central Bank Intervention rate performs the same function.

LONDON INTERBANK BID RATE

London Interbank Bid Rate (LIBID) is the rate of interest paid for funds in the London interbank market. This rate has been used as a reference for floating rate payments for especially strong borrowers.

LONDON INTERBANK MEAN RATE

London Interbank Mean Rate (LIMEAN) is the midpoint of the *LIBOR-LIBID* spread. LIMEAN has been used as a reference for floating rate payments.

LONDON INTERBANK OFFERED RATE (LIBOR)

London Interbank Offered Rate (LIBOR), the most prominent of the *interbank offered rates*, is the rate of interest at which banks in London lend funds to other prime banks in London. LIBOR is frequently used as a basis for determining the rate of interest payable on Eurodollars and other Eurocurrency loans. The effective rate of interest on these Eurocredits is LIBOR plus a markup negotiated between lender and borrower. The rate, however, varies according to circumstances at which funds can be borrowed in particular currencies, amounts, and maturities in the market.

See also INTERBANK OFFERED RATE.

LONDON STOCK EXCHANGE

The London Stock Exchange is the largest stock exchange in Europe. In 1997, it introduced a new electronic settlement system called CREST, in which all trades are settled within five days. In the same year, it also installed an order-driven electronic trading system called SETS (Stock Exchange Trading Service).

LONG POSITION

An investment is long if an investor makes money when the price of a security or the price of its underlying security goes up. Generally, a long position is a position taken when the investor buys something for future delivery. This may be done in the expectation that the security bought will rise in value. It may also be done to hedge a *currency risk.*

LONG-TERM NATIONAL FINANCIAL MARKETS

Long-term national financial markets are markets for long-term capital markets trading stocks and bonds in different nations. Many countries allow the MNC to issue its own securities (stocks and bonds) or invest in the securities of other firms.

See also SHORT-TERM NATIONAL FINANCIAL MARKETS.

M

MAASTRICHT TREATY

Treaty agreed to in 1991 and ratified in mid-1993 that committed the 12 member-states of the European Community to a closer economic and political union.

See also EURO; EUROPEAN CURRENCY UNIT.

MAINTENANCE MARGIN

Maintenance margin is the minimum amount of equity that investors must carry in their margin account at all times. It is used to protect the brokerage houses in margin transactions. If the balance falls below the maintenance margin, for example, because of losses on the foreign currency futures contract, a *margin call* is issued. Enough new money must be added to the account balance to avoid the margin call. The amount of margin is always measured in terms of its relative amount of equity, which is considered the investor's collateral. The formula for margin is

$$\text{Margin } (\%) = \frac{V - M}{V}$$

where V = value of securities and M = margin loan balance (the amount of money being borrowed).

EXAMPLE 75

Consider two cases showing how margin changes:

Case 1: The Original Margin Transaction

400 shares at $20 per share using 50% *initial margin*	Value of securities (V)	$8,000
	Margin loan balance (M)	4,000
	Equity	4,000

Case 2: Some Time Later

The price of the 400 shares drops to $13 per share	Value of securities (V)	$5,200
	Margin loan balance (M)	4,000
	Equity	1,200

$$\text{New margin } (\%) = \frac{V - M}{V} = \frac{\$5{,}200 - \$4{,}000}{\$5{,}200} = \frac{\$1{,}200}{\$5{,}200} = 23.08\%$$

If, for example, the maintenance margin is 25%, this case is subject to a margin call. The investor must put up an extra \$133.33 (\$5,333.33 – \$5,200) to meet the 25% requirement, as computed below.

$$25\% = .25 = \frac{V - \$4,000}{V}$$

$$.25V = V - \$4,000$$

$$-.75V = -\$4,000$$

$$V = \$5,333.33$$

See also INITIAL MARGIN; MARGIN; MARGIN TRADING.

MANAGED FLOAT

Also called *dirty float*, a managed float is a country's attempt to control its currency exchange rates by intervention.

See also DIRTY FLOAT.

MARGIN

1. Also called *performance bond*, a margin is the amount of equity used (down payment given) for buying a security or *futures contract* with the balance being on credit. If you use 25% margin, for example, it means that 25% of the investment position is being financed with your own capital and the balance (75%) with borrowed money.
   See INITIAL MARGIN; MAINTENANCE MARGIN; MARGIN TRADING.
2. The forward vs. spot rate for a foreign currency.

MARGIN TRADING

Margin trading refers to the use of borrowed funds (financial leverage or debt) to buy securities. It can magnify returns by reducing the amount of capital that must be put up by the investor. The following example illustrates how margin trading can magnify investment returns (ignoring dividends and brokerage commissions).

EXAMPLE 76

Invested capital (equity)	\$4,000
Borrowed funds (margin loan – 50%)	4,000
Total investment (to purchase 400 shares)	\$8,000
Cost to buy 400 shares (at \$20 per share)	\$8,000
6 months later—stock sold at \$30 per sale—proceeds at sale	12,000
Gross profit from transaction	\$4,000
Less: interest cost (10% per annum) on borrowed funds	200
\$4,000 × 10% × 6/12	
Net profit	\$3,800
Return on invested capital	95%
(Net profit/invested capital = \$3,800/\$4,000)	
Annualized return (95% × 12/6)	190%

Note: Margin trading is a double-edged sword. The risk of loss is also magnified (by the same rate) when the price of the security falls.

MARKED-TO-MARKET

Futures contracts are marked to market daily. Marked to market is the daily adjustment of margin accounts to reflect profits and losses. At the end of each day, the contracts are settled and the resulting profits or losses paid.

EXAMPLE 77

On Monday morning, you take a long position in a pound futures contract that matures on Wednesday afternoon. The agreed-upon price is $1.68/£ for £62,500. At the close of trading on Monday, the futures price has risen to $1.69. At Tuesday close, the price rises further to $1.70. At Wednesday close, the price falls to $1.685, and the contract matures. You take delivery of the pounds at the prevailing price of $1.685. The daily marked-to-market (settlement) process and your profit (loss) are determined as shown in Exhibit 79.

EXHIBIT 79
Daily Settlement Process

Time	Action	Cash Flow
Monday morning	You buy pound futures contract that matures in two days. Price is $1.68.	None
Monday close	Futures price rises to $1.69. Contract is marked-to-market.	You receive: 62,500 × (1.69 – 1.68) = $625
Tuesday close	Futures price rises to $1.70. Contract is marked-to-market.	You receive: 62,500 × (1.70 – 1.69) = $625
Wednesday close	Futures price falls to $1.685. (1) Contract is marked-to-market (2) Investor takes delivery of £62,500.	(1) You pay: 62,500 × (1.70 –1.685) = $937.50 (2) You pay: £62,500 × 1.685 = $105,312.50

Your net profit is: $1,250 – 937.50 = $312.50.

MARKET-BASED FORECASTING

Market-based forecasting is the process of developing currency forecasts from market indicators. While this approach is very simple, it is also very effective. The model relies basically on the spot rate or forward rate to forecast the spot rate in the future. The model assumes that the spot rate reflects the foreign exchange rate in the near future. Let us suppose that the Italian lire is expected to depreciate vs. the U.S. dollar. This would encourage speculators to sell lire and later purchase them back at the lower (future) price. This process if continued would drive down the prices of lire until the excess (arbitrage) profits were eliminated.

The model also suggests that the forward exchange rate equals the future spot price. Again, let us suppose that the 90-day forward rate is .987. The market forecasters believe that the exchange rate in 90 days is going to be .965. This provides an arbitrage opportunity. Markets will keep on selling the currency in the forward market until the opportunity for excess profit is eliminated. This model, however, relies heavily on market and *market efficiency.* It assumes that capital markets and currency markets are highly efficient and that there is perfect information in the marketplace. Under these circumstances, this model can provide accurate forecasts.

Indeed, many of the world currency markets such as the market for U.S. dollar, British pound, and Japanese yen are highly efficient and this model is well suited for such markets. However, market imperfections or lack of perfect information reduces the effectiveness of this model for most currency.

A. Long-Term Forecasting with Forward Rates

While forward rates are sometimes available for two to five years, such rates are rarely quoted. However, the quoted interest rates on risk-free instruments of various countries can be used to determine what the forward rates would be under conditions of interest parity. For example, assume that the U.S. five-year interest rate is 10% annualized, while the British five-year interest rate is 13%. The five-year compounded return on investments in each of these countries is computed as follows:

Country	Five-Year Compounded Return
U.S. (Home)	$(1.10)^5 - 1 = 61\%$
U.K. (Foreign)	$(1.13)^5 - 1 = 84\%$

Thus, the appropriate five-year forward rate premium (or discount) of the British pound would be

$$P = -\frac{(r_f - r_h)}{(1 + r_f)} = -(0.84 - 0.61/1.84) = -0.125 \quad \text{or} \quad -12.5\%$$

The results of our computation suggest that the five-year forward rate of the pound should contain a 12.5% discount. That is, the spot rate of the pound is expected to depreciate by 12.5% over the five-year period for which the forward rate is used to forecast.

B. Skepticism about the Forward Rate Forecast

One reason firms might not accept the forward rate as a predictor is because of a second market force. According to *interest rate parity*, the forward premium (or discount) is determined by the interest rate differential. If, for example, the 30-day interest rate of the British pound is above the 30-day U.S. interest rate, we expect the forward rate on the pound to exhibit a discount. The size of the discount will be such that U.S. investors cannot achieve abnormal returns by using *covered interest arbitrage* (converting dollars to pounds, investing pounds in a British 30-day account, and simultaneously selling pounds 30 days forward in exchange for dollars). Based on this information, consider that the forward rates of some currencies, such as the British pound and French franc, are usually below their current spot rates. That is, they almost always exhibit a forward discount (because their interest rates are typically above the U.S. rates). The discount within the forward rate suggests that the pound and French franc should depreciate, even when all other factors suggest that these currencies will appreciate.

See also COVERED INTEREST ARBITRAGE; FOREIGN EXCHANGE RATE FORECASTING; INTEREST RATE PARITY.

MARKET EFFICIENCY

Market efficiency requires that all relevant information is quickly reflected in both the spot and forward exchange markets.

See also EFFICIENT MARKET.

MARKET RISK

Market risk is the extent to which the possibility of financial loss can arise from any unfavorable market forces such as changes in interest rates, currency rates, equity prices, or commodity prices. Prices of all securities are correlated to some degree with broad swings in the stock market. Market risk refers to changes in a firm's security's price resulting from changes in the stock or bond market as a whole, regardless of the fundamental change in the firm's earning power.

MARKET-SERVING FDI

A form of *foreign direct investment (FDI)* by the MNC aimed at targeting the host country.

MARKKA

Finland's currency.

MIDAS

See MULTICURRENCY INTEREST-COMPENSATING DAILY ACCOUNT-MANAGEMENT SYSTEM.

MIXED FORECASTING

Mixed forecasting in not a unique technique for *foreign exchange rate (currency) forecasting* but rather a mixture of the three popular approaches—*market-based forecasting, fundamental forecasting*, and *technical forecasting*. In some cases, mixed forecasting is nothing but a weighted average of a variety of the forecasting techniques. The techniques can be weighted arbitrarily or by assigning a higher weight to the more reliable technique. Mixed forecasting may often lead to a better result than relying on one single forecast.

See also FOREIGN EXCHANGE RATE FORECASTING.

MODIFIED INTERNAL RATE OF RETURN (MIRR)

The modified internal rate of return (MIRR) is defined as the discount rate at which the present value of a project's cost is equal to the present value of its terminal value, where the terminal value is found as the sum of the future values of the cash flows, compounded at the firm's cost of capital. The MIRR is a modified version of the internal rate of return (IRR) and a better indicator of relative profitability, hence better for use in capital budgeting. The MIRR has significant advantages over the regular IRR.

- The MIRR assumes that the cash flows from each project are reinvested at the cost of capital rather than at the project's own IRR. Since reinvestment at the cost of capital is generally more correct and realistic, the MIRR is a better indicator of a project's true profitability.
- The MIRR solves the multiple IRR problem.
- If two mutually exclusive projects are of equal size and have the same life, NPV and MIRR will always lead to the same ranking. The kinds of conflicts encountered between NPV and the regular IRR will not occur. Example 78 illustrates this points.

EXAMPLE 78

Assume the following:

	Cash Flows Year					
Projects	0	1	2	3	4	5
A	($100)	$120				
B	($100)					$201.14

Computing IRR and NPV at 10% gives the following different rankings:

Projects	IRR	NPV at 10%
A	20%	$9.01
B	15	24.90

The difference in ranking between the two methods is caused by the methods' reinvestment rate assumptions. The IRR method assumes Project A's cash inflow of $120 is reinvested at 20% for the subsequent 4 years and the NPV method assumes $120 is reinvested at 10%. The correct decision is to select the project with the higher NPV (Project B), since the NPV method assumes a more realistic reinvestment rate, that is, the cost of capital (10% in this example).

To calculate Project A's MIRR, first, compute the project's terminal value at a 10% cost of capital.

$$120\ T1(10\%, 4\ \text{years}) = 120 \times 1.4641 = 175.69$$

Next, find the IRR by setting:

$$100 = 175.69\ T3(\text{MIRR}, 5\ \text{years})$$
$$T3 = 100/175.69 = 0.5692, \text{ which gives MIRR } = \text{ about } 12\%$$

Now we see the consistent ranking from both the NPV and MIRR methods as shown above.

Note: Microsoft Excel has a function MIRR(values, finance_rate, reinvest_rate).
See also INTERNAL RATE OF RETURN; NET PRESENT VALUE.

MONETARY APPROACH

See ASSET MARKET MODEL.

MONETARY ASSETS AND LIABILITIES

See MONETARY BALANCE.

MONETARY BALANCE

Monetary balance refers to minimizing *accounting exposure*. It involves avoiding either a net receivable or a net payable position. If an MNC had net *positive* exposure (more monetary assets than liabilities), it could use more financing from foreign monetary sources to balance things out. MNCs with assets and liabilities in more than one foreign currency may try to reduce risk by balancing off exposure in the different countries. Often, the monetary balance is practiced across *several* countries simultaneously. Monetary assets and liabilities are those items whose value, expressed in local currency, does not change with devaluation or revaluation. They are listed in Exhibit 80.

EXHIBIT 80
Monetary Assets and Liabilities

Monetary Assets	Monetary Liabilities
Cash	Accounts payable
Marketable securities	Notes payable
Accounts receivable	Tax liability reserve
Tax refunds receivable	Bonds
Notes receivable	Preferred stock
Prepaid insurance	

A firm's monetary balance can be looked at in terms of a firm's position with regard to real assets. For example, the basic balance sheet equation can be written as follows:

$$\text{Monetary assets} + \text{Real assets} = \text{Monetary liabilities} + \text{Equity}$$

EXAMPLE 79

Consider the following two cases:

	Monetary Assets	+	Real Assets	=	Monetary Liabilities	+	Equity (Net Worth)
Firm A: Monetary creditor	$7,000		$4,000		$5,000		$7,000
Firm B: Monetary debtor	5,000		7,000		7,000		5,000

Firm A is a monetary creditor because its monetary assets exceed its monetary liabilities; its net worth position is negative with respect to its investment coverage of net worth by real assets. In contrast, Firm B is a monetary debtor because it has monetary liabilities that exceed its monetary assets; its net worth coverage by investment in real assets is positive. Thus, the monetary creditor can be referred to as a firm with a negative position in real assets, and the monetary debtor as a firm with a positive position in real assets. Exhibit 81 summarizes these equivalent relationships.

EXHIBIT 81
Monetary Creditor versus Monetary Debitor

Firm A	(Long position in foreign currency)	Monetary creditor	Monetary assets exceed monetary liabilities	Negative position in real assets	Balance of receipts in foreign currency obligations in foreign currency is *positive*
Firm B	(Short position in foreign currency)	Monetary debtor	Monetary liabilities exceed monetary assets	Positive position in real assets	Balance of receipts in foreign currency less obligations in foreign currency is *negative*

Thus, if Firm A has a long position in a foreign currency, on balance it will be receiving more funds in foreign currency, or it will have a net monetary asset position that exceeds its monetary liabilities in that currency. The opposite holds for Firm B, which is in a short position with respect to a foreign currency. Hence the analysis with respect to a firm with net future receipts or net future obligations can be applied also to a firm's balance sheet position. A firm with net receipts is a net monetary creditor. Its foreign exchange rate risk exposure is vulnerable to a decline in value of the foreign currency. On the contrary, a firm with future net obligations in foreign currency is in a net monetary debtor position. The foreign exchange risk exposure it faces is the possibility of an increase in the value of the foreign currency.

In addition to the specific actions of hedging in the forward market or borrowing and lending through the money markets, other business policies can help the firm achieve a balance sheet position that minimizes the foreign exchange rate risk exposure to either currency devaluation or currency revaluation upward. Specifically, in countries whose currency values are likely to fall, local management of subsidiaries should be encouraged to follow these policies:

1. Never have excessive idle cash on hand. If cash accumulates, it should be used to purchase inventory or other real assets.
2. Attempt to avoid granting excessive trade credit or trade credit for extended periods. If accounts receivable cannot be avoided, an attempt should be made to charge interest high enough to compensate for the loss of purchasing power.
3. Wherever possible, avoid giving advances in connection with purchase orders unless a rate of interest is paid by the seller on these advances from the time the subsidiary—the buyer—pays them until the time of delivery, at a rate sufficient to cover the loss of purchasing power.
4. Borrow local currency funds from banks or other sources whenever these funds can be obtained at a rate of interest no higher than U.S. rates adjusted for the anticipated rate of devaluation in the foreign country.
5. Make an effort to purchase materials and supplies on a trade credit basis in the country in which the foreign subsidiary is operating, extending the final date of payment as long as possible.

The reverse polices should be followed in a country where a revaluation upward in foreign currency values is likely to transpire. All these policies are aimed at a monetary balance position in which the firm is neither a monetary debtor nor a monetary creditor. Some MNCs take a more aggressive position. They seek to have a net monetary debtor position in a country whose exchange rates are expected to fall and a net monetary creditor position in a country whose exchange rates are likely to rise.

See also CURRENCY RISK MANAGEMENT; TRANSLATION EXPOSURE.

MONETARY/NONMONETARY METHOD

The monetary/nonmonetary method is a translation method that applies the current exchange rate to all monetary assets and liabilities, both current and long term, while all other assets (physical, or nonmonetary, assets) are translated at historical rates. In contrast with the *current/noncurrent method*, this method rewards holding of physical assets under devaluation.

See also CURRENT RATE METHOD; CURRENT/NONCURRENT METHOD; TEMPORAL METHOD.

MONEY-MARKET HEDGE

Also called *credit-market hedge*, a money-market hedge is a *hedge* in which the exposed position in a foreign currency is offset by borrowing or lending in the money market. It basically calls for matching the exposed asset (accounts receivable) with a liability (loan payable) in the same currency. An MNC borrows in one currency, invests in the money market, and converts the proceeds into another currency. Funds to repay the loan may be generated from business operations, in which case the hedge is *covered*. Or funds to repay the loan may be purchased in the foreign exchange market at the *spot rate* when the loan matures, which is called an *uncovered* or open edge. The cost of the money-market hedge is determined by differential interest rates.

EXAMPLE 80

XYZ, an American importer enters into a contract with a British supplier to buy merchandise for £4,000. The amount is payable on the delivery of the good, 30 days from today. The company knows the exact amount of its pound liability in 30 days. However, it does not know the payable in dollars. Assume that the 30-day money-market rates for both lending and borrowing in the U.S. and U.K. are .5% and 1%, respectively. Assume further that today's foreign exchange rate is \$1.50/£. In a money-market hedge, XYZ can make any of the following choices:

1. Buy a one-month U.K. money-market security, worth £4,000/(1 +.005) = £3,980.00. This investment will compound to exactly £4,000 in one month.
2. Exchange dollars on today's spot (cash) market to obtain the £3,980. The dollar amount needed today is £3,980.00 × \$1.50/£ = \$5,970.00.
3. If XYZ does not have this amount, it can borrow it from the U.S. money market at the going rate of 1%. In 30 days XYZ will need to repay \$5,970.00 × (1 +.01) = \$6,029.70.

Note: XYZ need not wait for the future exchange rate to be available. On today's date, the future dollar amount of the contract is known with certainty. The British supplier will receive £4,000, and the cost of XYZ to make the payment is \$6,029.70.

MONEY MARKETS

Money markets are the markets for short-term (less than 1 year) debt securities. Examples of money-market securities include U.S. Treasury bills, federal agency securities, bankers' acceptances, commercial paper, and negotiable certificates of deposit issued by government, business, and financial institutions.

See FINANCIAL MARKETS.

MORGAN GUARANTY DOLLAR INDEX

See CURRENCY INDEXES; DOLLAR INDEXES.

MORGAN STANLEY CAPITAL INTERNATIONAL EUROPE, AUSTRALIA, FAR EAST INDEX

See EAFE INDEX.

MORGAN STANLEY EAFE INDEX

See EAFE INDEX.

MULTIBUYER POLICY

See EXPORT-IMPORT BANK.

MULTICURRENCY CROSS-BORDER CASH POOLING

Multicurrency cross-border cash pooling allows a facility to notionally offset debit balances in one currency against credit balances in another. For example, a corporation with credit balances in British pounds and debit balances in German marks and French francs can use pooling to offset the debit and credit balances without the administrative burden of physically moving or converting currencies. The concept of centralized cash pooling is to offset debit and credit balances within a currency and among different currencies without converting the funds physically. Without a centralized pooling system, local subsidiaries lose interest on credit balances or incur higher interest expense on debit balances due to the high margins on interest rates usually taken by local banks. In many cases, credit balances in foreign currency accounts do not earn interest. Through centralized pooling, cash-rich entities pledge their balances so that entities that need to overdraw their cash pool accounts can do so. Credits in one currency may be used to offset debits in another prior to interest calculations—a strategy that often decreases the net amounts borrowed and increases interest yields. The multicurrency system is managed per account on a daily basis. Pooling is based on a zero-balance concept—the volume of credit balances equals the volume of debit balances. When the overall position of all the cash pool accounts is zero or positive, the subsidiaries that are in an overdraft position will actually borrow at credit interest terms. *Note:* Cash pooling does not eliminate natural interest rate differences between currencies, but it does eliminate the margins on debit balances, thus reducing borrowing costs. Exhibit 82 summarizes goals of the system.

EXHIBIT 82
Reasons for Setting up Cross-Currency Cash Pooling Systems

Optimizing the use of excess cash
Reducing interest expense and maximizing interest yields
Reducing costly foreign exchange, swap transactions, and intercompany transfers
Minimizing administrative paper work
Centralizing and speeding information for tighter control and improved decision making

The following example illustrates both the advantages of cash pooling and the return edge provided by a multicurrency approach.

EXAMPLE 81

Assume that three subsidiaries operating in Australia, the United Kingdom, and the United States maintain multicurrency accounts in the pool. Each has signed an offset agreement with its Amsterdam-based pooling bank. The U.K. company has a local non-interest-bearing DM account. The interbank interest rates are 7.5% for Australian dollars, 4.25% for Deutsche marks, and 5.5% for U.S. dollars. The Australian company's excess funds in A$ are transferred to its pooling account. The U.K. company has a receivable in DM and has instructed the payor to make the payment directly to its DM pooling account. These pooled credit balances allow other pool members to overdraft their accounts in their preferred currency. For example, the U.S. pooling participant can overdraft its US$ account the countervalue of the available pool balance for investment. Because the overall pooled balance is positive, the pooling mechanism applies credit conditions to all balances in the pool, including debit balances. Consequently, borrowings from the pool are charged interest at credit rates. The positive effect of the pooling is apparent for the U.K. company, which earns interest on its DM balance at 4.25%. Without pooling, no interest would have been earned. Additionally, the U.S. company can borrow from the pool at a rate of 5.5%, which is a credit interest rate.

See also NETTING; MULTILATERAL NETTING.

MULTICURRENCY INTEREST-COMPENSATING DAILY ACCOUNT-MANAGEMENT SYSTEM

The multicurrency interest-compensating daily account-management system (MIDAS) works as follows: Each participating entity sets up its own account(s) at the bank—multicurrency accounts, in many cases, for units that conduct business in more than one currency. Once participating entities open accounts, they must sign offset agreements that permit credit balances in their accounts to be applied against debit balances in sister accounts without transaction approval. The overall net balance should be positive. The overall gain created may be credited to a separate treasury account or allocated among participants according to formulas that take into account participation incentives as well as tax criteria.

MULTILATERAL NETTING

Multilateral netting is an extension of *bilateral netting*. Under bilateral netting, if a Japanese subsidiary owes a British subsidiary $5 million and the British subsidiary simultaneously owes the Japanese subsidiary $3 million, a bilateral settlement will be made a single payment of $2 million from the Japanese subsidiary to the British subsidiary, the remaining debt being canceled out. Multilateral netting is extended to the transactions between multiple subsidiaries within an international business. It is the strategy used by some MNCs to reduce the number of transactions between subsidiaries of the firm, thereby reducing the total transaction costs arising from foreign exchange dealings with transfer fees. It attempts to maintain balance between receivables and payables denominated in a foreign currency. MNCs typically set up multilateral netting centers as a special department to settle the outstanding balances of affiliates of a multinational company with each other on a net basis. It is the development of a "clearing house" for payments by the firm's affiliates. If there are amounts due among affiliates they are offset insofar as possible. The net amount would be paid in the currency of the transaction. The total amounts owed need not be paid in the currency of the transaction; thus, a much lower quantity of the currency must be acquired. Note that the major advantage of the system is a reduction of the costs associated with a large number of separate foreign exchange transactions.

See also MULTICURRENCY CROSS-BORDER CASH POOLING.

MULTINATIONAL CAPITAL BUDGETING

See ANALYSIS OF FOREIGN INVESTMENTS.

MULTIPERIOD RETURNS

See ARITHEMATIC AVERAGE RETURN VS. COMPOUND (GEOMETRIC) AVERAGE RETURN.

N

NEAR MONEY

Liquid assets easily convertible into money as needed such as marketable securities, money-market funds, and time deposits.

NEGOTIABLE INSTRUMENT

Any financial instrument that can readily be converted into cash. It is a written *draft* or promissory note, which is signed by the maker or drawer, has an unconditional promise, and is an order to make payment of a certain sum of money on demand by the bearer or to the order of a named party at a determinable future date. A "holder in due course" of a negotiable instrument is entitled to payment despite any personal disagreements between drawee and drawer.

NEGOTIABLE LETTER OF CREDIT

A *letter of credit* issued in such form that it allows any bank to negotiate the documents. Negotiable credits incorporate the opening bank's engagement, stating that the drafts will be duly honored on presentation, provided they comply with all terms of the credit.

NET LIQUIDITY BALANCE

See OFFICIAL SETTLEMENTS BALANCE.

NET PRESENT VALUE

Net present value (NPV) is the excess of the present value (PV) of cash inflows generated by the project over the amount of the initial investment (I). The present value of future cash flows is computed using the so-called cost of capital (or minimum required rate of return) as the discount rate.

$$NPV = -I + \sum_{t=1}^{T} \frac{CF_t}{(1+k)^t}$$

where $-I$ = the initial investment or cash outlay, CF_t = estimated cash flows in t ($t = 1, \ldots T$), and k = the discount rate on those cash flows. When cash inflows are uniform, the present value would be $PV = CF \cdot T4\,(k, t)$ where CF is the amount of the annuity. The value of $T4$ is found in Exhibit 4 of the Appendix.

Decision rule: If *NPV* is positive, accept the project. Otherwise reject it.

EXAMPLE 82

Consider the following foreign investment project:

Initial investment (*I*)	$12,950,000
Estimated life	10 years
Annual cash inflows (*CF*)	$3,000,000
Cost of capital	12%

The net present value of the cash inflows is:	
$PV = CF \times T4(k, t)$	
$= \$3{,}000{,}000 \times T4(12\%, 10 \text{ years})$	
$= \$3{,}000{,}000\ (5.650)$	\$16,950,000
minus Initial investment (I)	−12,950,000
Net present value ($NPV = -I + PV$)	\$4,000,000

Since the *NPV* of the investment is positive, the investment should be accepted. The advantages of the *NPV* method are that it obviously recognizes the time value of money, and it is easy to compute whether the cash flows form an annuity or vary from period to period. Spreadsheet programs can be used in making *NPV* calculations. For example, the Excel formula for *NPV* is *NPV*(discount rate, cash inflow values) + *I*, where *I* is given as a negative number.

Year 0	1	2	3	4	5	6	7	8	9	10
2,950,000	3,000,000	3,000,000	3,000,000	3,000,000	3,000,000	3,000,000	3,000,000	3,000,000	3,000,000	3,000,000
NPV = \$4,000.67										

NETTING

Netting involves the consolidation of payables and receivables for one currency so that only the difference between them must be bought and sold. Centralization of cash management allows the MNC to offset subsidiary payments and receivables in a netting process.

See also MULTILATERAL NETTING.

NET TRANSACTION EXPOSURE

Net transaction exposure takes into account cash inflows and outflows in a given currency to determine the exposure after offsetting inflows against outflows.

NEW ECONOMY

See OLD ECONOMY VERSUS NEW ECONOMY.

NOMINAL EXCHANGE RATE

Actual spot rate of foreign exchange, in contrast to *real exchange rate*, which is adjusted for changes in purchasing power.

NONDELIVERABLE FORWARD CONTRACTS

Nondeliverable forward contracts (NDFs) are *forward contracts* that do not result in actual delivery of currencies. Instead, the agreement specifies that a payment is made by one party to the other party based on the exchange rate at the future date.

NONDIVERSIFIABLE RISK

Also called *unsystematic risk* or *uncontrollable risk*, nondiversifiable risk is that part of a security's risk that cannot be diversified away. It includes *market risk* that comes from factors systematically affecting most firms (such as inflation, recessions, political events, and high interest rates).

See also CAPITAL ASSET PRICING MODEL.

NONDOCUMENTARY LETTER OF CREDIT

Also called a *clean letter of credit*, this letter of credit for which no documents need to be attached to the draft is normally used in transactions other than commercial ones.
See also DOCUMENTARY LETTER OF CREDIT; LETTERS OF CREDIT.

NONSTERILIZED INTERVENTION

Unlike *sterilized intervention* in the foreign exchange market, nonsterilized intervention does not adjust for the change in money supply.
See also STERILIZED INTERVENTION.

NOSTRO ACCOUNT

A nostro account is working balances maintained with the correspondent to facilitate delivery and receipts of currencies.

NOTE ISSUANCE FACILITY

Note issuance facility (NIF) is a facility provided by a syndicate of banks that allows borrowers to issue short-term notes (typically of three- or six-months' maturity) in their own names. A group of underwriting banks guarantees the availability of funds to the borrower by purchasing any unsold notes or by providing standby credit. Borrowers usually have the right to sell their notes to the bank syndicate at a price that yields a prearranged spread over *LIBOR*.

O

OFFER

Also called *ask* or *sell*, the rate at which a trader is willing to sell a foreign currency or other securities.

OFFICIAL RESERVE TRANSACTIONS BALANCE

The official reserve transaction balance shows an adjustment to be made in official reserves for the *balance of payments* to balance.

OFFICIAL SETTLEMENTS BALANCE

Also called *overall balance* or *net liquidity balance*, the official settlements balance is the bottom line *balance of payments* when all private sector transactions have been accounted for and all that remain are official exchanges between central banks (and the *IMF*). It is equal to changes in short-term capital held by foreign monetary agencies and official reserve asset transactions. This balance is a comprehensive balance often used to judge a nation's overall competitive position in terms of all private transactions with the rest of the world. Exhibit 83 summarizes this and other commonly used balance of payments measures.

EXHIBIT 83
Commonly Used Balance of Payments Measures

Group	Category Component	Popular Name
A	Merchandise Trade	
	Other Current Items	
	Current Account	Current Balance (A)
B	Direct Investment	
	Portfolio Investment	
	Other Long-Term Items	
	Long-Term Capital	Basic Balance ($A + B$)
C	Short-Term Capital	
D	Errors and Omissions	Official Settlements Account ($A + B + C + D$)

See also BASIC BALANCE; CURRENT ACCOUNT BALANCE.

OFFSHORE BANKING

Offshore banking means accepting deposits and making loans in foreign currency, i.e., the *Eurocurrency market*, although the activity is not limited to Europe. The terms *offshore, overseas,* and *foreign* are frequently used interchangeably.

OFFSHORE MUTUAL FUND

A mutual fund that is managed and resides out of a foreign country, usually outside the U.S.

OLD ECONOMY VERSUS NEW ECONOMY

The new economy is the new, digital economy driven by industrial information technology, much of which is related to telecommunications such as the Internet—a technology that, many argue, has a huge potential to transform the engineering industry. In the new economy, production and distribution systems are automated, computer-based systems. The old economy, classical or traditional, is undergoing sweeping changes through the speed and efficiency brought by applications of information technology and the Internet.

OPEN INTEREST

Total number of *futures* or options on futures contracts that have not yet been offset or fulfilled by delivery. An indicator of the depth or liquidity of a market (the ability to buy or sell at or near a given price) and of the use of a market for risk- and/or asset-management.

See also FUTURES.

OPERATING EXPOSURE

See ECONOMIC EXPOSURE.

OPPORTUNITY COST

1. The difference between the forward rate and the (eventual) future spot rate. The forward contract does fix a "minimum loss" for the firm by setting the guaranteed price of exchange in the future.

EXAMPLE 83

Suppose that the 3-month forward rate for the French franc is $0.1457 per FFr. However, if the French franc devalues over the next three months to, say $0.1357/FFr, the forward contract holder will have an *opportunity cost* of the difference between the forward rate and the (eventual) future spot rate (a difference of $0.01/FFr, or 6.4%).

---

2. Net benefit lost by rejecting some alternative course of action. Its significance in decision making is that the best decision is always sought, as it considers the cost of the best available alternative not taken. The opportunity cost does not appear on formal accounting records.

EXAMPLE 84

If $1 million can be invested in a *Euro-commercial paper (Euro-P or EUP)* earning 9%, the opportunity cost of using that money for a particular business venture would be computed to be $90,000 ($1 million × .09).

OPTIMUM CURRENCY AREA

The optimum currency area is the best area within which exchange rates are fixed and between which exchange rates are flexible. It is the region characterized by relatively inexpensive mobility of the factors of production (capital and labor).

OPTION

An option is a contract to give the investor the right—but *not the obligation*—to buy or sell something. It has three main features. It allows you, as an investor to "lock in": (1) a specified number of shares of stock, (2) at a fixed price per share, called *strike* or *exercise price*, (3) for a limited length of time. For example, if you have purchased an option on a stock, you have the right to "exercise" the option at any time during the life of the option. This means that, regardless of the current market price of the stock, you have the right to buy or sell a specified number of shares of the stock at the strike price (rather than the current market price). Options possess their own inherent value and are traded in *secondary markets*. You may want to acquire an option so that you can take advantage of an expected rise in the price of the underlying stock. Option prices are directly related to the prices of the common stock to which they apply. Investing in options is very risky and requires specialized knowledge. Options may be American style (i.e., they can be exercised at any time up to the expiration date) or European style (i.e., exercisable only at maturity). Almost all exchange traded options are American style; over-the-counter may be either American or European style, but are often European.

All options are divided into two broad categories: calls and puts. A *call option* gives you the right (but not the obligation) to buy:

1. 100 shares of a specific stock,
2. at a fixed price per share, called the *strike* or *exercise price*,
3. for up to 9 months, depending on the expiration date of the option.

When you purchase a call, you are buying the right to purchase stock at a set price. You expect price appreciation to occur. You can make a sizable gain from a minimal investment, but you may lose all your money if the stock price does not go up.

EXAMPLE 85

You purchase a 3-month call option on Dow Chemical stock for $4 1/2 at an exercise price of $50 when the stock price is $53.

On the other hand, a *put option* gives you the right to sell (and thus force someone else to buy):

1. 100 shares of a specific stock,
2. at a fixed price, the strike price,
3. for up to 9 months, depending on the expiration date of the option.

Purchasing a put gives you the right to sell stock at a set price. You buy a put if you expect a stock price to fall. You have the chance to earn a considerable gain from a minimal investment, but you lose the whole investment if price depreciation does not materialize. The buyer of the contract (called the *holder*) pays the seller (called the *writer*) a premium for the contract. In return for the premium, the buyer obtains the right to buy securities from the writer or sell securities to the writer at a fixed price over a stated period of time.

Option Holder = Option Buyer = Long Position

Option Writer = Option Seller = Short Position

	Call Option	**Put Option**
Buy (long)	The right to call (buy) from the writer	The right to put (sell) to the writer
Sell (short)	Known as *writing a call,* being obligated to sell if called	Known as *writing a put,* if the stock or contract is put

Calls and puts are typically for widely held and actively traded securities on organized exchanges. With calls there are no voting privileges, ownership interest, or dividend income. However, option contracts are adjusted for stock splits and stock dividends.

Calls and puts are not issued by the company with the common stock but rather by option makers or option writers. The maker of the option receives the price paid for the call or put minus commission costs. The option trades on the open market. Calls and puts are written and can be acquired through brokers and dealers. The writer is required to purchase or deliver the stock when requested.

Holders of calls and puts do not have to exercise them to earn a return. They can trade them in the secondary market for whatever their value is. The value of a call increases as the underlying common stock goes up in price. The call can be sold on the market before its expiration date.

A. More on the Terms of an Option

There are three key terms with which you need to be familiar in connection with options: the exercise or strike price, expiration date, and option premium. The *exercise price* for a call is the price per share for 100 shares, at which you may buy. For a put, it is the price at which the stock may be sold. The purchase or sale of the stock is to the writer of the option. The striking price is set for the life of the option on the options exchange. When stock price changes, new exercise prices are introduced for trading purposes reflecting the new value.

In case of conventional calls, restrictions do not exist on what the striking price should be. However, it is usually close to the market price of the stock to which it relates. But in the case of listed calls, stocks having a price lower than $50 a share must have striking prices in $5 increments. Stocks between $50 and $100 must have striking prices in $20 increments. Striking prices are adjusted for material stock splits and stock dividends.

The *expiration date* of an option is the last day it can be exercised. For conventional options, the expiration date can be any business day; for a listed option there is a standardized expiration date.

The cost of an option is referred to as a *premium.* It is the price the buyer of the call or put has to pay the seller (writer). In other words, the option premium is what an option costs to you as a buyer. *Note:* With other securities, the premium is the excess of the purchase price over a determined theoretical value.

B. Why Do Investors Use Options?

Why use options? Reasons can vary from the conservative to the speculative. The most common reasons are:

1. You can earn large profits with *leverage*, that is, without having to tie up a lot of your own money. The leverage you can have with options typically runs 20:1 (each investor dollar controls the profit on twenty dollars of stock) as contrasted with the 2:1 leverage with stocks bought on margin or the 1:1 leverage with stocks bought outright with cash. *Note:* Leverage is a two-edge sword. You can lose a lot, too. That is why it is a risky derivative instrument.
2. Options may be purchased as "insurance or hedge" against large price drops in underlying stocks already held by the investor.
3. If you are neutral or slightly bullish in the short term on stocks you own, you can sell (or write) options on those stocks and realize extra profit.
4. Options offer a range of strategies that cannot be obtained with stocks. Thus, options are a flexible and complementary investment vehicle to stocks and bonds.

C. How Do You Trade Options?

Options are traded on listed option exchanges (secondary markets) such as the *Chicago Board Options Exchange, American Stock Exchange, Philadelphia Stock Exchange*, and *Pacific Stock Exchange*. They may also be exchanged in the *over-the counter (OTC)* market. Option exchanges are only for buying and selling call and put options. Listed options are traded on organized exchanges. Conventional options are traded in the OTC market. The *Options Clearing Corporation (OCC)* acts as principal in every options transaction for listed options contracts. As principal it issues all listed options, guarantees the contracts, and is the legal entity on the other side of every transaction. Orders are placed with this corporation, which then issues the calls or closes the position. Because certificates are not issued for options, a brokerage account is required. When an investor exercises a call, he goes through the Clearing Corporation, which randomly selects a writer from a member list. A call writer is obligated to sell 100 shares at the exercise price. Exchanges permit general orders (i.e., limit) and orders applicable only to the option (i.e., spread order).

D. How Do You Use Profit Diagrams?

In order to understand the risks and rewards associated with various option strategies, it is very helpful to understand how the profit diagram works. In fact, this is essential for understanding how an option works. The profit diagram is a visual portrayal of your profit in relation to the price of a stock at a single point in time.

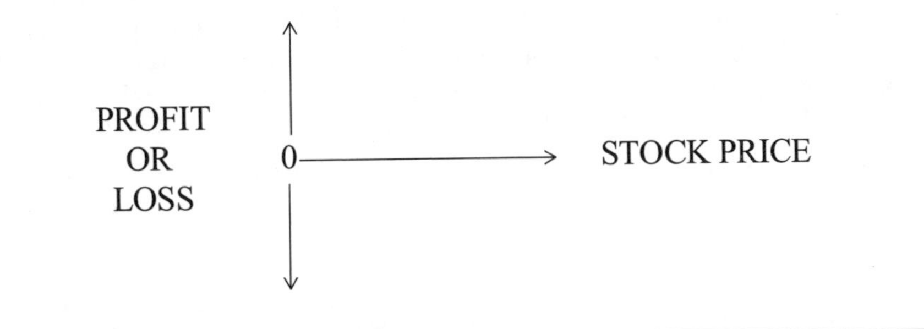

EXAMPLE 86

Nokia Stock Price in 3 months	Profit (Loss)
$60	–$2000
$70	–$1000
$80	0
$90	$1000
$100	$2000

The following shows the profit diagram for 100 shares of Nokia stock if you bought them today at $80 per share and sold them in 3 months. (Commissions are ignored in this example.)

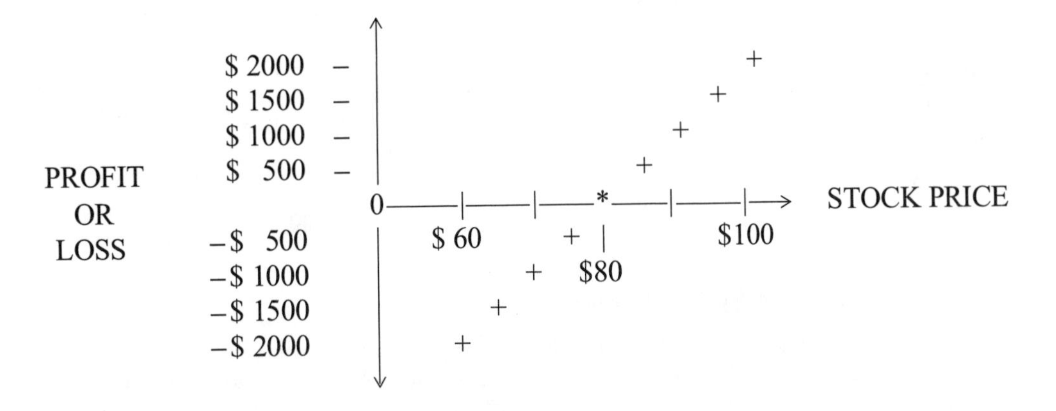

Note that all stocks have the same shape on the profit diagram at any point in the future. You will later see that this is *not* the case with options.

EXAMPLE 87

Assume that on April 7, you become convinced that Nokia stock which is trading at $80 a share will move considerably higher in the next few months. So, you buy one call option on Nokia stock with a premium of $2 a share. Since the call option involves a block of 100 shares of stock, it costs you a total of $2 times 100 shares or $200. Assume further that this call option has a striking price of $85 and an expiration date near the end of September. What this means is that for $200 you have the right to buy:

1. 100 shares of Nokia stock
2. at $85 a share
3. until near the end of September.

This may not sound like you are getting much for $200, but if Nokia stock goes up to $95 a share by the end of September, you would have the right to purchase 100 shares of Nokia stock for $8500 ($85 times 100 shares) and to turn right around and sell them for $9500, keeping the difference of $1000, an $800 profit. That works out to 400% profit in less than five months.

However, if you are wrong and Nokia stock goes down in price, the most you could lose would be the price of the option, $200. The following displays the profit table for this example.

If the Nokia Stock Price in Sep Turns Out to be:		The Value of the Call Option would be:		And Your Profit could be:
$75	→	$0	→	– $200
$80		$0		– $200
$85		$0		– $200
$87		$200		$0
$90		$500		$300
$95		$1000		$800

The profit diagram will look like this:

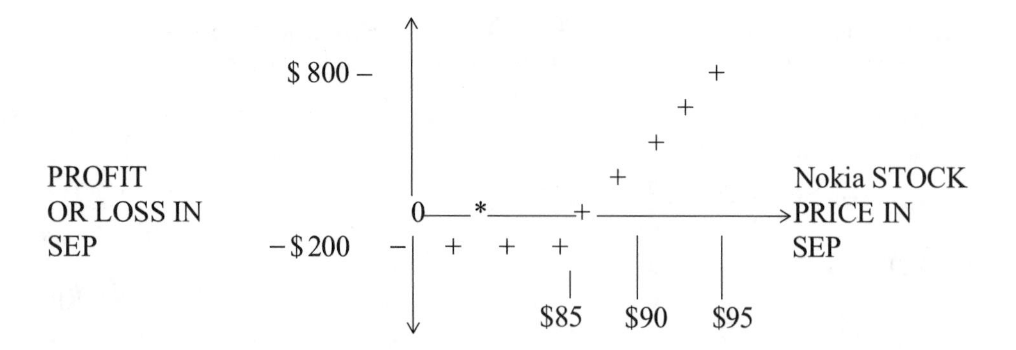

You are "long 1 Nokia Sep 85 call" option.

Notice where the profit line bends—at $85, unlike stocks that have the same shape on the profit diagram at any point in the future. This is *not* the case with options. You start making money after the price of Nokia stock goes higher than the $85 striking price of the call option. When this happens, the option is called "in-the-money."

On the other hand, the profit diagram for a put option looks like this:

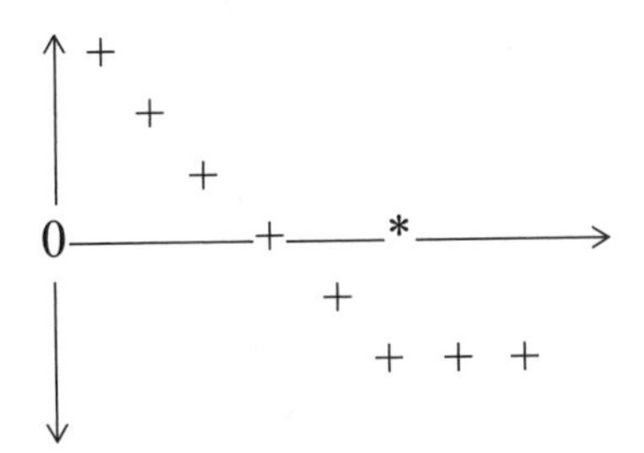

So, a put is typically used by an investor who is bearish on that particular stock. The put option can also be used as "insurance" against price drops for the investor with a long stock position.

E. How Much Does an Option Cost?

The premium for an option (or cost of an option) depends primarily on:

- Fluctuation in price of the underlying security (A higher variability means a higher premium because of the greater speculative appeal of the option.)
- Time period remaining before the option's expiration (The more time there is until the expiration, the greater the premium you must pay the seller.)
- Price spread between the stock compared to the option's strike price (A wider difference translates to a higher price.)

EXAMPLE 88

ABC stock is selling at \$32 a share today. Consider two options: (1) Option X gives you the right to buy the stock at \$25 per share and (2) Option Y gives you the right to buy the stock at \$40 per share. Because you would rather have an option to pay \$25 for a \$32 stock instead of \$32, Option X is more valuable than Option Y. Thus, it will cost you more to buy Option X than to buy Option Y.

Other factors that determine the cost of an option are:

- The dividend trend of the underlying security
- The volume of trading in the option
- The exchange the option is listed on
- "Going" interest rates
- The market price of the underlying stock

F. In-the-Money and Out-of-the-Money Call Options

Options may or may not be exercised, depending on the difference between the market price of the stock and the exercise price.

Let P = the price of the underlying stock and S = the exercise price.

There are three possible situations:

1. If $P > X$ or $P - X > 0$, then the call option is said to be *in the money.* (By exercising the call option, you, as a holder, realize a positive profit, $P - X$.) The value of the call in this case is:

$$\text{Value of call} = (\text{market price of stock} - \text{exercise price of call}) \times 100$$

2. If $P - X = 0$, then the option is said to be *at the money.*
3. If $P - X < 0$, then the option is said to be *out of the money.* It is unprofitable. The option holder can purchase the stock at the cheaper price in the market rather than exercising the option and thus the option is thrown away. Out-of-the-money call options have no intrinsic value.

If the total premium (option price) of an option is \$14 and the intrinsic value is \$6, there is an additional premium of \$8 arising from other factors. Total premium is composed of the intrinsic value and time value (speculative premium) based on variables such as risk, expected future prices, maturity, leverage, dividend, and fluctuation in price.

$$\text{Total premium} = \text{intrinsic value} + \text{time value (speculative premium)}$$

Intrinsic value = In-the-money option (i.e., $P - S > 0$ for a call and $S - P > 0$ for a put option). For in-the-money options, time value is the difference between premium and intrinsic value. For other options all value is time value. Exhibit 83 shows the time value and intrinsic value associated with a call option.

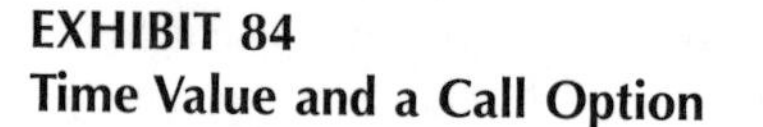

EXHIBIT 84
Time Value and a Call Option

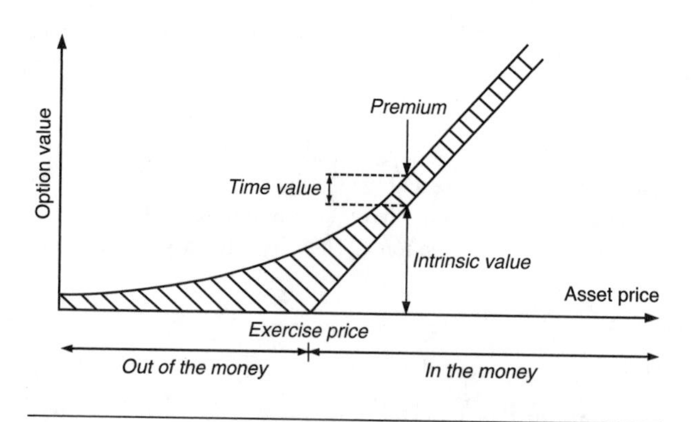

G. In-the-Money and Out-of-the-Money Put Options

A put option on a common stock allows the holder of the option to sell (put) a share of the underlying stock at an exercise price until an expiration date. The definition of *in-the-money* and *out-of-the-money* are different for puts because the owner may sell stock at the strike price. For a put option, the option is in the money if $P - X < 0$.

Its value is determined as follows:

$$\text{Value of put} = (\text{exercise price of put} - \text{market price of stock}) \times 100$$

And the option is out of the money when $P - X > 0$ and has no value.

EXAMPLE 89

Assume a stock has a market price of \$100 and a strike price of the put is \$116. The value of the put is \$1,600. If market price of stock exceeds strike price, an out-of-the money put exists.

Because a stock owner can sell it for a greater amount in the market relative to exercising the put, no intrinsic value exists of the out-of-money put.

	ABC Calls at 60 Strike Price Stock Price	ABC Puts at 60 Strike Price Stock Price
In-the-money	Over 60	Under 60
At-the-money	60	60
Out-of-the-money	Under 60	Over 60

The theoretical value for calls and puts reflects the price the options should be traded. But usually they are traded at prices exceeding true value when options have a long period to go. This difference is referred to as investment premium.

$$\text{Investment premium} = \frac{\text{option premium} - \text{option value}}{\text{option value}}$$

EXAMPLE 90

Assume a put with a theoretical value of $2,500 and a price of $3,000. It is therefore traded at an investment premium of 20% [($3,000 – $2,500)/$2,500].

H. What Are the Risks and Rewards of Options?

Your risk in buying options is limited to the premium you paid. That is the downside risk for option investing. For example, assume you own a two-month call option to acquire 500 shares of ABC Company at $20 per share. Within that time period, you exercise the option when the market price is $38. You make a gain of $9,000 ($18 × 500 shares) except for the brokerage commission. Of course, the higher the stock's price goes, the more you can profit. However, if the market price had declined from $20 you would not have exercised the call option, and you would have lost the cost of the option. *Note:* If you owned the stock whose price fell $10 per share, you would have lost $10 a share. But if you had an option to buy that stock, you could have lost only the cost (premium) of that option, no matter how far the stock price fell.

I. How Do Calls Work?

By buying a call you can own common stock for a low percentage of the cost of buying regular shares. Leverage is obtained since a small change in common stock price can magnify a major move in the call option's price. An element of the percentage gain in the price of the call is the speculative premium related to the remaining time left on the call. Calls can also be viewed as a way of controlling 100 shares of stock without a large monetary commitment.

EXAMPLE 91

Assume that a security has a present market price of $70. A call can be bought for $600 permitting the purchase of 100 shares at $70 per share. If the stock price goes up, the call increases in value. Assume the stock goes to $95 by the call's expiration date. The profit is $25 per share in the call, or a total of $2,500 on an investment of $600. There is a return of 417%. When you exercise the call for 100 shares at $70 each, you can immediately sell them at $95 per share. *Note:* You could have earned the same amount by investing directly in the common stock. However, you would have needed to invest $7,000 resulting in a much lower return rate.

J. How Do Puts Work?

The put holder may sell 100 shares at the exercise price for a specified time period to a put writer. A put is bought when a price decline is expected. As with a call option, the entire premium cost (investment) would be lost if the price does not drop.

EXAMPLE 92

Assume that a stock has a market price of \$80. You buy a put to sell 100 shares of stock at \$80 per share. The put cost is \$500. At the exercise date, the price of the stock goes to \$70 a share. The profit is \$10 per share, or \$1,000. You can buy on the market 100 shares at \$70 each and then sell them to the writer of the put for \$80 each. The net gain is \$500 (\$1,000 – \$500).

Exhibit 85 summarizes payoffs, risks, and break-even stock prices for various option participants.

EXHIBIT 85
Option Payoffs, Risks, and Break-Even Points

	Call Buyer	Call Seller (Writer)
Payoff	$-c + (P - S)$ where c = the call premium For a break-even, $-c + (P - S) = 0$ or $P = S + c$	$+c - (P - S)$
Risk	Maximum risk is to lose the premium because investor throws away the out-of-the-money option	No risk limit as the stock price rises above the exercise price—Uncovered (naked) option To be covered, investor should own the underlying stock or hold a long call on the same stock
	Put Buyer	**Put Seller (Writer)**
Payoff	$-c + (S - P)$ where c = the put premium For a break-even, $-c + (S - P) = 0$ or $P = S - c$	$+c - (S - P)$
Risk	Maximum risk is to lose the premium	Maximum risk is the strike price when the stock price is zero—Uncovered (naked) option To be covered, investor should sell the the underlying stock short or hold a long put on the same stock

Option Parties	Break-Even Market Price
A call-holder	The strike price + the premium
A put-holder	The strike price – the premium
A call-writer	The strike price + the premium
A put-writer	The strike price – the premium
A covered call-writer	The original cost of the security – the premium
A covered put-writer (short the stock)	The strike price + the premium

---

K. Are There Call and Put Investment Strategies You May Use?

Investment possibilities with calls and puts include (1) hedging, (2) speculation, (3) straddles, and (4) spreads. If you own call and put options, you can *hedge* by holding two or more securities to reduce risk and earn a profit. You may purchase a stock and subsequently buy an option on it. For instance, you may buy a stock and write a call on it. Further, if you own a stock that has appreciated, you may buy a put to insulate from downside risk.

EXAMPLE 93

You bought 100 shares of XYZ at $52 per share and a put for $300 on the 100 shares at an exercise price of $52. If the stock does not move, you lose $300 on the put. If the price falls, your loss offsets your gain on the put. If stock price goes up, you have a capital gain on the stock but lose your investment in the put. To obtain the advantage of a hedge, you incur a loss on the put. Note that at the expiration date, you have a loss with no hedge any longer.

You may employ calls and puts to *speculate*. You may buy options when you believe you will make a higher return compared to investing in the underlying stock. You can earn a higher return at lower risk with out-of-the-money options. However, with such an option, the price is composed of only the investment premium, which may be lost if the stock does not increase in price.

EXAMPLE 94

You speculate by buying an option contract to purchase 100 shares at $55 a share. The option costs $250. The stock price increases to $63 a share. You exercise the option and sell the shares in the market, recognizing a gain of $550 ($63 – $55 – $2.50 = $5.50 × 100 shares). You, as a speculator, can sell the option and earn a profit due to the appreciated value. But if stock price drops, your loss is limited to $250 (the option's cost). Obviously, there will also be commissions. In sum, this call option allowed you to buy 100 shares worth $5,500 for $250 up to the option's expiration date.

Straddling combines a put and a call on the identical security with the same strike price and expiration date. It allows you to trade on both sides of the market. You hope for a substantial change in stock price either way so as to earn a gain exceeding the cost of both options. If the price change does materialize, the loss is the cost of the both options. You may increase risk and earning potential by closing one option prior to the other.

EXAMPLE 95

You buy a call and put for $8 each on October 31 when the stock price is $82. There is a three month expiration date. Your investment is $16, or $1,600 in total. If the stock increases to $150 at expiration of the options, the call generates a profit of $60 ($68 – $8) and the loss on the put is $8. Your net gain is $52, or $5,200 in total.

In a *spread,* you buy a call option (long position) and write a call option (short position) in the identical stock. A sophisticated investor may write many spreads to profit from the spread in option premiums. There is substantial return potential but high risk. Different kinds of spreads exist such as a *bull call spread* (two calls having the same expiration date) and *horizontal spread* (initiated with either two call options or two put options on the identical underlying stock). These two options must be with the same strike price but different expiration dates.

You may purchase straddles and spreads to maximize return or reduce risk. You may buy them through dealers belonging to the *Put and Call Brokers and Dealers Association.*

L. How Does Option Writing Work?

The writer of a call contracts to sell shares at the strike price for the price incurred for the call option. Call option writers do the opposite of buyers. Investors write options expecting price appreciation in the stock to be less than what the call buyer anticipates. They may even anticipate the price of the stock to be stable or decrease. Option writers receive the option premium less applied transaction costs. If the option is not exercised, the writer earns the price he paid for it. If the option is exercised, the writer incurs a loss, possibly significant.

If the writer of an option elects to sell, he must give the stock at the contracted price if the option is exercised. In either instance, the option writer receives income from the premium. (Shares are in denominations of 100.) An investor typically sells an option when he anticipates it will not exercised. The risk of option writing is that the writer, if uncovered, must purchase stock or, if covered, loses the gain. As the writer, you can purchase back an option to end your exposure.

EXAMPLE 96

Assume a strike price of $50 and a premium for the call option of $7. If the stock is below $50, the call would not be exercised, and you earn the $7 premium. If the stock is above $50, the call may be exercised, and you must furnish 100 shares at $50. The call writer loses money only if the stock price was above $57.

M. Can You Sell (or Write) an Option on Something You Do Not Own?

Naked (uncovered) and *covered* options exist. Naked options are on stock the writer does not own. There is much risk because you have to buy the stock and then immediately sell it to the option buyer on demand, irrespective of how much you lose. The investor writes the call or put for the premium and will retain it if the price change is beneficial to him or insignificant. The writer has unlimited loss possibilities.

To eliminate this risk, you may write *covered options* (options written on stocks you own). For instance, a call can be written for stock the writer owns or a put can be written for stock sold short. This is a conservative strategy to generate positive returns. The objective is to write an out-of-the-money option, retain the premium paid, and have the stock price equal but not exceed the option exercise price. The writing of a covered call option is like hedging a position because if stock price drops, the writer's loss on the security is partly offset against the option premium.

N. What Are Some Option Strategies?

Currently, about 90% of the option strategies implemented by investors are long calls and long puts only. These are the most basic strategies and are the easiest to implement. However, they are usually the riskiest in terms of a traditional measure of risk: variability (uncertainty) of outcomes. A variety of other strategies can offer better returns at less risk.

N.1. Long Call

This strategy is implemented simply by purchasing a call option on a stock. This strategy is good for a very bullish stock assessment.

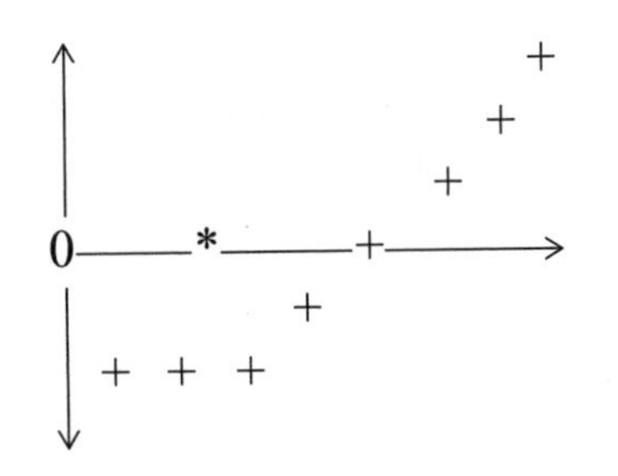

N.2. Bull Call Spread

This strategy requires two calls, both with the same expiration date. It is good for a mildly bullish assessment of the underlying stock.

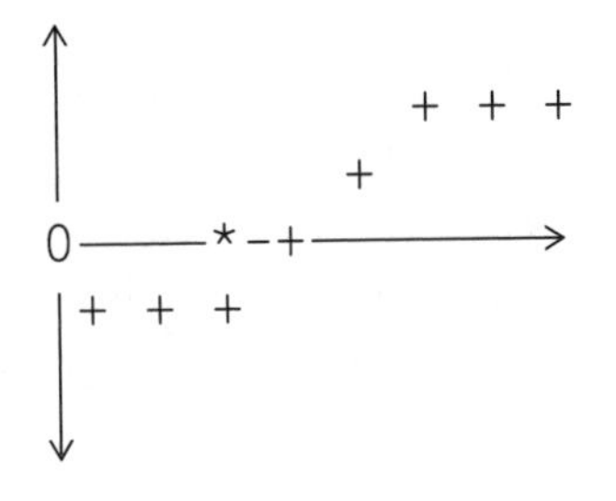

N.3. Naked Put Write

This strategy is implemented by writing a put and is appropriate for a neutral or mildly bullish projection on the underlying stock.

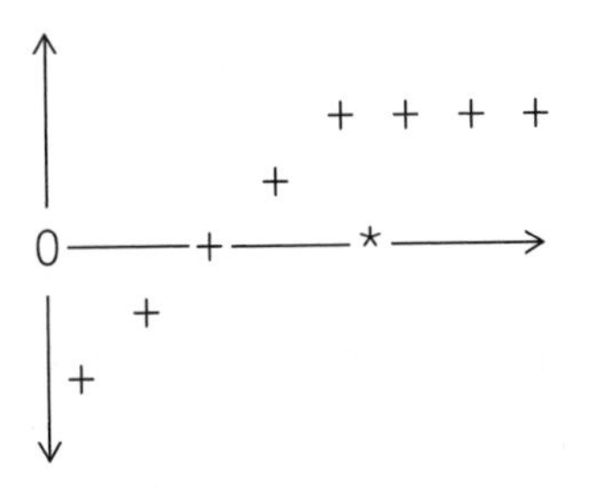

N.4. Covered Call Write

This strategy is equivalent to the *naked put write* and is good as a neutral or mildly bullish assessment of the underlying stock.

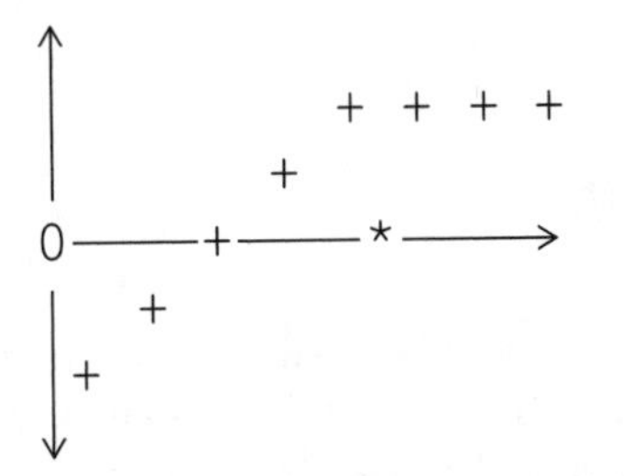

N.5. Straddle

This strategy is implemented by purchasing both a call and a put option on the same underlying stock. This strategy is good when the underlying stock is likely to make a big move but there is uncertainty as to its direction.

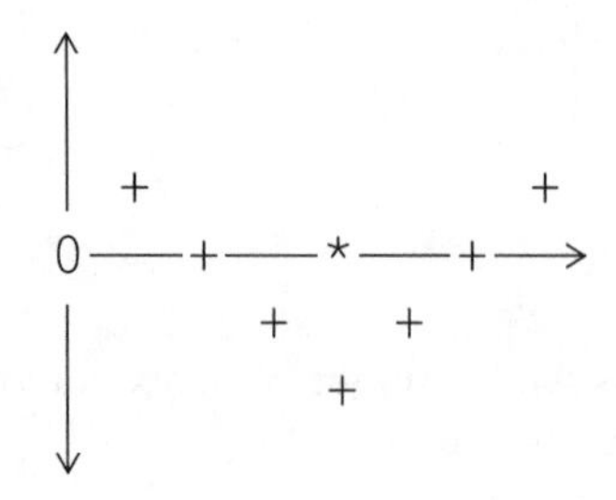

N.6. Inverse Straddle

This strategy is implemented by writing both a call and a put on the same underlying stock and is appropriate for a neutral assessment of the underlying stock. A substantial amount of collateral is required for this strategy due to the open-ended risk should the underlying stock make a big move.

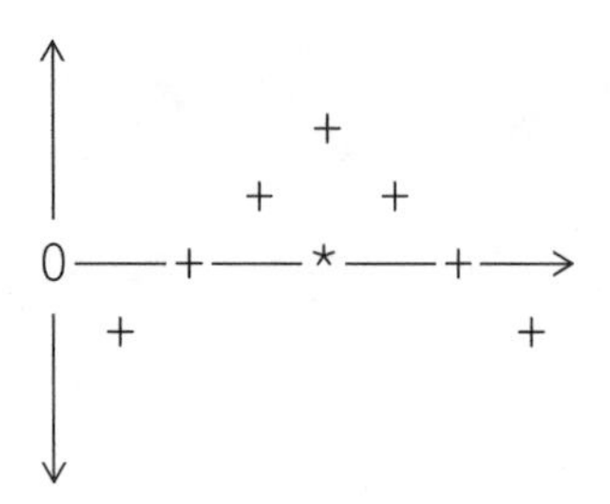

N.7. Horizontal Spread

This strategy is implemented with either two call options or two put options on the same underlying stock. These two options must have the same striking price but have different expiration dates.

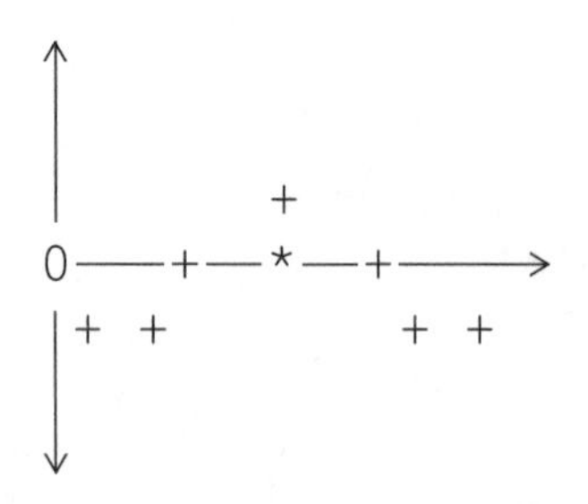

N.8. Naked Call Write

This strategy is implemented by writing a call and is appropriate for a neutral or mildly bearish assessment on the underlying stock. A substantial amount of collateral is required for this strategy due to the open-ended risk should the underlying stock rise in value.

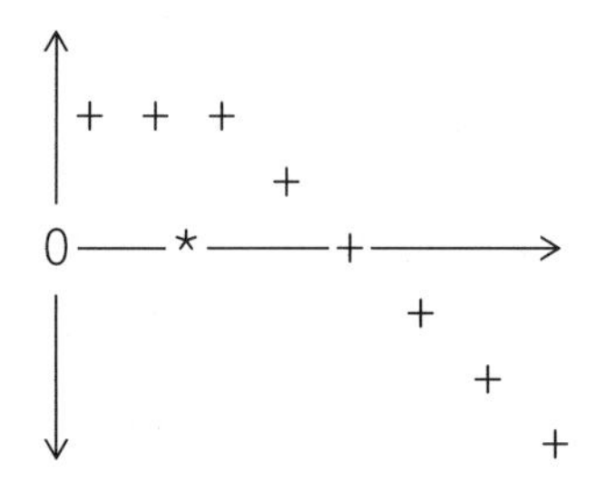

N.9. Bear Put Spread

This strategy is the opposite of the bull call spread. It is implemented with two puts, both with the same expiration date, and is appropriate for a mildly bearish assessment of the underlying stock.

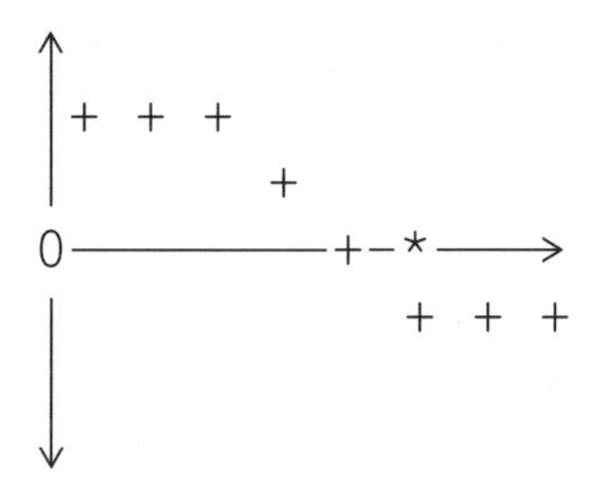

N.10. Long Put

This strategy is implemented simply by purchasing a put option on a stock. It is good for a very bearish stock assessment.

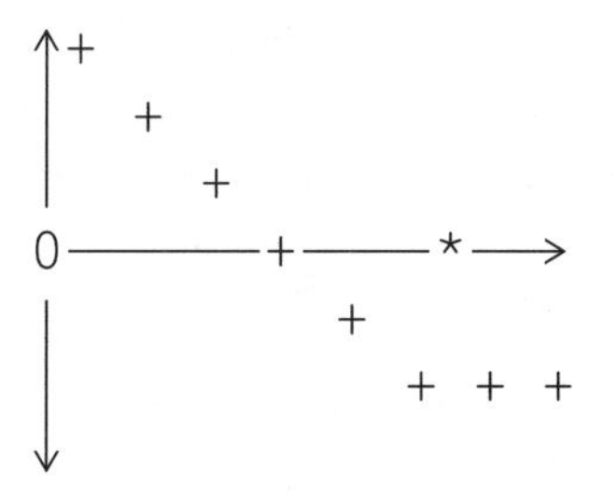

Note: Computer software such as *OptionVue* plots profit tables and diagrams and helps you evaluate large numbers of options for minimum risk and maximum reward.

O. How Do You Choose an Option Strategy?

The key question remains: Which option strategy should you choose? What factors should you consider? What would be a typical decision process? There are three major steps in the decision process:

O.1. Select the Underlying Stock

First, you should decide which stock to consider and do a thorough analysis on the stock, including the effects of current market trends.

O.2. Choose the Strategy

You then determine the risk involved in the stock based on its volatility. Computer software can be of great help. Based on the assessment on the stock (bullish or bearish) and its volatility, a strategy is chosen. For example, a strongly bullish, high volatility stock would indicate a long call strategy, because the underlying stock is likely to rise a substantial amount.

The ranking of strategies so far discussed, from bullish to bearish, is as follows:

Bullish	Long Call Bull Call Spread Naked Put Write (Covered Call Write)
Neutral	Straddle Inverse Straddle Horizontal Spread
Bearish	Naked Call Write Bear Put Spread Long Put

Note: The key to choosing the specific option contracts to implement a strategy is to accurately forecast both the price of the underlying stock and the amount of time it will take to get to that price. This will facilitate choosing the striking price and expiration date of the options to be used.

O.3. Assess the Risk

Option strategies have some interesting risk/reward trade-offs. Some strategies have a small chance of a very large profit while other strategies have a large chance of making a small profit. You have to decide exactly how much to risk for how much reward.

P. Are Index Options Safer?

Options on stock indexes rather than on individual stocks have been popular among investors. Index options include ones on S&P 100, S&P OTC 250, S&P 500, Gold/Silver Index, and Computer Technology Index.

Index options offer advantages over stock options in several ways:

1. There is greater stability in a stock index due to *diversification*. Because an index is a composite of stocks, the effects of mergers, announcements, and reports are much milder in an index than with an individual stock.
2. Index options provide a wider selection of striking prices and expiration dates than stock options.
3. It appears easier to predict the behavior of the market than of an individual stock.
4. More liquidity exists with index options. Due to the high volume of activity, it is easier to buy and sell index options for the price you want. This is especially helpful for far out-of-the-money or deep in-the-money options.
5. Index options are always settled in cash, never in shares of the underlying stock. This settlement is automatic at expiration and the cash settlement prevents unintended stock assignment.

A disadvantage of index options is that *no* covered writing is possible on index options.

Q. Software for Options Analysis

The Value Line Options Survey (800-535-9643 ext. 2854—Dept. 414M10) recommends the few dozen buying and covered writing candidates (out of more than 10,000 options listed on the several exchanges), based on their computerized model. The following is a list of popular options software:

1. Stock Option Analysis Program and Stock Options Scanner, H&H Scientific, (301) 292-2958
2. An Option Valuator/An Option Writer, Revenge Software, (516) 271-9556
3. Strategist, Iotinomics Corp., (800) 255-3374 or (801) 466-2111

4. Option-80, (508) 369-1589
5. Optionvue IV, Optionvue Systems International, Inc., (800) 733-6610 or (708) 816-6610
6. Option Pro, Essex Trading Co., (800) 726-2140 or (708) 416-3530
7. Options and Arbitrage Software Package, Programmed Press, (516) 599-6527

OPTION COLLAR

The option collar is constructed by simultaneously buying a put and selling a call.

EXAMPLE 97

XYZ will receive 2,000,000 pounds in 90 days. It wishes to maximize the dollar receipts. Assume the following data:

Accounts receivable	2,000,000 pounds		
Forward rate	1.5540 $/pnd		
Call strike price	1.5500 $/pnd	Premium	0.0350 $/pnd
Put strike price	1.5500 $/pnd	Premium	0.0250 $/pnd

It is possible to determine the sensitivity of the option collar to strike prices and premiums, and compare the results with the *forward contract.* Exhibits 86 and 87 present valuation of an option collar.

EXHIBIT 86
Valuation of an Option Collar

Ending Spot Rate	1.53	1.54	1.55	1.56	1.57	1.58	1.59	1.60
				Buy a Put				
Premium cost	−0.0250	−0.0250	−0.0250	−0.0250	−0.0250	−0.0250	−0.0250	−0.0250
Put receipts	1.5500	1.5500	0.0000	0.0000	0.0000	0.0000	0.0000	0.0000
Spot receipts	0.0000	0.0000	1.5500	1.5600	1.5700	1.5800	1.5900	1.6000
Put Total	1.5250	1.5250	1.5250	1.5350	1.5450	1.5550	1.5650	1.5750
				Write a Call				
Premium receipts	0.0350	0.0350	0.0350	0.0350	0.0350	0.0350	0.0350	0.0350
Covering call	0.0000	0.0000	0.0000	−1.5600	−1.5700	−1.5800	−1.5900	−1.6000
Call receipts	0.0000	0.0000	0.0000	1.5500	1.5500	1.5500	1.5500	1.5500
Call Total	0.0350	0.0350	0.0350	0.0250	0.0150	0.0050	−0.0050	−0.0150
Combined "Collar" (put + call)	1.5600	1.5600	1.5600	1.5600	1.5600	1.5600	1.5600	1.5600
Forward Contract	1.5540	1.5540	1.5540	1.5540	1.5540	1.5540	1.5540	1.5540

EXHIBIT 87

Option Collar Valuation
(Buying a Put and Selling a Call)

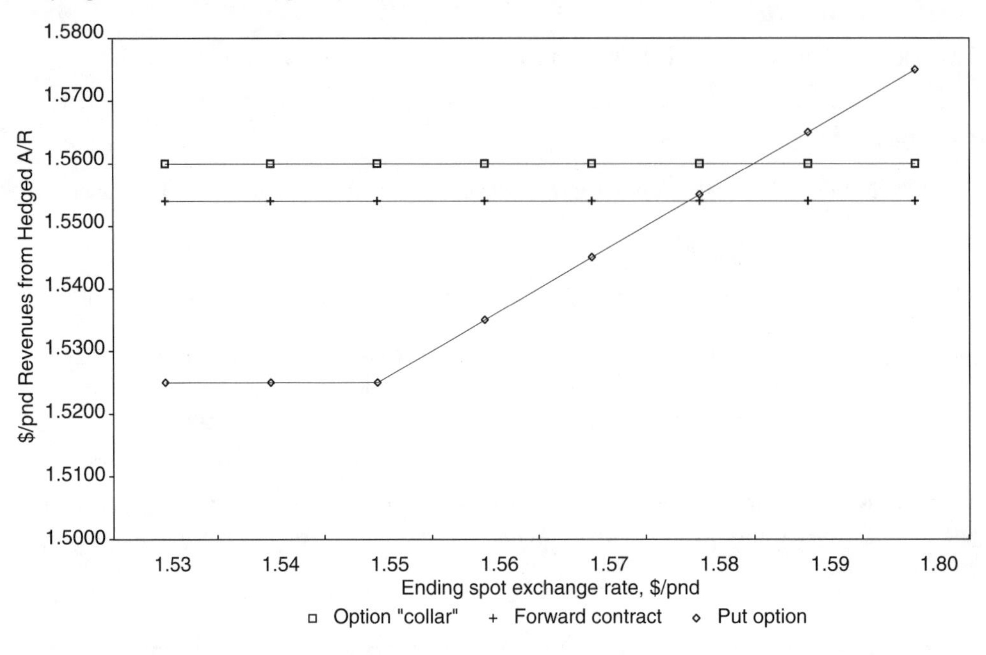

OUTRIGHT QUOTATION

Exchange rate quotation in terms of the full price, in one currency, per unit of another currency.

OUTRIGHT RATE

Actual forward rate expressed in dollars per unit of a currency, or vice versa. Commercial customers are quoted this way.

See FORWARD EXCHANGE RATE; SWAP RATE.

OVERALL BALANCE

See OFFICIAL SETTLEMENTS BALANCE.

OVERSEAS PRIVATE INVESTMENT CORPORATION

Overseas Private Investment Corporation (OPIC), established in 1971, is a self-sustaining federal agency responsible for insuring direct U.S. investments in foreign countries against the risks of expropriation, currency inconvertibility, and other political uncertainties (such as damages from wars and revolutions).

See also PRIVATE EXPORT FUNDING CORPORATION.

P

PARALLEL LOAN

See BACK-TO-BACK LOANS.

PARALLEL MARKETS

A government-tolerated alternative to the official foreign exchange market, such as a *black market*. Some countries unofficially acknowledge the benefits of such a market by allowing the market to exist openly.

PARITY CONDITIONS

More exactly called *international exchange rate parity conditions*, parity conditions are fundamental economic relationships that help to explain exchange rate movements. There are five key relationships among spot exchange rates, forward rates, inflations rates, and interest rates.

Parity Condition	Relationship in Exhibit 87
Purchasing Power Parity	A
Fisher Effect	B
International Fisher Effect	C
Interest Rate Parity	D
Forward Rates as Unbiased Predictor of Future Spot Rates	E

Exhibit 88 displays that under a freely floating exchange rate system, future spot exchange rates are theoretically determined by differing national rates of inflation, interest rates, and the forward premium or discount on each currency. Parity conditions are for any percentage change, or difference in rates. Hypothetical data used in the exhibit are as follows:

1. Currency rates:
   a. Spot Rate (*indirect*) S_1 = ¥104/$
   b. Forward Rate (one year) F = ¥100/$
   c. Expected Spot Rate S_2 = ¥100/$
   d. Expected change in S: $\frac{S_1 - S_2}{S_2} = \frac{104 - 100}{100} = 4\%$
   e. Forward premium on yen: $\frac{104 - 100}{100} = 4\%$
2. Expected rate of inflation:
   a. Japan 2%
   b. United States 6%
   c. Difference 4%
3. Interest on 1-year Government Security
   a. Japan 1%
   b. United States 5%
   c. Deference 4%

See also FISHER EFFECT; FORWARD RATES AS UNBIASED PREDICTORS OF FUTURE SPOT RATES; INTEREST RATE PARITY; INTERNATIONAL FISHER EFFECT; PURCHASING POWER PARITY.

EXHIBIT 88
International Parity Conditions

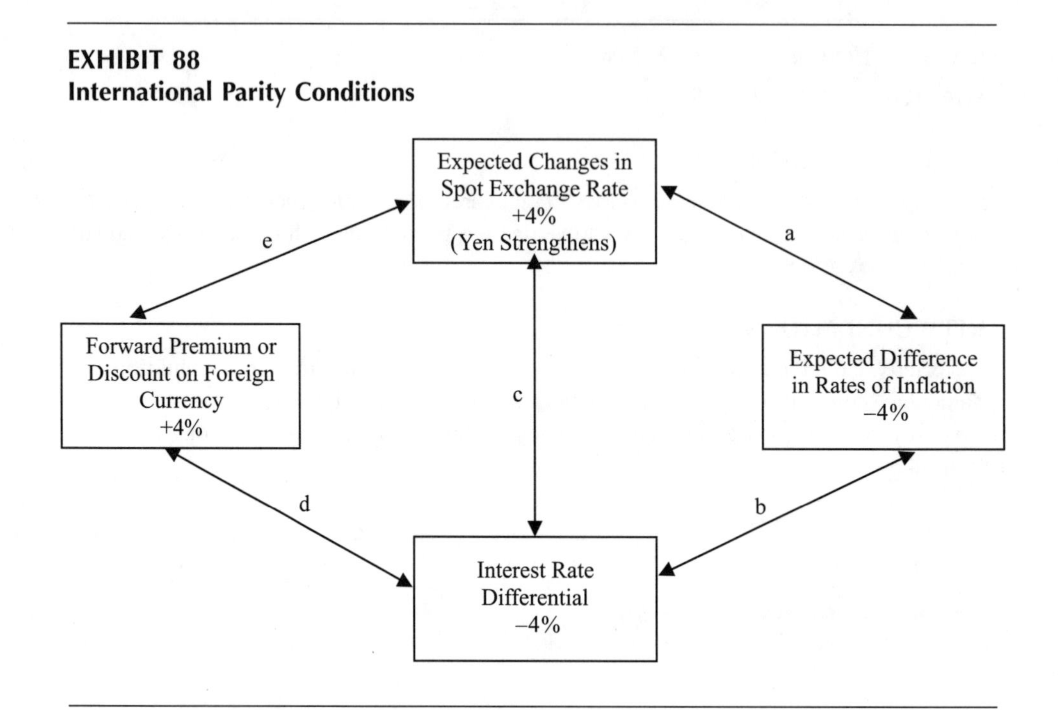

PARTICIPATING FORWARD

See FORWARD WITH OPTION EXIT.

PAR VALUE

1. For a country's currency, the value set by a government, by agreement or regulation, on its currency in terms of other currencies. At *Bretton Woods*, other currencies were assigned par values in terms of the U.S. dollar.
2. For a stock, the dollar amount assigned each share of stock in the company's charter. For preferred issues and bonds, the value on which the issuer promises to pay dividends.

PEGGED EXCHANGE RATES

An exchange rate system under which a country establishes a fixed exchange rate with another major currency and, as a result, values of pegged currencies move together over time. For example, Kuwait pegs its currency to a mix of currencies that roughly represent the composite of currencies used by its trading partners to buy its oil.

PERFORMANCE BOND

Previously referred to as *margin*, funds that must be deposited as a collateral by a customer with his or her broker, by a broker with a clearing member, or by a clearing member with the Clearing House. The performance bond helps to ensure the financial integrity of brokers, clearing members, and the Exchange as a whole.
See MARGIN.

PERFORMANCE BOND CALL

Previously referred to as *margin call*, a demand for additional funds because of adverse price movement.

See also MAINTENANCE MARGIN.

PESETA

Monetary unit of Spain, Andorra, Balearic Island, and Canary Islands.

PESO

The currency of the following countries: Argentina, Chile, Columbia, Cuba, Dominican Republic, Mexico, Republic of Philippines, and Uruguay.

PHI

See CURRENCY OPTION PRICING SENSITIVITY.

PIP

One-hundredth of a *basis* point.

PLAIN-VANILLA

Often called *generic*, a security, especially a bond or a swap, issued with standard features.

PLAIN-VANILLA SWAPS

A plain-vanilla *swap* is a bank-intermediated *interest rate swap*. Typically, in this type of swap, a BBB-rated U.S. corporation is paired with an AAA-rated foreign bank. The BBB corporation wants to borrow at fixed rates but is discouraged from doing so by the high rates attached to its low rating. The company can borrow in the floating-rate market, however. The AAA foreign bank wants to borrow at floating rates, as its assets are tied to the *London Interbank Offer Rate (LIBOR)*. The bank normally pays at LIBOR to obtain lendable funds. It is able to borrow at fixed rates, which are better than those the corporation could get. When the swap is planned, the bank borrows at fixed rates in the Eurobond market, and the BBB corporation borrows at floating rates. With this done, the two then swap rates on their respective loans. The corporation entices the bank into doing this by discounting the floating rate to the bank by LIBOR minus 0.25%. In return, the bank gives the BBB corporation a fixed rate that is better than the one the corporation could acquire on its own.

EXAMPLE 98

Assume the BBB firm can borrow at a fixed rate of 12.5%.

1. AAA bank issues 11% fixed-rate debt.
2. BBB firm taps floating-rate market at LIBOR plus 0.5%.
3. BBB firm swaps its floating rate to bank at LIBOR minus 0.25%.
4. AAA bank swaps its 11% fixed rate to BBB firm.

The net result is the BBB firm saves 150 basis points and passes 75 of them on to the AAA bank; the bank saves 25 basis points on the normal floating rate.
See also SWAPS.

POINT

One percent (%).

POINTS QUOTATION

A *forward rate* quotation expressed as only the number of decimal points.

POLITICAL RISK

Also called *sovereign risk*, political risk implies unwanted consequence of political activity. It is commonly interpreted as (usually host) government interference with business operations. But political changes that are gradual and progressive and neither unexpected nor difficult to anticipate do not constitute political risk. Political risk means potential actions by a host government that would threaten the value of an MNC's investment. It is the risk associated with political or sovereign uncertainty. Clearly, political factors are a major determinant of the attractiveness for investment in any country. Examples of political risk are government expropriation of property, currency conversion restrictions, and import barriers. Countries viewed as likely candidates for internal political upheaval or with a pronounced trend toward elimination of the private sector will be unattractive to all investors, foreign and domestic alike. There is no reason to believe that local investors will be systematically optimistic regarding their country's future. When political risks increase significantly, such investors will attempt to diversify from the home market as rapidly as will foreigners. As a result, prices will fall until someone will be satisfied to hold the securities of a risky country. Political instability, limited track records, poor statistics all make gauging risk a risky business. Several companies try to evaluate the risk in some of the countries that are receiving the most attention from foreign investors:

- *Euromoney* Magazine's *Country Risk Ratings*, published twice a year, are based on a measure of different countries' access to international credit, trade finance, and political risk and the country's payment record. The rankings are generally confirmed by political risk insurers and top syndicate managers in the Euromarkets. See Exhibit 89.
- *International Country Risk Guide*, published by *International Business Communications, Ltd.*, a London company, offers a composite risk rating, as well as individual ratings for political, financial, and economic risk. The political variable—which makes up half of the composite index—includes factors such as government corruption and how economic expectations diverge from reality. The financial rating looks at such things as the likelihood of losses from exchange controls and loan defaults. Finally, economic ratings consider such factors as inflation and debt-service costs.
- Rating by *Economist Intelligence Unit*, a New York-based subsidiary of the *Economist Group*, London, is based on such factors as external debt and trends in the current account, the consistency of the government policy, foreign-exchange reserves, and the quality of economic management.
- *Data Book* (quarterly), published by Thompson BankWatch, Inc., 61 Broadway, New York, NY 10006, provides a Thompson country rating assessing overall political and economic stability of a country in which a bank is domiciled.

EXHIBIT 89
Euromoney Magazine's Country Risk Ratings, September 2000

Sep-00	Mar-00			Total Score	Political Risk	Economic Performance	Debt Indicators	Debt in Default or Rescheduled	Credit Ratings
			Weighting:	100	25	25	10	10	10
1	1	Luxembourg		99.02	24.38	25.00	10.00	10.00	10.00
2	2	Switzerland		96.89	25.00	21.89	10.00	10.00	10.00
3	4	United States		94.25	24.94	19.30	10.00	10.00	10.00
4	3	Norway		94.24	23.90	20.42	10.00	10.00	10.00
5	6	Netherlands		92.90	24.66	18.27	10.00	10.00	10.00
6	5	Denmark		92.77	23.09	19.95	10.00	10.00	9.79
7	11	Germany		92.77	24.52	18.27	10.00	10.00	10.00
8	8	France		92.33	24.34	18.01	10.00	10.00	10.00
9	7	Austria		92.29	23.69	18.63	10.00	10.00	10.00
10	12	United Kingdom		91.54	24.63	16.92	10.00	10.00	10.00
11	9	Finland		91.38	23.52	18.23	10.00	10.00	9.69
12	10	Sweden		91.12	23.65	18.65	9.50	10.00	9.38
13	14	Japan		90.70	23.51	18.41	10.00	10.00	9.58
14	15	Singapore		90.04	22.39	19.07	10.00	10.00	9.58
15	16	Ireland		89.71	23.42	16.53	10.00	10.00	9.79
16	13	Belgium		89.63	23.07	17.84	10.00	10.00	8.75
17	17	Canada		89.10	23.75	16.24	10.00	10.00	9.17
18	18	Australia		88.01	22.69	16.60	10.00	10.00	9.17
19	20	Spain		87.29	23.03	14.93	10.00	10.00	9.38
20	19	Italy		87.11	22.61	16.14	10.00	10.00	8.44
21	21	Iceland		86.72	19.91	18.71	10.00	10.00	8.54

(Continued)

EXHIBIT 89
Euromoney Magazine's Country Risk Ratings, September 2000 (continued)

Sep-00	Mar-00		Total Score	Political Risk	Economic Performance	Debt Indicators	Debt in Default or Rescheduled	Credit Ratings
		Weighting:	100	25	25	10	10	10
22	22	New Zealand	85.27	21.42	14.54	10.00	10.00	9.38
23	23	Portugal	83.25	22.21	13.72	10.00	10.00	8.75
24	24	Taiwan	80.65	20.46	14.95	9.87	10.00	8.75
25	25	Greece	78.66	20.13	13.35	10.00	10.00	6.25
26	27	Hong Kong	77.42	18.22	15.95	10.00	10.00	7.08
27	26	Cyprus	76.62	18.18	13.49	9.76	10.00	7.50
28	29	United Arab Emirates	75.59	18.07	14.59	9.98	10.00	6.88
29	30	Bermuda	73.67	18.67	13.71	10.00	10.00	8.96
30	28	Kuwait	73.28	16.35	15.45	9.64	10.00	6.67
31	33	Malta	72.89	18.40	14.20	9.95	10.00	7.08
32	31	Israel	72.68	16.56	13.70	9.74	10.00	6.88
33	32	Slovenia	68.93	17.43	12.31	5.66	10.00	7.29
34	38	Saudi Arabia	68.27	16.74	10.85	9.79	10.00	4.38
35	34	Qatar	68.14	16.10	12.20	8.50	10.00	5.31
36	35	Brunei	66.84	16.40	18.44	10.00	10.00	0.00
37	40	Korea South	66.28	17.50	12.43	9.05	10.00	5.83
38	36	Oman	66.17	15.57	10.27	9.49	10.00	5.00
39	41	Chile	65.83	18.24	10.39	8.67	10.00	6.88
40	42	Hungary	65.24	17.06	10.78	8.35	10.00	6.04
41	39	Bahrain	65.20	15.22	10.56	10.00	10.00	3.75
42	43	Poland	63.59	16.56	9.92	9.26	10.00	6.25
43	44	Czech Republic	63.14	17.11	9.96	8.94	10.00	6.46
44	45	Malaysia	61.11	15.86	10.08	8.88	10.00	5.42
45	48	China	59.75	15.64	9.46	9.58	10.00	5.83
46	49	Mexico	59.68	15.68	9.28	8.81	10.00	4.58

47	54	Thailand	59.55	14.89	8.55	8.44	10.00	4.79
48	37	Bahamas	59.46	16.50	6.03	10.00	10.00	5.31
49	47	Mauritius	59.08	14.16	11.44	8.90	10.00	5.00
50	51	South Africa	57.71	13.85	9.51	9.41	10.00	4.79
51	46	Tunisia	57.48	13.87	10.00	8.80	10.00	5.21
52	75	Barbados	57.32	15.62	10.71	9.43	10.00	6.56
53	50	Uruguay	56.79	14.19	9.41	8.83	10.00	4.79
54	52	Egypt	56.36	14.48	9.21	9.19	9.90	4.79
55	55	Estonia	55.69	13.85	9.31	9.61	10.00	5.63
56	53	Morocco	55.11	12.76	9.27	8.55	10.00	4.38
57	57	Argentina	54.97	12.81	9.58	7.68	10.00	3.13
58	60	Trinidad & Tobago	54.18	14.10	10.95	9.10	10.00	5.00
59	58	India	53.75	13.79	7.90	9.11	10.00	3.96
60	67	Costa Rica	53.42	13.07	10.88	9.20	10.00	3.54
61	59	Latvia	53.11	12.84	8.56	9.62	10.00	5.42
62	66	Slovak Republic	52.95	12.86	8.57	8.76	10.00	4.38
63	56	Philippines	52.81	13.23	7.61	8.75	10.00	4.38
64	65	Turkey	52.74	14.07	8.54	8.74	10.00	2.08
65	64	Panama	52.20	12.02	9.34	8.54	10.00	4.38
66	62	Botswana	51.83	15.14	9.93	9.82	10.00	0.00
67	69	Brazil	51.31	12.49	8.66	7.51	9.96	2.29
68	63	El Salvador	51.12	10.98	9.63	9.29	10.00	4.58
69	61	Lithuania	50.79	12.54	8.28	9.49	10.00	3.96
70	70	Croatia	49.67	11.69	8.49	9.14	7.67	4.58
71	68	Colombia	48.88	11.26	7.13	8.64	10.00	3.96
72	71	Fiji	47.38	10.12	9.18	9.68	10.00	3.75
73	76	Guatemala	47.31	9.77	8.84	9.33	10.00	3.13
74	73	Lebanon	46.85	10.33	7.29	8.62	10.00	2.50
75	74	Jordan	46.32	11.22	7.96	7.56	8.49	3.44
76	78	Venezuela	43.85	10.71	6.85	8.72	10.00	1.67
77	77	Dominican Republic	43.84	10.25	8.81	9.45	9.19	1.25
78	90	Jamaica	42.70	7.73	8.76	8.74	10.00	2.19

(Continued)

EXHIBIT 89
Euromoney Magazine's Country Risk Ratings, September 2000 (continued)

Sep-00	Mar-00			Total Score	Political Risk	Economic Performance	Debt Indicators	Debt in Default or Rescheduled	Credit Ratings
			Weighting:	100	25	25	10	10	10
79	85	Bolivia		42.52	8.54	7.10	8.08	10.00	2.81
80	84	Bulgaria		42.51	10.46	6.69	8.23	10.00	1.88
81	88	Kazakhstan		42.46	9.28	6.65	9.21	10.00	2.29
82	86	Paraguay		41.31	9.74	7.01	9.45	10.00	1.88
83	80	Iran		40.64	8.99	7.09	9.25	10.00	1.25
84	82	Belize		40.61	10.81	5.30	8.82	10.00	3.13
85	79	Sri Lanka		39.81	9.37	5.82	9.08	10.00	0.00
86	91	Seychelles		39.71	9.14	6.32	9.14	10.00	0.00
87	94	Macau		39.59	14.15	12.75	0.00	0.00	5.63
88	125	Maldives		39.22	9.64	8.23	9.03	10.00	0.00
89	92	Peru		39.12	10.63	7.50	9.15	1.25	3.33
90	93	Syria		38.95	9.67	6.62	7.96	10.00	0.00
91	116	Honduras		38.77	8.00	7.12	8.13	10.00	1.25
92	136	Dominica		38.60	7.39	9.98	9.08	10.00	0.00
93	115	Indonesia		38.48	7.98	5.96	6.65	8.65	0.63
94	87	Vietnam		38.36	9.82	6.01	8.64	9.96	1.88
95	133	Russia		37.88	8.02	6.65	8.62	8.26	0.42
96	99	Algeria		37.71	8.27	6.77	7.94	8.90	0.00
97	72	Ghana		37.64	8.57	6.61	7.92	10.00	0.00
98	89	Kenya		37.64	7.41	7.37	8.61	10.00	0.00
99	102	Gambia		37.63	7.85	7.67	9.43	10.00	0.00
100	120	Macedonia (FYR)		37.37	6.30	7.67	8.18	10.00	0.00
101	83	Papua New Guinea		37.17	8.49	5.33	8.74	10.00	2.29
102	103	Azerbaijan		36.94	8.42	6.72	9.22	10.00	0.00

103	107	Romania	36.62	8.17	5.32	8.89	10.00	0.83
104	100	Lesotho	36.42	8.47	6.74	8.54	10.00	0.00
105	98	St Lucia	35.91	9.98	4.57	8.69	10.00	0.00
106	118	Kyrgyz Republic	35.76	8.05	8.13	8.59	10.00	0.00
107	110	Equatorial Guinea	35.69	4.19	9.88	8.93	10.00	0.00
108	95	Bangladesh	34.96	7.73	5.48	9.24	10.00	0.00
109	96	Senegal	34.28	6.28	7.03	8.22	10.00	0.00
110	164	Uzbekistan	34.15	6.76	6.35	9.43	10.00	0.00
111	137	Yemen	33.99	7.57	5.86	8.38	9.95	0.00
112	101	Uganda	33.73	6.77	6.68	8.95	7.72	0.00
113	112	Zimbabwe	33.43	4.22	5.04	7.88	10.00	0.00
114	105	Cape Verde	33.06	6.33	4.95	8.89	10.00	0.00
115	134	Ukraine	33.06	6.05	5.68	9.26	9.61	0.00
116	97	Gabon	33.03	6.86	8.42	8.42	8.26	0.00
117	81	Swaziland	32.92	9.09	8.96	0.00	10.00	0.00
118	131	Cambodia	32.90	3.81	9.40	8.80	10.00	0.00
119	106	Nepal	32.72	6.24	5.38	8.96	10.00	0.00
120	123	Côte d'Ivoire	32.47	5.84	6.56	7.29	8.70	0.00
121	114	St Vincent & the Grenadines	32.14	7.53	4.27	7.70	10.00	0.00
122	108	Nigeria	32.09	4.88	6.20	8.48	10.00	0.00
123	129	Pakistan	31.99	6.36	4.87	8.61	10.00	0.94
124	119	Burkina Faso	31.95	6.27	4.65	8.85	10.00	0.00
125	132	Turkmenistan	31.81	5.98	5.96	7.32	10.00	0.94
126	109	Malawi	31.66	4.99	5.55	7.73	10.00	0.00
127	147	Samoa	31.28	7.75	0.87	8.52	10.00	0.00
128	130	Ethiopia	30.95	4.79	7.62	7.41	9.80	0.00
129	140	Belarus	30.74	5.77	3.24	9.82	10.00	0.00
130	139	Mongolia	30.64	6.10	3.68	8.71	10.00	1.25
131	144	Armenia	30.47	6.22	3.44	8.91	10.00	0.00
132	121	Grenada	30.41	8.09	1.46	8.71	10.00	0.00
133	155	Georgia	30.40	4.19	6.05	9.26	10.00	0.00
134	135	Solomon Islands	30.39	6.86	0.79	9.25	10.00	0.00

(Continued)

EXHIBIT 89
Euromoney Magazine's Country Risk Ratings, September 2000 (continued)

Sep-00	Mar-00		Total Score	Political Risk	Economic Performance	Debt Indicators	Debt in Default or Rescheduled	Credit Ratings
		Weighting:	100	25	25	10	10	10
135	146	Albania	30.38	5.40	4.56	9.49	10.00	0.00
136	127	Vanuatu	30.02	6.19	0.92	9.58	10.00	0.00
137	126	Cameroon	29.74	5.64	5.52	7.73	7.72	0.00
138	117	Madagascar	29.48	3.90	5.66	7.87	9.37	0.00
139	153	Ecuador	29.28	3.90	5.34	7.94	10.00	0.00
140	142	Moldova	29.24	4.90	2.42	8.37	10.00	0.94
141	113	Tanzania	28.89	4.96	6.12	6.63	8.79	0.00
142	145	Mozambique	28.63	4.60	5.48	6.31	8.84	0.00
143	111	Togo	28.63	5.46	3.61	7.69	9.19	0.00
144	157	Bhutan	28.53	6.19	0.71	9.22	10.00	0.00
145	141	Benin	28.51	4.00	3.69	8.64	10.00	0.00
146	152	Guyana	28.27	6.69	3.79	6.10	10.00	0.00
147	104	Mali	28.15	4.66	3.52	8.02	9.99	0.00
148	150	Chad	27.79	3.04	3.51	8.73	9.82	0.00
149	149	Mauritania	27.19	3.52	5.33	5.66	10.00	0.00
150	138	Zambia	27.04	3.86	6.35	6.57	9.32	0.00
151	122	Guinea	26.77	5.02	3.19	8.04	9.62	0.00
152	171	Myanmar	26.35	5.03	3.24	7.19	10.00	0.00
153	158	Nicaragua	26.34	4.71	5.76	4.29	8.72	1.25
154	154	Sudan	25.45	2.41	3.33	8.82	10.00	0.00
155	124	Niger	25.43	2.67	3.17	8.26	9.89	0.00
156	—	Micronesia (Fed. States)	25.24	13.06	0.93	0.00	10.00	0.00
157	156	Central African Republic	25.13	2.86	4.27	8.11	9.00	0.00
158	168	Djibouti	24.63	3.30	0.91	9.17	10.00	0.00
159	148	Haiti	24.32	2.71	0.69	9.40	10.00	0.00

160	128	Tonga	24.01	8.64	1.06	0.00	10.00	0.00
161	163	Laos	23.99	6.06	0.00	7.04	10.00	0.00
162	143	Namibia	23.65	10.29	7.58	0.00	0.00	0.00
163	176	Suriname	23.11	7.36	3.22	0.00	10.00	1.25
164	162	Sierra Leone	23.06	2.62	2.96	6.62	9.97	0.00
165	161	Dem. Rep. of the Congo (Zaire)	22.87	2.35	2.61	7.01	10.00	0.00
166	—	Eritrea	22.18	1.33	0.64	9.31	10.00	0.00
167	159	Congo	22.03	3.66	3.44	5.75	9.19	0.00
168	151	Rwanda	21.13	1.52	0.77	9.55	8.40	0.00
169	167	Angola	20.97	3.30	3.29	4.44	8.42	0.00
170	—	Burundi	20.81	2.29	0.62	7.01	10.00	0.00
171	166	Guinea-Bissau	20.04	4.19	2.56	2.47	9.93	0.00
172	165	New Caledonia	19.96	12.86	3.67	0.00	0.00	0.00
173	—	Marshall Islands	19.44	12.19	1.00	0.00	0.00	0.00
174	174	Antigua & Barbuda	19.43	4.61	2.87	0.00	10.00	0.00
175	169	Libya	19.30	9.20	8.19	0.00	0.00	0.00
176	172	Tajikistan	17.76	3.09	4.42	8.99	0.00	0.00
177	160	Sao Tome & Principe	16.67	2.41	0.65	0.00	10.00	0.00
178	—	Bosnia-Herzegovina	15.82	3.38	3.25	8.16	0.00	0.00
179	173	Liberia	15.30	3.91	0.85	0.00	10.00	0.00
180	175	Yugoslavia (Fed. Republic)	14.81	1.99	1.73	0.00	10.00	0.00
181	170	Somalia	14.76	2.29	0.74	0.00	10.00	0.00
182	177	Cuba	10.67	3.80	5.81	0.00	0.00	0.00
183	178	Iraq	9.04	2.36	5.80	0.00	0.00	0.00
184	179	Korea North	4.72	2.98	0.85	0.00	0.00	0.00
185	180	Afghanistan	2.81	0.00	1.56	0.00	0.00	0.00

Source: **www.euromoney.com.**

A. Methods for Dealing with Political Risk

To the extent that forecasting political risks is a formidable task, what can an MNC do to cope with them? There are several methods suggested.

- *Avoidance*—Try to avoid political risk by minimizing activities in or with countries that are considered to be of high risk and by using a higher discount rate for projects in riskier countries.
- *Adaptation*—Try to reduce such risk by adapting the activities (for example, by using hedging techniques).
- *Diversification*—Diversity across national borders, so that problems in one country do not risk the company.
- *Risk transfer*—Buy insurance policies for political risks. Most developed nations offer insurance for political risk to their exporters. Examples include: in the U.S., the *Eximbank* offers policies to exporters that cover such political risks as war, currency inconvertibility, and civil unrest. Furthermore, the *Overseas Private Investment Corporation (OPIC)* offers policies to U.S. foreign investors to cover such risks as currency inconvertibility, civil or foreign war damages, or expropriation. In the U.K., similar policies are offered by the *Export Credit Guarantee Department (ECGD)*; in Canada, by the *Export Development Council (EDC)*; and in Germany, by an agency called *Hermes*.

PORTFOLIO-BALANCE APPROACH

See ASSET MARKET MODEL.

PORTFOLIO DIVERSIFICATION

The rationale behind portfolio diversification is the reduction of risk. The main method of reducing the risk of a portfolio is the combining of assets which are not perfectly positively correlated in their returns.

EXAMPLE 99

Consider the following two-asset portfolio:

Stock	Return	Risk	Correlation
Wal-mart (US)	18.60%	22.80%	0.20
Smith-Kline (UK)	16.00%	24.00%	

By changing the correlation coefficient, the benefits of risk reduction can be clearly observed, as shown in Exhibits 90 and 91.

EXHIBIT 90
Portfolio Analysis

Weight of Wal-mart in Portfolio	Weight of Smith-Kline in Portfolio	Expected Return (percent)	Expected Risk (percent)
1.00	0.00	18.60%	22.80%
0.95	0.05	18.47%	21.93%
0.90	0.10	18.34%	21.13%
0.85	0.15	18.21%	20.41%
0.80	0.20	18.08%	19.77%
0.75	0.25	17.95%	19.22%
0.70	0.30	17.82%	18.78%

0.65	0.35	17.69%	18.44%
0.60	0.40	17.56%	18.22%
0.55	0.45	17.43%	18.11%
0.50	0.50	17.30%	18.13%
0.45	0.55	17.17%	18.27%
0.40	0.60	17.04%	18.52%
0.35	0.65	16.91%	18.89%
0.30	0.70	16.78%	19.36%
0.20	0.80	16.52%	20.60%
0.25	0.75	16.65%	19.94%
0.20	0.80	16.52%	20.60%
0.15	0.85	16.39%	21.35%
0.10	0.90	16.26%	22.17%
0.05	0.95	16.13%	23.06%
0.00	1.00	16.00%	24.00%

EXHBIT 91
Portfolio Analysis: Risk and Return Two Asset Portfolio

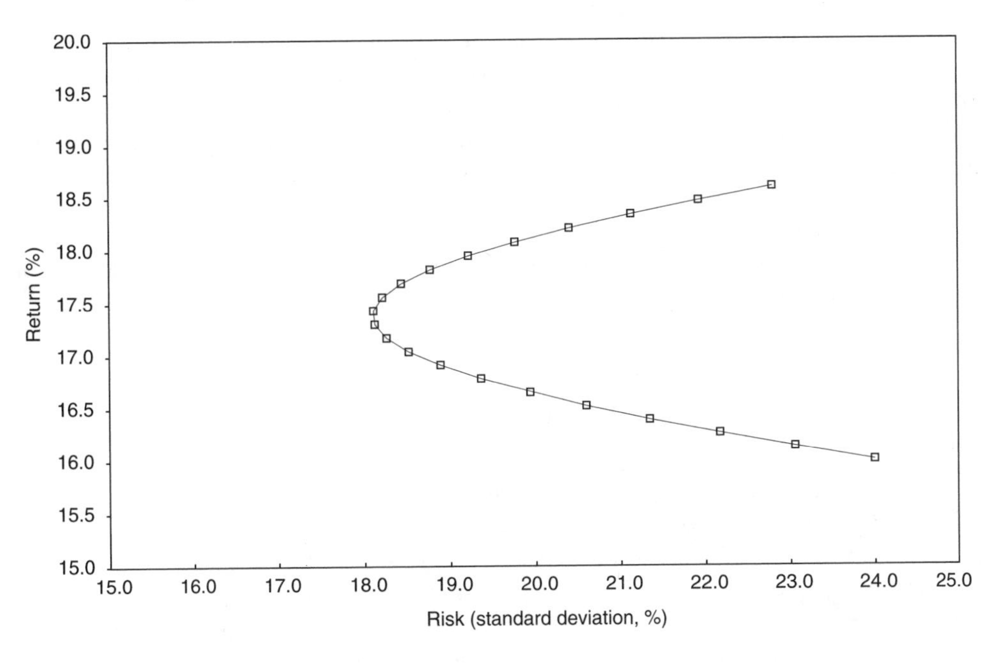

See also DIVERSIFICATION; EFFICIENT PORTFOLIO; INTERNATIONAL DIVERSIFICATION; PORTFOLIO INVESTMENTS; PORTFOLIO THEORY.

PORTFOLIO INVESTMENTS

1. Investing in a variety of assets to reduce risk by diversification. An example of a portfolio is a mutual fund that consists of a mix of assets which are professionally managed and that seeks to reduce risk by diversification. Investors can own a variety of securities with a minimal capital investment. Since mutual funds are professionally managed, they tend to involve less risk. To reduce risk, securities in a portfolio should have negative or no correlations to each other.

See also DIVERSIFICATION; EFFICIENT PORTFOLIO; INTERNATIONAL DIVERSIFICATION; PORTFOLIO THEORY.

2. Investments that are undertaken for the sake of obtaining investment income or capital gains rather than entrepreneurial income which is the case *with foreign direct investments (FDI)*. This typically involves the ownership of stocks and/or bonds issued by public or private agencies of a foreign country. The investors are not interested in assuming control of the firm.

PORTFOLIO THEORY

Theory advanced by H. Markowitz in attempting a well-diversified portfolio. The central theme of the theory is that rational investors behave in a way that reflects their aversion to taking increased risk without being compensated by an adequate increase in expected return. Also, for any given expected return, most investors will prefer a lower risk, and for any given level of risk, they will prefer a higher return to a lower return. Markowitz showed how quadratic programming could be used to calculate a set of "efficient" portfolios. An investor then will choose among a set of efficient portfolios the best that is consistent with the risk profile of the investor.

Most financial assets are not held in isolation but rather as part of a portfolio. Therefore, the risk–return analysis should not be confined to single assets only. What is important is the expected return on the portfolio (not just the return on one asset) and the portfolio's risk.

Most financial assets are not held in isolation; rather, they are held as parts of portfolios. Therefore, risk–return analysis should not be confined to single assets only. It is important to look at portfolios and the gains from diversification. What is important is the return on the portfolio (not just the return on one asset) and the portfolio's risk.

A. Portfolio Return

The expected return on a portfolio (r_p) is simply the weighted average return of the individual sets in the portfolio, the weights being the fraction of the total funds invested in each asset:

$$r_p = w_1 r_1 + w_2 r_2 + \cdots + w_n r_n = \sum_{j=1}^{n} w_j r_j$$

where

r_j = expected return on each individual asset
w_j = fraction for each respective asset investment
n = number of assets in the portfolio

$$\sum_{j=1}^{n} w_j = 1.0$$

EXAMPLE 100

A portfolio consists of assets A and B. Asset A makes up one-third of the portfolio and has an expected return of 18%. Asset B makes up the other two-thirds of the portfolio and is expected to earn 9%. The expected return on the portfolio is:

Asset	Return (r_j)	Fraction (w_j)	$w_j r_j$
A	18%	1/3	1/3 × 18% = 6%
B	9%	2/3	2/3 × 9% = 6%
			r_p = 12%

B. Portfolio Risk

Unlike returns, the risk of a portfolio (σ_p) is not simply the weighted average of the standard deviations of the individual assets in the contribution, for a portfolio's risk is also dependent on the correlation coefficients of its assets. The correlation coefficient (ρ) is a measure of the degree to which two variables "move" together. It has a numerical value that ranges from −1.0 to 1.0. In a two-asset (A and B) portfolio, the portfolio risk is defined as:

$$\sigma_p = \sqrt{w_A^2\sigma_A^2 + w_B^2\sigma_B^2 + 2\rho_{AB}w_Aw_B\sigma_A\sigma_B}$$

where

σ_A and σ_B = standard deviations of assets A and B
w_A and w_B = weights, or fractions, of total funds invested in assets A and B
ρ_{AB} = the correlation coefficient between assets A and B.

Incidentally, the correlation coefficient is the measurement of joint movement between two securities.

C. Diversification

As can be seen in the above formula, the portfolio risk, measured in terms of σ is not the weighted average of the individual asset risks in the portfolio. We have in the formula a third term (ρ), which makes a significant contribution to the overall portfolio risk. What the formula basically shows is that portfolio risk can be minimized or completely eliminated by diversification. The degree of reduction in portfolio risk depends upon the correlation between the assets being combined. Generally speaking, by combining two perfectly negatively correlated assets ($\rho = -1.0$), we are able to eliminate the risk completely. In the real world, however, most securities are negatively, but not perfectly correlated. In fact, most assets are positively correlated. We could still reduce the portfolio risk by combining even positively correlated assets. An example of the latter might be ownership of two automobile stocks or two housing stocks.

EXAMPLE 101

Assume the following:

Asset	σ	w
A	20%	1/3
B	10%	2/3

The portfolio risk then is:

$$\begin{aligned}\sigma_p &= \sqrt{w_A^2\sigma_A^2 + w_B^2\sigma_B^2 + 2\rho_{AB}w_Aw_B\sigma_A\sigma_B}\\ &= [(1/3)^2(0.2)^2 + (2/3)^2(0.1)^2 + 2\rho_{AB}(1/3)(2/3)(0.2)(0.1)]^{1/2}\\ &= 0.0089 + 0.0089\rho_{AB}\end{aligned}$$

(a) Now assume that the correlation coefficient between A and B is +1 (a perfectly positive correlation). This means that when the value of asset A increases in response to market conditions,

(Continued)

so does the value of asset B, and it does so at exactly the same rate as A. The portfolio risk when $\rho_{AB} = +1$ then becomes:

$$\sigma_p = 0.0089 + 0.0089\rho_{AB} = 0.0089 + 0.0089(+1) = 0.1334 = 13.34\%$$

(b) If $\rho_{AB} = 0$, the assets lack correlation and the portfolio risk is simply the risk of the expected returns on the assets, i.e., the weighted average of the standard deviations of the individual assets in the portfolio. Therefore, when $\rho_{AB} = 0$, the portfolio risk for this example is:

$$\sigma_p = 0.0089 + 0.0089\rho_{AB} = 0.0089 + 0.0089(0) = 0.0089 = 8.9\%$$

(c) If $\rho_{AB} = -1$ (a perfectly negative correlation coefficient), then as the price of A rises, the price of B declines at the very same rate. In such a case, risk would be completely eliminated. Therefore, when $\rho_{AB} = -1$, the portfolio risk is

$$\sigma_p = 0.0089 + 0.0089\rho_{AB} = 0.0089 + 0.0089(-1) = 0.0089 - 0.0089 = 0 = 0$$

When we compare the results of (a), (b), and (c), we see that a positive correlation between assets increases a portfolio's risk above the level found at zero correlation, while a perfectly negative correlation eliminates that risk.

EXAMPLE 102

To illustrate the point of diversification, assume data on the following three securities are as follows:

Year	Security X (%)	Security Y (%)	Security Z (%)
20×1	10	50	10
20×2	20	40	20
20×3	30	30	30
20×4	40	20	40
20×5	50	10	50
r_j	30	30	30
σ_p	14.14	14.14	14.14

Note here that securities X and Y have a perfectly negative correlation, and securities X and Z have a perfectly positive correlation. Notice what happens to the portfolio risk when X and Y, and X and Z are combined. Assume that funds are split equally between the two securities in each portfolio.

Year	Portfolio XY (50% − 50%)	Portfolio XZ (50% − 50%)
20 × 1	30	10
20 × 2	30	20
20 × 3	30	30
20 × 4	30	40
20 × 5	30	50
r_p	30	30
σ_p	0	14.14

Again, see that the two perfectly negative correlated securities (XY) result in a zero overall risk.

D. Markowitz's Efficient Portfolio

Dr. Harry Markowitz, in the early 1950s, provided a theoretical framework for the systematic composition of optimum portfolios. Using a technique called *quadratic programming*, he attempted to select from among hundreds of individual securities, given certain basic information supplied by portfolio managers and security analysts. He also weighted these selections in composing portfolios. The central theme of Markowitz's work is that rational investors behave in a way reflecting their aversion to taking increased risk without being compensated by an adequate increase in expected return. Also, for any given expected return, most investors will prefer a lower risk and, for any given level of risk, prefer a higher return to a lower return. Markowitz showed how quadratic programming could be used to calculate a set of "efficient" portfolios such as illustrated by the curve in Exhibit 92.

EXHIBIT 92
Efficient Frontier

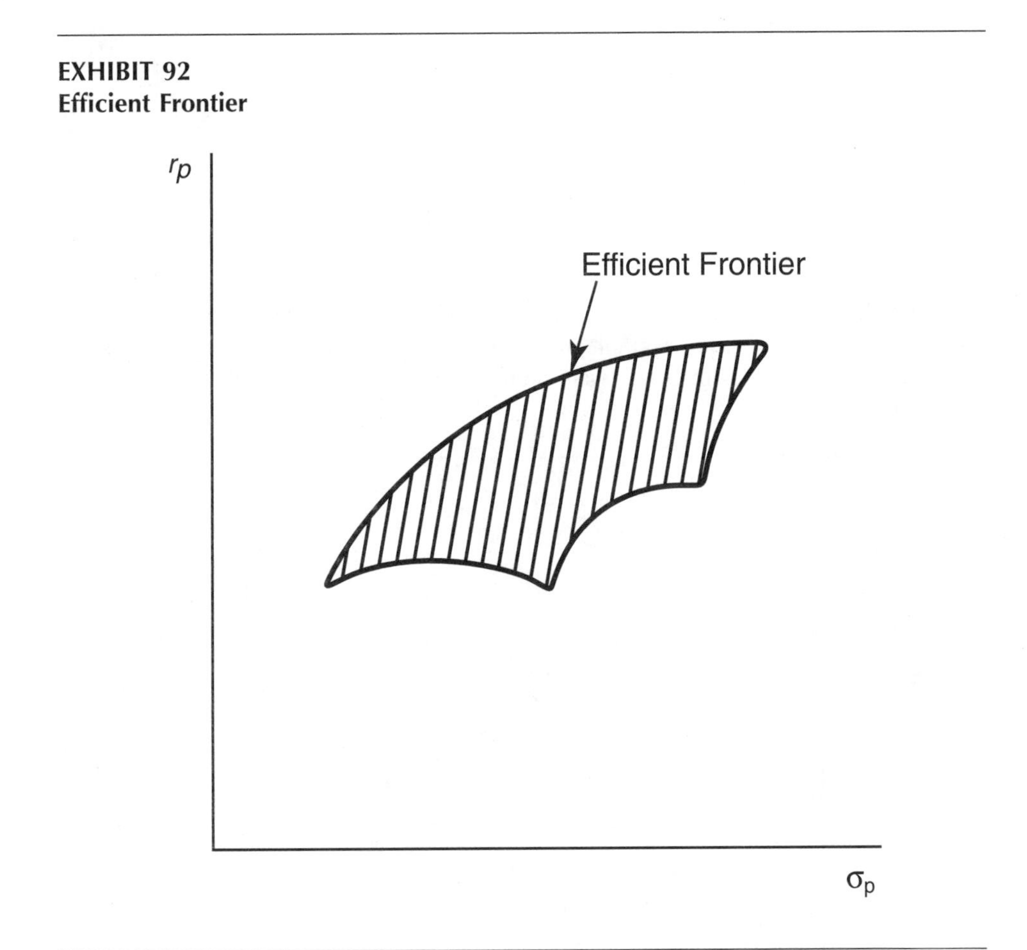

In Exhibit 93, an efficient set of portfolios that lie along the ABC line, called "efficient frontier," is noted. Along this frontier, the investor can receive a maximum return for a given level of risk or a minimum risk for a given level of return. Specifically, comparing three portfolios A, B, and D, portfolios A and B are clearly more efficient than D, because portfolio A could produce the same expected return but at a lower risk level, while portfolio B would have the same degree of risk as D but would afford a higher return.

To see how the investor tries to find the optimum portfolio, we first introduce the indifference curve, which shows the investor's trade-off between risk and return. Exhibit 94 shows the two

EXHIBIT 93
Efficient Portfolio

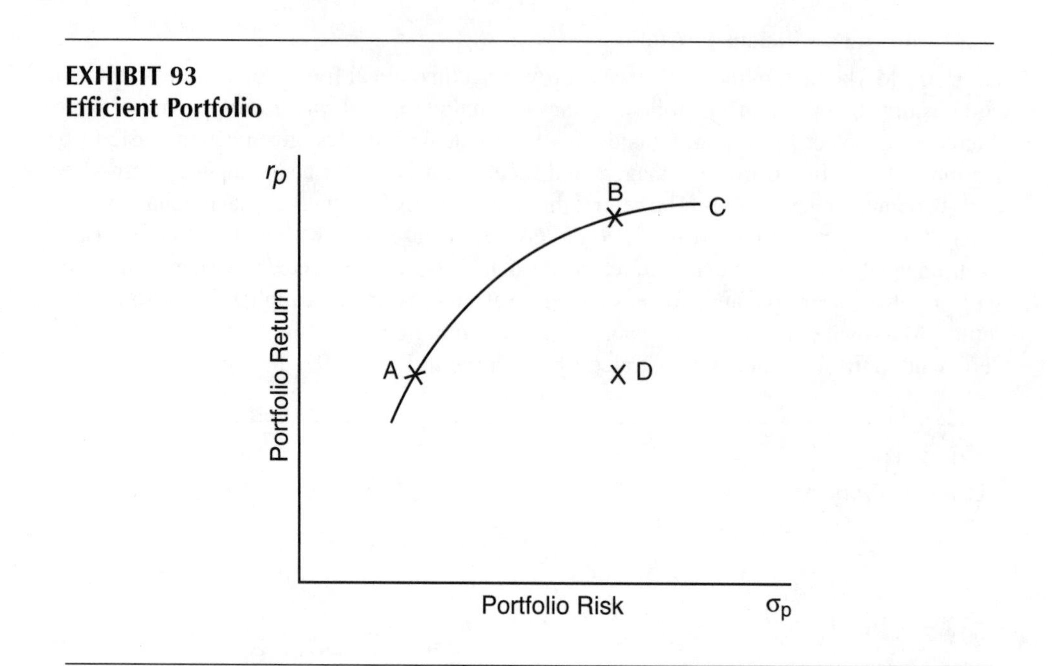

different indifference curves for two investors. The steeper the slope of the curve, the more risk averse the investor is. For example, investor B's curve has a steeper slope than investor A's. This means that investor B will want more incremental return for each additional unit of risk.

EXHIBIT 94
Risk–Return Indifference Curves

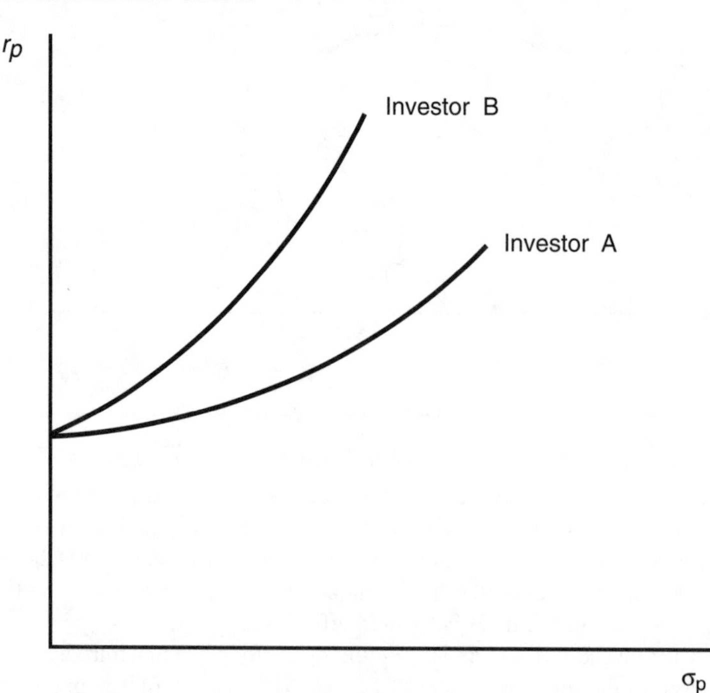

Exhibit 95 depicts a family of indifference curves for investor A. The objective is to maximize his satisfaction by attaining the highest curve possible.

EXHIBIT 95
Matching the Efficient Frontier and Indifference Curve

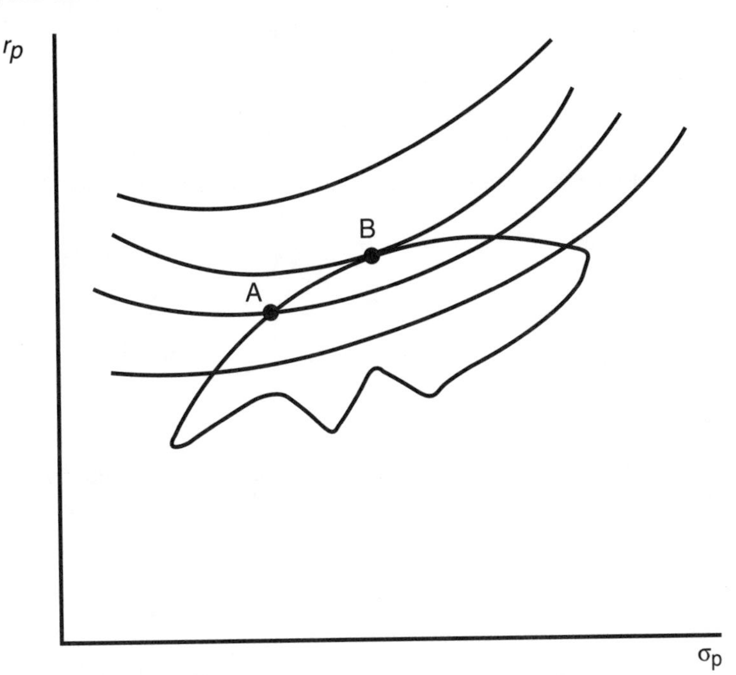

By matching the indifference curve showing the risk–return trade-off with the best investments available in the market as represented by points on the efficient frontier, investors are able to find an optimum portfolio. According to Markowitz, investor A will achieve the highest possible curve at point B along the efficient frontier. Point B is thus the optimum portfolio for this investor.

E. Portfolio Selection as a Quadratic Programming Problem

A portfolio selection problem was formulated by Markowitz as a quadratic programming model as follows:

$$\text{Minimize } E(r_p) - \lambda V(r_p)$$

subject to

$\Sigma x_i = 1, \quad (i = 1, 2, \ldots n)$
$x_i \geq 0$

where

$E(r_p)$ = the expected return
$V(r_p)$ = the variance or covariance of any given portfolio
x_i = proportion of the investor's total investment in security i

n = number of securities
λ (Lamda) = coefficient of risk aversion.

λ represents the rate at which a particular investor is just willing to exchange expected rate of return for risk. $\lambda = 0$ indicates the investor is a risk lover, while $\lambda = 1$ means he is a risk averter. The resulting solution to the problem would identify a portfolio that lies on the efficient portfolio. If one knows the coefficient of risk aversion, λ, for a particular investor, the model will be able to find the optimal portfolio for that investor.

F. The Market Index Model

For even a moderately sized portfolio, the formulas for portfolio return and risk require estimation of a large number of input data. Concern for the computational burden in deriving these estimates led to the development of the following market index model:

$$r_j = a + br_m$$

where

r_j = return on security j
r_m = return on the market portfolio
b = the beta or systematic risk of a security.

What this model attempts to do is measure the systematic or uncontrollable risk of a security. The beta is measured as follows:

$$b = \text{Cov}(r_j, r_m)/\sigma_m^2$$

where

$\text{Cov}(r_j, r_m)$ = the covariance of the returns of the security with the market return.
σ_m^2 = the variance (standard deviation squared) of the market return, which is the return on the Standard & Poor's 500 or Dow Jones 30 Industrials.

An easier way to compute beta is to determine the slope of the least-squares linear regression line $(r_j - r_f)$, where the excess return of the security $(r_j - r_f)$ is regressed against the excess return of the market portfolio $(r_m - r_f)$. The formula for beta is:

$$b = \left(\sum MK - n\overline{M}\overline{K}\right) / \left(\sum M^2 - n\overline{M}^2\right)$$

where $M = (r_m - r_f)$, $K = (r_j - r_f)$, n = the number of periods, $\overline{M}$ = the average of M, and $\overline{K}$ = the average of K.

The market index model was initially proposed to reduce the number of inputs required in portfolio analysis. It can also be justified in the context of the capital asset pricing model.

G. The Capital Asset Pricing Model (CAPM)

The capital asset pricing model (CAPM) takes off where the efficient frontier concluded with an assumption that there exists a risk-free security with a single rate at which investors can borrow and lend. By combining the risk-free asset and the efficient frontier, we create a whole new set of investment opportunities which will allow us to reach higher indifference curves than would be possible simply along the efficient frontier. The r_f mx line in Exhibit 96 shows this possibility. This line is called the capital market line (CML) and the formula for this line is:

$$r_p = r_f + [(r_m - r_f)/(\sigma_m - 0)]\sigma_p$$

which indicates the expected return on any portfolio (r_p) is equal to the risk-free return (r_f) plus the slope of the line times a value along the horizontal axis (σ_p) indicating the amount of risk undertaken.

EXHIBIT 96
Graph of CAPM

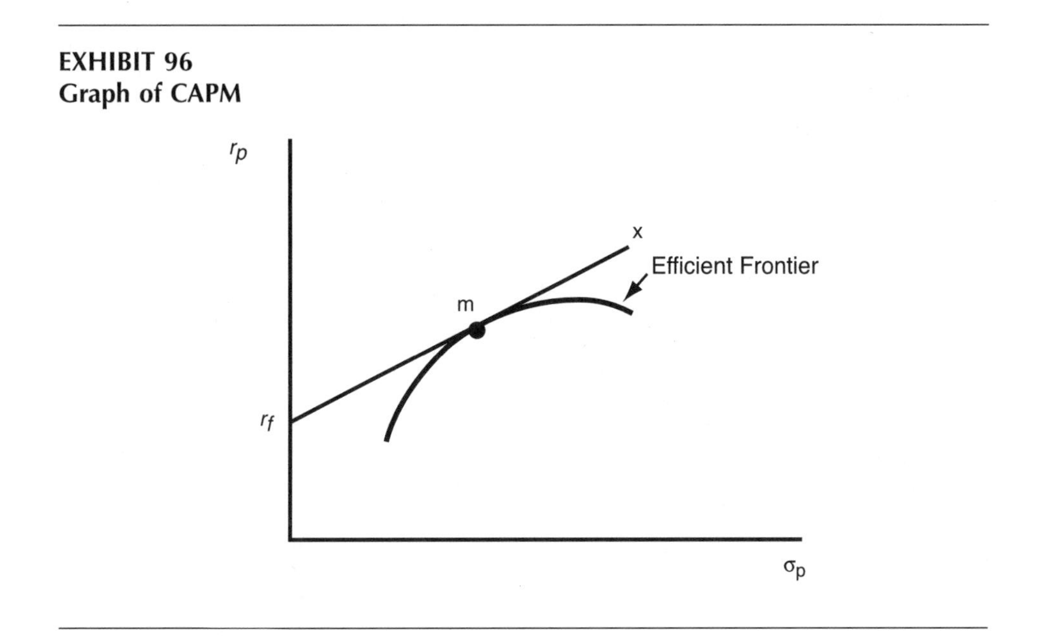

H. The Security Market Line

We can establish the trade-off between risk and return for an individual security through the security market line (SML) in Exhibit 97. SML is a general relationship to show the risk–return trade-off for an individual security, whereas CML achieves the same objective for a portfolio.

The formula for SML is:

$$r_j = r_f + b(r_m - r_f)$$

where

r_j = the expected (or required) return on security j
r_f = the risk-free security (such as a T-bill)
r_m = the expected return on the market portfolio (such as Standard & Poor's 500 Stock Composite Index or Dow Jones 30 Industrials)
b = beta, an index of nondiversifiable (noncontrollable, systematic) risk

This formula is called the *Capital Asset Pricing Model (CAPM)*. The model shows that investors in individual securities are only assumed to be rewarded for systematic, uncontrollable, market-related risk, known as the beta (b) risk. All other risk is assumed to be diversified away and thus is not rewarded.

The key component in the CAPM, beta (b), is a measure of the security's volatility relative to that of an average security. For example, $b = 0.5$ means the security is only half as volatile, or risky, as the average security; $b = 1.0$ means the security is of average risk; and $b = 2.0$ means the security is twice as risky as the average risk. The whole term $b(r_m - r_f)$ represents the risk premium, the additional return required to compensate investors for assuming a given level of risk.

Thus, in words, the CAPM (or SML) equation shows that the required (expected) rate of return on a given security (r_j) is equal to the return required for securities that have no risk

(r_f) plus a risk premium required by investors for assuming a given level of risk. The higher the degree of systematic risk (b), the higher the return on a given security demanded by investors. Exhibit 97 shows the graph of the equation, known as *the security market line (SML)*.

EXHIBIT 97
The Security Market Line (SML)

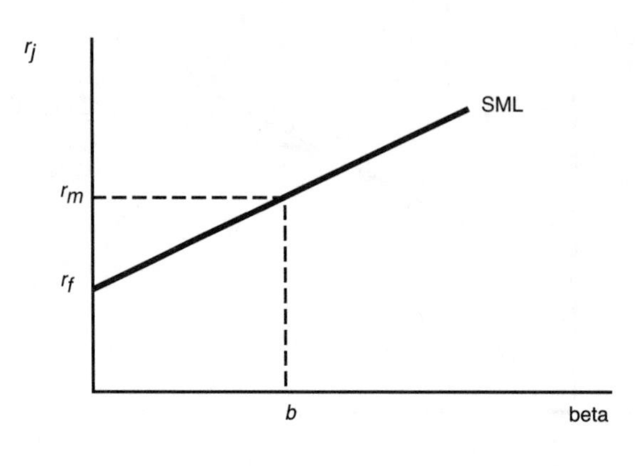

EXAMPLE 103

Assuming that the risk-free rate (r_f) is 8%, and the expected return for the market (r_m) is 12%, then if

$b = 0$ (risk-free security)	$r_j = 8\% + 0(12\% - 8\%)$	=	8%
$b = 0.5$	$r_j = 8\% + 0.5(12\% - 8\%)$	=	10%
$b = 1.0$ (market portfolio)	$r_j = 8\% + 1.0(12\% - 8\%)$	=	12%
$b = 2.0$	$r_j = 8\% + 2.0(12\% - 8\%)$	=	16%

See also BETA; DIVERSIFICATION; EFFICIENT PORTFOLIO; INTERNATIONAL DIVERSIFICATION; PORTFOLIO INVESTMENTS.

POUND

Monetary unit of Great Britain, Cyprus, Egypt, Gibraltar, Republic of Ireland, Lebanon, Malta, Sudan, and Syria.

PREMIUM

1. The price agreed upon between the purchaser and seller for the purchase or sale of an *option*—purchasers pay the premium and sellers (writers) receive the premium.
2. The excess of one futures contract price over that of another, or over the cash market price.

PRIVATE EXPORT FUNDING CORPORATION

Private Export Funding Corporation (PEFCO) is a private U.S. corporation, established with government support, which helps finance U.S. exports of big-ticket items from private sources. PEFCO purchases at fixed interest rates the medium- to long-term debt obligations of importers of U.S. products. Foreign importer loans are financed through the sale of PEFCO's own securities. Guarantees of repayment on all of PEFCO's foreign obligations are provided by the *Eximbank*.

PROJECT FINANCE LOAN PROGRAM

See EXPORT-IMPORT BANK.

PUNT

Ireland's currency.

PURCHASING POWER PARITY (PPP)

Purchasing Power Parity (PPP) states that spot currency rates among countries will change to the differential in inflation rates between countries. There are two versions of this theory.

Absolute PPP: The price of internationally traded commodities should be the same in every country, that is, one unit of home currency should have the same purchasing power worldwide. The absolute version, popularly called the *law of one price*, is written as

$$S = \frac{P_h}{P_f} \quad \text{or} \quad P_h = SP_f$$

where S = spot exchange rate in *direct* quotes (i.e., the number of units of home currency that can be purchased for one unit of foreign currency), P_h = the price of the good in the home country, and P_f = the price of the good in the foreign country.

EXAMPLE 104

Suppose that a shoe is selling for 30 pounds in the U.K. and \$50 in the U.S. (home). If the exchange rate (direct quotes) is \$1.667 per pound, then

$$S = \frac{P_h}{P_f}$$

$$1.667/£ = \frac{P_h}{£30}$$

or

$$P_h = SP_f = (\$1.667/£)(£30) = \$50$$

Thus, the price of the shoe in the U.K. is the same as the U.S. price once we use the exchange rate to the convert the dollar into pounds and compare prices in a common currency.

Relative PPP: The relative version of purchasing power parity says that the exchange rate of one currency against another will adjust to reflect changes in the price levels of the two countries.

Purchasing power parity can be summarized as follows:

Expected spot rate = current spot rate × expected difference in inflation rate

Mathematically,

$$\frac{S_2}{S_1} = \frac{1 + I_h}{1 + I_f} \qquad \text{(Equation 1)}$$

where S_1 and S_2 = the spot exchange rate (*direct quote*) at the beginning of the period and the end of the period, I_f = foreign inflation rate, measured by price indexes, and I_h = home (domestic) inflation rate.

EXAMPLE 105

If the home currency experiences a 5% rate of inflation, and the foreign currency experiences a 2% rate of inflation, then the foreign currency will adjust by 2.94% (1.05/1.02 = 1.0294). In fact, the foreign currency is expected to appreciate by 2.94% in response to the higher rate of inflation of the home country relative to the foreign country.

If purchasing power parity is expected to hold, then the best prediction for the one-period spot rate, called the *purchasing power parity (PPP) rate*, should be:

$$S_2 = S_1 \frac{1 + I_h}{1 + I_f} \qquad \text{(Equation 2)}$$

A more simplified but less precise relationship of purchasing power parity is shown as:

$$\frac{S_2 - S_1}{S_1} = \text{\% change in the foreign currency} = I_h - I_f \qquad \text{(Equation 3)}$$

Note: Dividing both sides of Equation 2 by S_1 and then subtracting 1 from both sides yields

$$\frac{S_2 - S_1}{S_1} = \frac{I_h - I_f}{1 + I_h}$$

Equation 3 follows if I_h is relatively small.

Equation 3 indicates that the exchange rate change during a period should equal the inflation differential for that same time period. In effect, PPP says that currencies with high rates of inflation should devalue relative to currencies with lower rates of inflation.

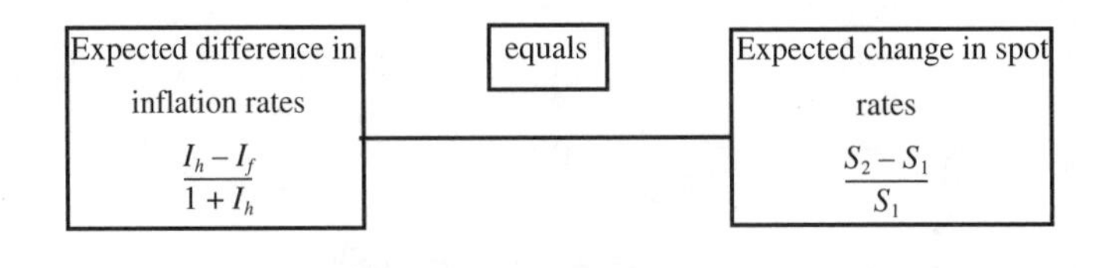

EXAMPLE 106

If the home currency experiences a 5% rate of inflation, and the foreign currency experiences a 2% rate of inflation, the foreign currency should adjust by about 3% (5% – 2% = 3%).

Equation 3 is illustrated in Exhibit 98. The vertical axis shows the percentage appreciation of the foreign currency relative to the home currency, and the horizontal axis measures the percentage higher or lower rate of inflation in the foreign country relative to the home country. Equilibrium is reached on the parity line, which contains all those points at which these two differentials are equal. At point A, for example, the 3% inflation differential is exactly offset by the 3% appreciation of the foreign currency relative to the home currency. Point B, on the other hand, portrays a situation of disparity, where the inflation differential of 3% is greater than the appreciation of 1% in the home currency value of the foreign currency.

EXHIBIT 98
Purchasing Power Parity

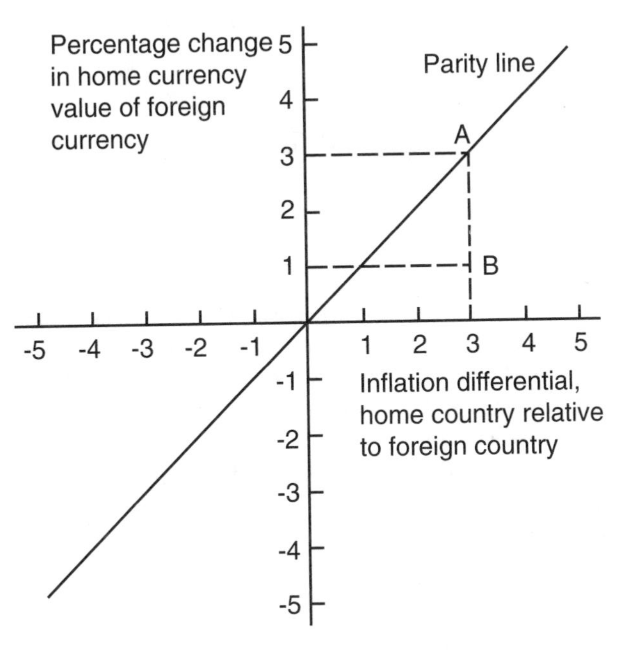

Note: (1) If absolute PPP holds, then relative PPP will also hold. But if absolute PPP does not hold, relative PPP still may. This is because the level of S may not equal I_f/I_h, but the change in S could still equal the inflation differential. (2) Empirical evidence has indicated that purchasing power parity holds up well over the long run, but not so well over shorter time periods.

PURCHASING POWER RISK

Also called *inflation risk.*

1. The failure of assets (financial and real) to earn a return to keep up with increasing price levels. Bonds are exposed to this risk because the issuer will be paying back in cheaper dollars in inflationary times.
2. The domestic counterpart to exchange risk. It involves uncertain changes in the exchange rate between domestic currency and domestic goods and services.

PUT

1. A right (not the obligation) to sell a specific security at a specified price within a designated period for which the option buyer pays the seller (writer) a premium or option price. Contracts on listed puts (and calls) have been standardized at date of issue for periods of three, six, and nine months, although as these contracts approach expiration, they may be purchased with a much shorter life.
2. Bondholder's right to redeem a bond prior to maturity.

PUT OPTION

See PUT.

Q

QUANTO

An option in which the foreign exchange risk in the underlying asset has been removed.

QUOTATION

Also *called, quote.*

1. The highest bid to buy and the lowest offer (ask) to sell a security in a particular market at a given time. For example, a quotation on a stock, say "40 1/4 to 40 3/4," means that $40.25 was the highest price any buyer wished to pay (bid) at the time the quotation was given on the exchange and that $40.75 was the lowest price at which any holder of the stock offered (asked) to sell.
2. In foreign exchange trading, the pair of prices (bid and ask) at which the dealer is willing to buy or sell foreign exchange.

See also CURRENCY QUOTATIONS.

R

RAND

Monetary unit of Lesotho, South Africa, and South West Africa.

REAL COST OF HEDGING

The real cost of *hedging* is the extra cost of hedging as opposed to not hedging. A *negative* real cost would signify that hedging was more favorable than not hedging.

REAL EXCHANGE RATE

In contrast to *nominal exchange rate*, the rate adjusted for inflation is roughly nominal exchange rate minus inflation rate.

REGISTERED BOND

A bond registered in the owner's name and listed on the records of the issuer as well as with the registrar. Transfer, as at the time of sale or if the bond is being possessed as collateral for a loan in default, requires power of attorney and the return of the physical bond to a transfer agent. The transfer agent replaces the old bond with a new one registered to the new owner. A *bearer bond* is an example of an unregistered bond.

REGRESSION ANALYSIS

Regression analysis is a statistical procedure for estimating mathematically the average relationship between the dependent variable and the independent variable(*s*). *Simple regression* involves one independent factor, such as inflation or interest rate differentials, in forecasting currency rate, whereas *multiple regression* involves two or more explanatory variables, e.g., inflation, interest rate differentials, and economic growth together. An example of simple regression is: A global fund's return is a function of the return on a world market portfolio, i.e., $r_j = a + br_m$, where b = beta, a measure of uncontrollable risk.

EXAMPLE 107

To explain how beta can be computed, using regression analysis, the data presented in Exhibit 99 is used for an illustrative purpose.

EXHIBIT 99
ABC Global Fund Returns versus EAFE Index Returns

EAFE Index Returns (%)	ABC Global Fund Returns (%)
9	15
19	20
11	14
14	16
23	25

(*Continued*)

12	20
12	20
22	23
7	14
13	22
15	18
17	18

Exhibit 100 shows MSExcel output for regression analysis.

EXHIBIT 100

Summary Output

Regression Statistics	
Multiple R	0.7800
R Square	0.6084
Adjusted R Square	0.5692
Standard Error	2.3436
Observations	12.0000

Anova

	df	SS	MS	F	Significance F
Regression	1	85.3243	85.3243	15.5345	0.0028
Residual	10	54.9257	5.4926		
Total	11	140.25			

	Coefficients	Standard Error	t Stat	P-value	Lower 95%	Upper 95%
Intercept	10.5836	2.1796	4.8558	0.0007	5.7272	15.4401
EAFE Index Returns	0.5632	0.1429	3.9414	0.0028	0.2448	0.8816

From the Excel's regression output, we see:

$$r_j = 10.5836 + 0.5632r_m$$

which indicates the beta for the particular global fund is 0.5632. It appears that ABC Global Fund is less risky than the world market, measured by the *EAFE Index*.

See also FUNDAMENTAL FORECASTING.

REINVOICING CENTER

A central financial facility designated by an MNC to centralize all payments and invoicing charges subsidiaries fees for its function. This way, it attempts to reduce *transaction exposure* by having all home country exports billed in the home currency and then reinvoiced to each operating affiliate in that affiliate's local currency. The reinvoicing center determines which currencies should be used and where, how, and when. This reinvoicing activity can effectively shift profits to subsidiaries where tax rates are low.

RELATIVE PURCHASING POWER PARITY

See PURCHASING POWER PARITY.

REMBRANDT BONDS

Dutch guilder-denominated bonds issued within the Netherlands by a foreign issuer (borrower).

REPATRIATION

Repatriation is the process of sending cash flows from a foreign subsidiary to the parent company. Broadly, foreign affiliates make the following payments to the parent company: (1) dividends, (2) interest and repayment of parent company loans, (3) royalties for use of trade names and services, (4) management fees for central services, and (5) payments for goods supplied by the parent. A foreign government may restrict the amount of the cash that may be repatriated to the parent company. For example, the restriction may be in terms of a ceiling stated as a percentage of the company's net worth. Such restrictions are intended to force multinational companies to reinvest earnings in the foreign country or/and to prevent large currency outflows, which might disrupt the exchange rate.

REPORTING CURRENCY

The parent company's currency used in preparing and translating its own financial statements (e.g., U.S. dollars for a U.S. firm).

REPRESENTATIVE OFFICES

Representative nonbanking offices are established in a foreign country primarily to assist the parent bank's customers in that country. Representative offices cannot accept deposits, make loans, transfer funds, accept drafts, transact in the international money market, or commit the parent bank to loans. In fact, they cannot cash a traveler's check drawn on the parent bank. What they may do, however, is provide information and assist the parent bank's clients in their banking and business contacts in the foreign country. For example, a representative office may introduce business people to local bankers or it may act as an intermediary between U.S. firms and firms located in the country of the representative office. Of course, while acting as an intermediary, it provides information about the parent bank's services. A representative's office is also a primary vehicle by which an initial presence is established in a country before setting up formal banking operations.

RETURN RELATIVE

It is often necessary to measure returns on a slightly different basis than *total returns (TRs)*. This is particularly true when calculating a *compound* (or *geometric*) *average return*, because negative returns cannot be used in the calculation. The return relative (RR) solves this problem by adding 1.0 to the total return. Although return relatives may be less than 1.0, they will be greater than zero, thereby eliminating negative numbers.

$$\text{RR} = \frac{C + (P_1 - P_0)}{P_0} + 1$$

This can be reduced simply to the following formula (by using the price at the end of the holding period in the numerator, rather than the *change in* price):

$$\text{RR} = \frac{C + P_1}{P_0}$$

EXAMPLE 108

A TR of 0.10 for some holding period is equivalent to a return relative of 1.10, and a TR of –0.15 is equivalent to a return relative of 0.85.

See also ARITHEMATIC AVERAGE RETURN VS. COMPOUND (GEOMETRIC) AVERAGE RETURN; TOTAL RETURN.

REVALUATION

See CURRENCY REVALUATION.

REVOCABLE LETTER OF CREDIT

A *letter of credit* that the opening bank may revoke at any time without the approval of the beneficiary. Neither the issuing bank nor the advising and paying bank guarantees payment.

REVOLVING LETTER OF CREDIT

A *letter of credit* which contains a provision for reinstating its face value after being drawn within a stated period of time facilitates the financing of ongoing regular purchases.

RHO

See CURRENCY OPTION PRICING SENSITIVITY.

RINGGIT

Malaysia's currency.

RISK

Risk refers to the variation in earnings. It includes the chance of losing money on an investment. As a measure of risk, we use the standard deviation, which is a statistical measure of dispersion of the probability distribution of possible returns of an investment. The smaller the deviation, the tighter the distribution, and thus, the lower the riskiness of the investment. Mathematically,

$$\sigma = \sqrt{\sum (r_i - \bar{r})^2 p_i}$$

To calculate σ, we proceed as follows:

Step 1. First compute the expected rate of return ($\bar{r}$)
Step 2. Subtract each possible return from $\bar{r}$ to obtain a set of deviations $(r_i - \bar{r})$
Step 3. Square each deviation, multiply the squared deviation by the probability of occurrence for its respective return, and sum these products to obtain the variance $(\sigma)^2$:

$$\sigma^2 = \sum (r_1 - \bar{r})^2 p_i$$

Step 4. Finally, take the square root of the variance to obtain the standard deviation (σ).

EXAMPLE 109

To follow this step-by-step approach, it is convenient to set up a table, as follows:

Return (r_i)	Probability (p_i)	(step 1) $r_i p_i$	(step 2) $(r_i - \bar{r})$	(step 3) $(r_i - \bar{r})^2$	(step 4) $(r_i - \bar{r})^2 p_i$
		Stock A			
–5%	.2	–1%	–24%	576	115.2
20	.6	12	1	1	.6
40	.2	8	21	441	88.2
		$\bar{r}$ = 19%			$\sigma^2 = 204$
					$\sigma = \sqrt{204}$
					= 14.18%
		Stock B			
10%	.2	2%	–5%	25	5
15	.6	9	0	0	0
20	.2	4	5	25	5
		$\bar{r}$ = 15%			$\sigma^2 = 10$
					$\sigma^2 = \sqrt{10}$
					σ = 3.16%

Exhibit 103 shows average return rates and standard deviations for selected types of investments.

EXHIBIT 103
Risk and Return 1926–1995

Series	Geometric Average	Arithmetic Average	Standard Deviation
Common stocks	8.81%	10.20%	16.89%
Long-term corporate bonds	5.03	5.58	11.26
Long-term government bonds	4.70	5.11	9.70
U.S. Treasury bills	6.25	6.29	3.10

Source: Ibbotson, R. and Rex A. Sinquefield, *Stocks, Bonds, Bills and Inflation: 1996 Yearbook* (Chicago: Dow Jones–Irwin, 1996) (www.ibbotson.com/).

The financial manager must be careful in using the standard deviation to compare risk, as it is only an absolute measure of dispersion (risk). In other words, it does not consider the risk in relationship to an expected return. In comparisons of securities with differing expected returns, we commonly use the *coefficient of variation*. The coefficient of variation is computed simply by dividing the standard deviation for a security by its expected value, i.e.,

$$\sigma / \bar{r}$$

The higher the coefficient, the more risky the security.

EXAMPLE 110

Based on the following data, we can compute the coefficient of variation for each stock as follows:

	Stock A	Stock B
$\bar{r}$	19%	15%
σ	14.28%	3.16%

The coefficient of variation is computed as follows:

For stock A,

$$\sigma/\bar{r} = 14.18/19 = .75$$

For stock B,

$$\sigma/\bar{r} = 3.16/15 = .21$$

Although stock A produces a considerably higher return than stock B, stock A is overall more risky than stock B, based on the computed coefficient of variation. Note, however, that if investments have the same expected returns there is no need for the calculation of the coefficient of determination.

A. Sources of Risk

Different sources of risk are involved in investment and financial decisions. Investors and decision makers must take into account the type of risk underlying an asset.

1. *Financial risk.* This is a type of investment risk associated with excessive debt.
2. *Industry risk.* This risk concerns the uncertainty of the inherent nature of the industry such as high-technology, product liability, and accidents.
3. *International and political risks.* These risks stem from foreign operations in politically unstable foreign countries. An example is a U.S. company having a location and operations in a hostile country.
4. *Economic risk.* This is the negative impact of a company from economic slowdowns. For example, airlines have lower business volume in recession.
5. *Currency exchange risk.* This risk arises from the fluctuation in foreign exchange rates.
6. *Social risk.* This is caused by problems facing the company due to ethnic boycott, discrimination cases, and environmental concerns.
7. *Business risk.* This is caused by fluctuations of earnings before interest and taxes (operating income). Business risk depends on variability in demand, sales price, input prices, and amount of operating leverage.
8. *Liquidity risk.* It represents the possibility that an asset may not be sold on short notice for its market value. If an investment must be sold at a high discount, then it is said to have a substantial amount of liquidity risk.
9. *Default risk.* It is the risk that a borrower will be unable to make interest payments or principal repayments on a debt. For example, there is a great amount of default risk inherent in the bonds of a company experiencing financial difficulty.
10. *Market risk.* Prices of all stocks are correlated to some degree with broad swings in the stock market. Market risk refers to changes in a stock's price that result from changes in the stock market as a whole, regardless of the fundamental change in a firm's earning power.

11. *Interest rate risk.* This refers to the fluctuations in the value of an asset as the interest rates and conditions of the money and capital markets change. Interest rate risk relates to fixed income securities such as bonds. For example, if interest rates rise (fall), bond prices fall (rise).
12. *Inflation (purchasing power) risk.* This risk relates to the possibility that an investor will receive a lesser amount of purchasing power than was originally invested. Bonds are most affected by this risk because the issuer will be paying back in cheaper dollars during an inflationary period.
13. *Systematic and unsystematic risk.* Many investors hold more than one financial asset. A portion of a security's risk, called unsystematic risk, can be controlled through diversification. This type of risk is unique to a given security. Business, liquidity, and default risks fall in this category. Nondiversifiable risk, more commonly referred to as systematic risk, results from forces outside of the firm's control and is therefore not unique to the given security. Purchasing power, interest rate, and market risks fall into this category. This type of risk is measured by the *beta coefficient.*

RISK ANALYSIS

Risk analysis is the process of measuring and analyzing the risks associated with financial and investment decisions. It is important especially in making *foreign direct investment* decisions because of the large amount of capital involved and the long-term nature of the investment being considered. The higher the risk associated with a proposed project, the greater the return that must be earned to compensate for that risk. There are several methods for the analysis of risk, including risk-adjusted discount rate, certainty equivalent, Monte Carlo simulation, sensitivity analysis, and decision trees.

See also BETA; CAPITAL ASSET PRICING MODEL; RISK–RETURN TRADE-OFF.

RISK ARBITRAGE

See SPECULATION.

RISK-FREE RATE

Risk-free rate is the rate of return earned on a riskless security if no inflation were expected. A proxy for a risk-free return is a rate of interest on short-term U.S. Treasury securities in an inflation free world.

RISK PREMIUM

1. The difference in the expected future spot rate versus forward rate for a currency.
2. The amount by which the required return on an asset or security exceeds the risk-free rate, r_f. In terms of the *capital asset pricing model (CAPM)*, it can be expressed as $b(r_m - r_f)$, where b is the security's beta coefficient, a measure of systematic risk, and r_m is the required return on the market portfolio. The risk premium is the additional return required to compensate investors for assuming a given level of risk. The higher this premium, the more risky the security and vice versa.

See BETA; CAPITAL ASSET PRICING MODEL.

RISK–RETURN TRADE-OFF

Integral to the theory of finance and investment is the concept of a risk–return trade-off. All financial decisions involve some sort of risk–return trade-off. The greater the risk associated with any financial decision, the greater the return expected from it. Risk, along with the return, is a major consideration in investment decisions. The investor must compare the expected return from a given investment with the risk associated with it. Generally speaking, the higher the risk undertaken, the more ample the return, and conversely, the lower the risk, the more modest the return. Proper assessment and balance of the various risk–return trade-offs available is part of creating a sound financial and investment plan. In the case of investment in stock, you, as an investor, would demand higher return from a speculative stock to compensate for the higher level of risk. In the case of working capital management, the fewer inventories you keep, the higher the expected return (because less of your current assets is tied up), but also the greater the risk of running out of stock and thus losing potential revenue. Exhibit 101 depicts the risk–return trade-off, where the risk-free rate is the rate of return commonly required on a risk-free security such as a U.S. Treasury bill. Exhibit 102 illustrates risk–return trade-off of internationally diversified portfolios over the period 1976–1999.

EXHIBIT 101
Risk–Return Trade-off

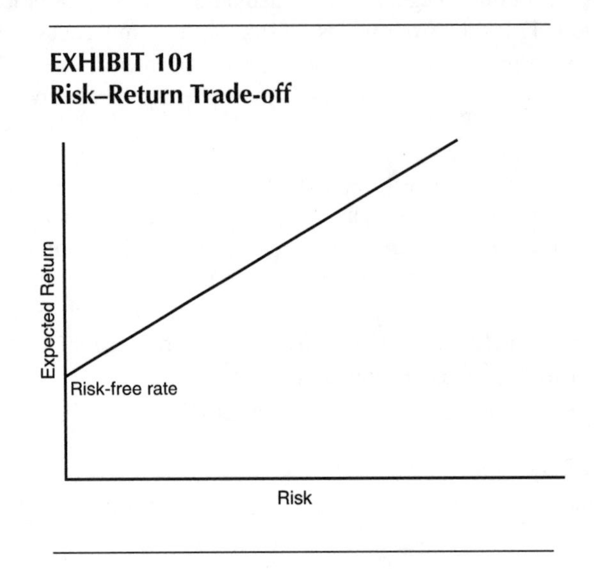

EXHIBIT 102
Risk–Return Trade-off of an Internationally Diversified Portfolios over the Period 1976–1999

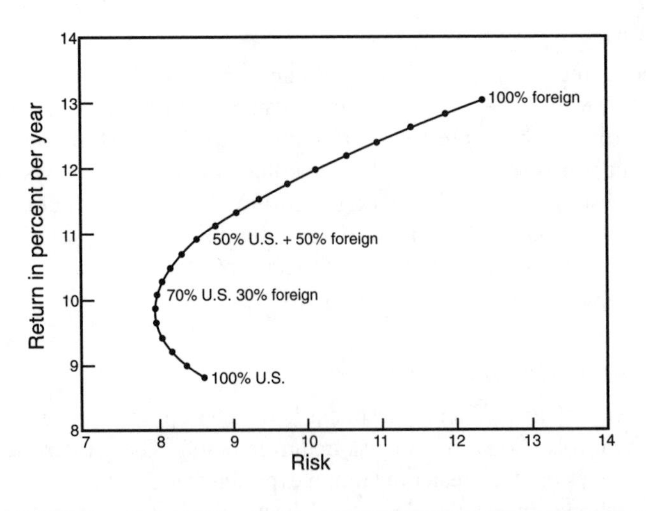

See also INTERNATIONAL DIVERSIFICATION.

ROUNDING RULES

When a conversion between a national currency unit and the *euro* results in a figure that is 0.005 or above, each amount is rounded up (100.445 and FFr 45.675 become 100.45 and FFr 45.68). Intermediate products cannot be rounded.

See also EURO.

ROUND TURN

Procedure by which a long or short position is offset by an opposite transaction or by accepting or making delivery of the actual financial instrument or physical commodity. This complete buy/sell or sell/buy is a *round turn*.

RUBLE

Soviet Union's currency.

RUPEE

Monetary unit of India, Maldives, Mauritius, Nepal, Pakistan, Seychelles, and Sri Lanka.

RUPIAH

Indonesia's currency.

S

SAMURAI BONDS

Yen-denominated bonds issued in Japan by a foreign borrower. This contrasts with *Shogun bonds*.

See also FOREIGN BOND; SHOGUN BONDS.

SCHILLING

Austria's currency.

SCRIPTURAL MONEY

The form in which the *euro* can be used during the transition period—before it is available in notes and coins. It includes all forms of noncash money such as checks, money orders, credit cards, payment cards, and electronic purses.

See also EURO.

S/D-B/L

Short form for sight draft and bill of lading attached.

SECURITIZATION

Securitization is the process of matching up borrowers and lenders wholly or partly by way of the financial markets. Using this process, traditional bank assets, mainly loans or mortgages, are converted into negotiable securities. More broadly, it refers to the development of markets for a variety of new negotiable instruments.

SEIGNIORAGE

Seigniorage is the difference between the cost to the reserve country of creating new balances and the real resources the reserve country is able to acquire. It then is a financial reward accruing to the reserve currency as a result of the use of the currency as a world money.

SELLING SHORT

Short selling means selling a security (e.g., stocks, foreign currencies) that is not owned by the seller. This method has been used as a method for making profit from a fall in stock price. The rationale behind short selling is as follows. We all know that the simplest way to make money in the stock market is to buy a stock at a low price and sell it later at a higher one. In a short selling situation, investors are reversing the sequence; they are selling high, promising to buy back the stock later at what they hope will be a lower price. If the stock price falls, they make money. If it rises and they have to buy back their stocks for more than they sold it, they lose money. An investor who wants to sell short has to set up a margin account with a stockbroker and comply with rules established by the federal government, the SEC, and the brokerage house.

SETTLEMENT DATE

See VALUE DATE.

SETTLEMENT RISK

Also called *Hersatt risk*, settlement risk, a form of *credit risk*, is the risk that a bank's counterpart will be unable to deliver money in a foreign exchange deal.

SHARPE'S RISK-ADJUSTED RETURN

Sharpe's risk-adjusted return is a risk-adjusted grade that compares five-year, risk-adjusted returns. This measure, developed by Nobel Laureate William Sharpe, is excess returns per unit of risk. The index concentrates on total risk as measured by the standard deviation of returns (noted with the Greek letter σ, read as sigma).

$$\text{Sharpe measure} = \frac{\text{Excess returns}}{\text{Fund standard deviation}} = \frac{\text{Total fund return} - \text{Risk-free rate}}{\text{Fund standard deviation}}$$

EXAMPLE 111

If a mutual fund has a return of 10%, the risk-free rate of 6%, and the fund and standard deviation of 18%, the Sharpe measure is .22, as shown below.

$$\text{Sharpe measure} = \frac{10\% - 6\%}{18\%} = \frac{4\%}{18\%} = .22$$

Mutual fund analysis by *Morningstar Inc.* (www.morningstar.net) and others use the Sharpe measure. An investor should rank the performance of his mutual funds based on Sharpe's index of portfolio performance. The funds would be ranked from high to low return. For example, a fund with an index of .6 would be far superior to one with an index of .3. Sharpe's index should be compared with other trends as well as with the average market. The larger the index, the better the performance. *Note:* The index should be used by investors with some mathematical knowledge. And remember, the portfolio with the best risk-adjusted performance will likely not produce the greatest profits. Formulas such as Sharpe's take into account the risk undertaken to earn profits.
See also TREYNOR'S PERFORMANCE MEASURE.

SHEKEL

Israel's currency.

SHILLING

Monetary unit of Kenya, Somalia, Tanzania, and Uganda.

SHOGUN BONDS

Foreign currency-denominated bonds issued within Japan by Japanese companies. This contrasts with *Samurai bonds*.
See also FOREIGN BOND; SAMURAI BONDS.

SHORT

An investment is short if the investor makes money when its price or the price of the underlying security drops.
See also LONG POSITION.

SHORT-TERM NATIONAL FINANCIAL MARKETS

Also called *national money markets*, short-term national financial markets are markets, in different countries, for short-term instruments such as bank deposits and government bills in different nations. Typically, local bank deposits and short-term investments in government securities give the firm an interest-earning opportunity for its locally available funds. This contrasts with *long-term national financial markets*.
See also LONG-TERM NATIONAL FINANCIAL MARKETS.

SIMEX

Singapore International Monetary Exchange.

SIMPLE ARBITRAGE

Also called a *two-way arbitrage* or *locational arbitrage*, simple arbitrage is the one that eliminates exchange rate differentials across the markets for a single currency. If the exchange rate quotations between two markets are out of line, then an arbitrage trader could make a profit buying in the market where the currency was cheaper and selling in the other.

EXAMPLE 112

Suppose that the indirect rates are £0.6603/$ prevailing in New York and £0.6700/$ prevailing in London. If you simultaneously bought a pound in New York for £0.6603/$ and sold a pound in London for £0.6700/$, you would have (1) taken a zero investment position because you bought £1 and sold £1, (2) locked in a sure profit of £0.0097/$ no matter which way the pound subsequently moves, and (3) eliminated the different quotes in New York and London.

See also ARBITRAGE; COVERED INTEREST ARBITRAGE; FOREIGN EXCHANGE ARBITRAGE; TRIANGULAR ARBITRAGE.

SINGLE-BUYER POLICY

See EXPORT-IMPORT BANK.

SINGLE-COUNTRY FUNDS

Mutual funds invested in securities of a single country, single-country funds are the most focused and by far the most aggressive foreign stock funds. Almost all single-country funds are *closed-end funds* (exceptions are the Japan Fund and French Fund). Because of their aggressive investment style, single-country closed-end funds have been recognized to sell at both large discounts and premiums to their *net asset value*.

SMALL BUSINESS POLICY

See EXPORT-IMPORT BANK.

SOFT CURRENCY

Also called *weak currency*, a currency for which there is not much demand and whose values often fluctuate. The Nepal Rupee would be an example.
See also HARD CURRENCY.

SOFT LANDING VS HARD LANDING

Soft landing means, in Fed speak, that the economy is slowing enough to eliminate the need for the Fed to further raise interest rates to dampen activity—but not enough to threaten a recession, which is what results when the economy contracts instead of expands. Hard landing, on the hand, could mean a recession.

SOFT LOANS

Soft loans are loans that have grace periods during which no payments need be made; they may bear low or no interest. Loans granted by the International Development Agency (IDA) are an example of soft loans.

SOVEREIGN RISK

See POLITICAL RISK.

SPECIAL DRAWING RIGHTS

Also called, *paper gold*, special drawing rights (SDRs) are the official currency of the *International Monetary Fund (IMF)*. The SDR is the unit of account for all purposes of the Fund. Created in 1967, SDRs are a new form of international reserve assets. Outside the Fund, the SDR is widely used as a unit of account in private contracts such as SDR-denominated deposits with commercial banks. A number of the Fund's member countries peg their currency to the SDR. Its value is based on a weighted basket of five currencies: U.S dollar, West German mark, U.K. pound, French franc, and Japanese yen. Unlike gold, SDRs have no tangible life of their own and take the form of bookkeeping entries in a special account managed by the Fund. They are used as the instruments for financing international trade.

SPECULATION

Also called *risk arbitrage*, speculation is the process that ensures the equality of returns of a risk-adjusted basis on different securities, unless *market efficiency* prevents this from happening.
See also EFFICIENT MARKET.

SPECULATORS

Speculators are individuals who engage in *speculation*, that is, ones that seek to profit from differences in risk-adjusted returns on different securities. Speculators actively expose themselves to currency risk by buying or selling currencies forward in order to profit from currency movements, while *arbitrageurs*, traders, and *hedgers* seek to reduce or eliminate their exchange risks by "locking in" the exchange rate on future trade or financial operations. Their degree of participation is based on prevailing forward rates and their expectations for spot exchange rates in the future.
See also ARBITRAGEUR; HEDGER.

SPEED LIMITS

The economy's speed limit is the rate at which it can grow without triggering inflation.

SPOT EXCHANGE RATE

Also called *current exchange* or *cash exchange rate*, the spot exchange rate of one currency for another for immediate delivery can be defined as the rate that exists in today's market. A typical listing of foreign exchange rates is found in the business section of daily newspapers, and *The Wall Street Journal*. For example, the British pound is quoted at 1.5685 per dollar. This rate is the spot rate. It means you can go to the bank today and exchange $1.5685 for £1.00. In reality, for example, if you need £10,000 for paying off an import transaction on a given day, you would ask your bank to purchase £10,000. The bank would not hand you the money, but instead it would instruct its English subsidiary to pay £10,000 to your English supplier and it would debit you account by (10,000 × 1.5685) $15,685.
See also FORWARD EXCHANGE RATE.

SPOT TRANSACTION

Also called *cash transaction*, a spot transaction involves the purchase and sale of commodities, currency, and financial instruments for immediate delivery. This is settled (paid for) on the second following business day. A spot transaction contrasts with a *forward transaction* which provides the delivery at a future date.

STANDARD & POOR'S 500 STOCK COMPOSIITE INDEX

The 500 Stock Composite Index computed by Standard & Poor's is used as a broad measure of market direction. It is different from the *Dow Jones Industrial Average (DJIA)* in several respects. First, it is value-weighted, not price-weighted. The index thus considers not only the price of a stock but also the number of outstanding shares. It is based on the aggregate market value of the stock, i.e., price times number of shares. A benefit of the index over the DJIA is that *stock splits* and *stock dividends* do not impact the index value. A drawback is that large capitalization stocks—those with a large number of shares outstanding—significantly influence the index value. The S&P 500 consists of four separate indexes: the 400 industrials, the 40 utilities, the 20 transportation, and the 40 financial. They are also frequently used as proxies for market return when computing the systematic risk measure (beta) of individual stocks and portfolios. The S&P 500 Stock Index is one of the U.S. Commerce Department's 11 leading economic indicators. The purpose of the S&P 500 Stock Price Index is to portray the pattern of common stock price movement. The total market value of the S&P 500 represents nearly 90% of the aggregate market value of common stocks traded on the New York Stock Exchange. For this reason, many investors use the S&P 500 as a yardstick to help evaluate the performance of mutual funds.

STANDARD & POOR'S GUIDE TO INTERNATIONAL RATINGS

Standard & Poor's (S&P) debt rating is a current assessment of the creditworthiness of an obligor with respect to a specific obligation. The S&P ratings are based, in varying degrees, on the following considerations: (1) likelihood of default, (2) nature and provisions of the obligation, (3) protection afforded by, and relative position of, the obligation in the event of bankruptcy, reorganization, or other arrangements under the laws of bankruptcy and other laws affecting creditor's rights. Debt obligations of issuers *outside* the U.S. and its territories are rated on the same basis as domestic corporate and municipal issues. The ratings measure the creditworthiness of the obligor to repay in the currency of denomination of the issue. However, S&P does not assess the foreign exchange risk that the investor may bear. Exhibit 104 is a listing of the designations used by S&P (and the other well-known independent agency, Mergent, F.I.S., Inc.). Descriptions on ratings are summarized. Mergent now issues the Moody's Bond Ratings. For original versions of descriptions, see Mergent Bond Record and Standard & Poor's *Bond Guide*.

EXHIBIT 104
Description of Bond Ratings

Moody's	Standard & Poor's	Quality Indication
Aaa	AAA	Highest quality
Aa	AA	High quality
A	A	Upper medium grade
Baa	BBB	Medium grade

Ba	BB	Contains speculative elements
B	B	Outright speculative
Caa	CCC & CC	Default definitely possible
Ca	C	Default, only partial recovery likely
C	D	Default, little recovery likely
	r	Assigned to derivative products

Note: Ratings may also have a + or – sign to show relative standings in class.

You should pay careful attention to ratings since they can affect not only potential market behavior but relative yields as well. Specifically, the higher the rating, the lower the yield of a bond, other things being equal. It should be noted that the ratings do change over time and the rating agencies have "credit watch lists" of various types. Try to select only those bonds rated Baa or above by Moody's or BBB or above by Standard & Poor's, even though doing so means giving up about 3/4 of a percentage point in yield.

STATEMENT OF FINANCIAL ACCOUNTING STANDARDS NO. 8

Statement of Financial Accounting Standards No. 8. (FASB No. 8) is the currency translation standard previously in use by U.S. firms. This standard, effective on January 1, 1976, was based on the temporal method of translating into dollars foreign currency-denominated financial statements and transactions of U.S.-based MNCs.

STATEMENT OF FINANCIAL ACCOUNTING STANDARDS NO. 52

Statement of Financial Accounting Standards No. 52 (FASB No. 52), commonly called *SFAS 52*, was issued by the Financial Accounting Standards Board (FASB) and deals with the translation of foreign currency changes on the balance sheet and income statement. In recording foreign exchange translations, the Statement adopted the two-transaction approach. Under this approach, the foreign currency transaction has two components: the purchase/sale of the asset and the financing of this purchase/sale. Each component will be treated separately and not netted with the other. The purchase/sale is recorded at the exchange rate on the day of the transaction and is not adjusted for subsequent changes in that rate. Subsequent fluctuations in exchange rates will give rise to foreign exchange gains and losses. They are considered financing income or expense and are recognized separately in the income statement in the period the foreign exchange fluctuations happen. Thus, exchange gains and losses arising from foreign currency transactions have a direct effect on net income.

See also CURRENT RATE METHOD; FUNCTIONAL CURRENCY; TEMPORAL METHOD.

STERILIZED INTERVENTION

A government intervention in the foreign exchange market, with simultaneous interference in the Treasury securities market, made to offset any effects on the U.S. dollar money supply; thus, the intervention in the foreign currency market is accomplished without affecting the existing dollar money supply. It contrasts with *nonsterilized intervention.*

STERLING

Great Britain's currency. The monetary unit is the pound sterling.

See also BRITISH POUND.

STRIKE PRICE

See EXERCISE PRICE.

STRIPPED BONDS

Bonds created by stripping the coupons from a bond and selling them separately from the principal.

STRONG DOLLAR

See APPRECIATION OF THE DOLLAR.

SUBPART F INCOME

A type of foreign income, as defined in the U.S. tax code, which under certain conditions is taxed by the IRS in the United States whether or not it is remitted back to the United States.

SUCRE

Ecuador's currency.

SUSHI BONDS

Eurodollars-, or other non-yen-denominated bonds issued by a Japanese firm for sale to Japanese investors.

SWAP CONTRACT

In the context of the forward market, a swap contract is a spot contract immediately combined with a forward contract.

See also SWAP RATE.

SWAP FUNDS

Also known as *exchange funds*, swap funds are not the same as ordinary mutual funds. They are highly specialized types of fixed investment pools, typically set up as a limited partnership or as a limited-liability company. They appeal to very wealthy investors with large holdings in a single stock who want diversification without having to pay capital taxes.

Suppose you own $5 million of stock in one company that you bought a long time ago at prices far below today's values. Instead of selling these shares outright and paying taxes, you swap them for units of a swap fund, tax-free. Swap funds usually have stiff early-redemption penalties and very high minimum investment requirements. In one fund, for example, the minimum investment is $500,000 of stock.

SWAP RATE

A forward exchange rate quotation expressed in terms of the number of points by which the forward rate differs from the spot rate (i.e., as a discount from, or a premium on, the *spot rate*). The interbank market quotes the forward rate this way.

EXAMPLE 113

Suppose a French investor buys $100,000 at FFr 140/$. In order to reduce the currency risk, she immediately sells forward $100,000 for 90 days, at FFr 145/$. The combined spot and forward contract is a swap contract. The swap rate, FFr 5/$, is the difference between the rate at which the investor buys and the rate at which she sells.

See also FORWARD RATE QUOTATIONS; OUTRIGHT RATE.

SWAPS

A swap is the exchange of assets or payments. It is a simultaneous purchase and sale of a given amount of securities, with the purchase being effected at once and the sale back to the same party to be carried out at a price agreed upon today but to be completed at a specified future date. Swaps are basically of two types: *interest rate swaps* and *currency swaps*. Interest rate swaps typically involve exchanging fixed interest payments for floating interest payments. Currency swaps are the exchange of one currency into another at an agreed rate, combining a spot and forward contract in one deal.

See also BANK SWAPS; CURRENCY SWAP; INTEREST RATE SWAPS; PLAIN-VANILLA SWAPS.

SWAP TRANSACTION

A swap transaction is a combination of a spot deal with a reversal deal at some future date. A common type of *swap* is "spot against forward." For example, a bank in the interbank market buys a currency in the spot market and simultaneously sells the same amount in the forward market to the same bank. The difference between the spot and the forward rates, called the *swap rate*, is known and fixed.

See also SWAP RATE.

SYNTHETIC CROSS RATES

Synthetic cross rates are cross bid and ask rates that result from a combination of two or more other exchange transactions.

EXAMPLE 114

Given:

$$\text{DM/\$} \quad 2.4520 - 2.4530$$
$$\text{\$/£} \quad 1.3840 - 1.3850$$

The synthetic bid and ask DM/£ rates can be determined as follows:

First, find the right dimension of the rate. The dimension of the rate we are looking for is DM/£. Because the dimensions of the two quotes given to us are DM/$ and $/£. The way to obtain the synthetic rate is to multiply the rates, as follows:

$$\text{Synthetic DM/£} = \text{DM/\$} \times \text{\$/£}$$

Second, let us now think about bid and ask synthetic quotes. To synthetically buy £ against DM, we first buy $ against DM, that is, at the higher rate (ask); then we buy £ against $, again at the higher rate (ask).

$$\begin{aligned}\text{Synthetic DM/£}_{ask} &= \text{DM/\$}_{ask} \times \text{\$/£}_{ask} \\ &= 2.4530 \times 1.3850 = 3.397405.\end{aligned}$$

Thus, we can synthetically buy £1 at DM 3.397405. By a similar argument, we can obtain the rate at which we can synthetically sell £ against DM.

$$\begin{aligned}\text{Synthetic DM/£}_{bid} &= \text{DM/\$}_{bid} \times \text{\$/£}_{bid} \\ &= 2.4520 \times 1.3840 = 3.393568.\end{aligned}$$

Thus, the synthetic rates are DM/£ 3.393568—3.397405.

Note: This example is the first instance of the *Law of the Worst Possible Combination* or the *Rip-Off Rule*. For any single transaction, the bank gives you the worst rate from your point of view (this is how the bank makes money). It follows that if you make a sequence of transactions, you will inevitably get the worst possible cumulative outcome. This law is the first fundamental law of real-world capital markets.

EXAMPLE 115

Given:

$$\text{DM/\$} \quad 2.3697 - 2.3725$$
$$\text{£/\$} \quad 0.64371 - 0.64412$$

This example differs from Example 114 because it involves a quotient rather than a product. However, in this case, too, we end up with the worst possible outcome.

The synthetic bid and ask DM/£ rates can be determined as follows:

First, from the dimensions of the quote we are looking for and the dimensions of the two quotes that are given to us, we need to divide DM/$ by £/$:

$$\text{Synthetic DM/£} = \frac{\text{DM/\$}}{\text{\$/£}}$$

To identify where to use the bid and where to use the ask rate, we could explicitly go through the two transactions. The simpler way is to ask the bank to convert the £/$ quote into $/£. This transforms the problem into the problem we have already solved. The bank will gladly oblige and quote:

$$\text{Synthetic \$/£}_{\text{bid}} = 1/\text{£/\$}_{\text{ask}}$$
$$\text{Synthetic \$/£}_{\text{ask}} = 1/\text{£/\$}_{\text{bid}}$$

We can then simply feed these formulas into the solutions of Example 114, and obtain:

$$\text{Synthetic DM/£}_{\text{ask}} = \frac{\text{DM/\$}_{\text{ask}}}{\text{\$/£}_{\text{bid}}} = \frac{2.3725}{0.64371} = 3.6857$$

$$\text{Synthetic DM/£}_{\text{bid}} = \frac{\text{DM/\$}_{\text{bid}}}{\text{\$/£}_{\text{ask}}} = \frac{2.3697}{0.64412} = 3.6790$$

Thus, the synthetic rates are DM/£ 3.6790 – 3.6857.

Note: In this example, to get the correct DM/£ quote, we need to divide the DM/$ quote by the £/$ quote. Thus, to obtain the largest possible outcome (the synthetic DM/£ ask rate), we divide the larger number by the smaller; and to obtain the smallest possible outcome (the DM/£ bid rate), we divide the smaller number by the larger. This illustrates the *Law of the Worst Possible Combination.*

SYSTEMATIC RISK

Also called *nondiversifiable*, or *noncontrollable risk*, this risk that cannot be diversified away results from forces outside a firm's control. Purchasing power, interest rate, and market risks fall in this category. This type of risk is assessed relative to the risk of a diversified portfolio of securities or the market portfolio. It is measured by the *beta coefficient* used in the *Capital Asset Pricing Model* (*CAPM*). The systematic risk is simply a measure of a security's volatility relative to that of an average security. For example, b = 0.5 means the security is only half as volatile, or risky, as the average security; b = 1.0 means the security is of average risk; and b = 2.0 means the security is twice as risky as the average risk. The higher the beta, the higher the return required.

T

TARGET-ZONE ARRANGEMENT

Target-zone arrangement is an international monetary arrangement in which countries vow to maintain their exchange rates within a specific band around agreed-upon, fixed, central exchange rates.

TAX ARBITRAGE

Tax arbitrage is a form of arbitrage that involves the shifting of gains or losses from one tax authority to another to profit from tax rate differences.

TAX EXPOSURE

Tax exposure is the extent to which an MNC's tax liability is affected by fluctuations in foreign exchange values. As a general rule, only realized gains or losses affect the income tax liability of a company. Translation losses or gains are normally not realized and are not taken into account in tax liability. Some steps taken to reduce exposure, such as entering into forward exchange contracts, can create losses or gains that enter into tax liability. Other measures that can be taken have no income tax implications.

TECHNICAL ANALYSIS

As the antithesis of *fundamental analysis*, technical analysis concentrates on past price and volume movements—while totally disregarding economic fundamentals—to forecast a security price or currency rates. The two primary tools of technical analysts are charting and key indicators. Charting means plotting on a graph the stock's price movement over time. For example, the security may have moved up and down in price, but remained within a band bounded by the lower limit (support level) and the higher limit (resistance level). Key indicators of market and security performance include trading volume, market breadth, mutual fund cash position, short selling, odd-lot theory, and the Index of Bearish Sentiment.

See also FUNDAMENTAL ANALYSIS; TECHNICAL FORECASTING.

TECHNICAL FORECASTING

Technical forecasting involves the use of historical exchange rates to predict future values. For example, the fact that a given currency has increased in value over four consecutive days may provide an indication of how the currency will move tomorrow. It is sometimes conducted in a judgmental manner, without statistical analysis. Often, however, statistical analysis is applied in technical forecasting to detect historical trends. For example, a computer program can be developed to detect particular historical trends. There are also time series models that examine moving averages. Some develop a rule, such as, "The currency tends to decline in value after a rise in moving average over three consecutive periods."

Technical forecasting of exchange rates is similar to technical forecasting of stock prices. If the pattern of currency values over time appears random then technical forecasting is not appropriate. Unless historical trends in exchange rate movements can be identified, examination of past movements will not be useful for indicating future movements. Technical factors have sometimes been cited as the main reason for changing speculative positions that cause an adjustment in the dollar's value. For example, the *Wall Street Journal* frequently summarizes the dollar movements on particular days as shown below.

Date	Status of Dollar	Explanation
Oct. 14, 1999	Weakened	Technical factors overwhelmed economic news
Nov. 18, 1999	Weakened	Technical factors triggered sales of dollars
Dec. 16, 1999	Weakened	Technical factors triggered sales of dollars
Apr. 14, 2000	Strengthened	Technical factors indicated that dollars had been recently oversold, triggering purchase of dollars

These examples suggest that technical forecasting appears to be widely used by speculators who frequently attempt to capitalize on day-to-day exchange rate movements. Technical forecasting models have helped some speculators in the foreign exchange market at various times. However, a model that has worked well in one particular period will not necessarily work well in another. With the abundance of technical models existing today, some are bound to generate speculative profits in any given period.

Most technical models rely on the past to predict the future. They try to identify a historical pattern that seems to repeat and then try to forecast it. The models range from a simple moving average to a complex auto regressive integrated moving average (ARIMA). Most models try to break down the historical series. They try to identify and remove the random element. Then they try to forecast the overall trend with cyclical and seasonal variations. A moving average is useful to remove minor random fluctuations. A trend analysis is useful to forecast a long-term linear or exponential trend. Winter's seasonal smoothing and Census XII decomposition are useful to forecast long-term cycles with additive seasonal variations. ARIMA is useful to predict cycles with multiplicative seasonality. Many forecasting and statistical packages such as *Forecast Pro, Sibyl/Runner, Minitab, SPSS*, and *SAS* can handle these computations.

See also FOREIGN EXCHANGE RATE FORECASTING; FUNDAMENTAL FORECASTING.

TED SPREAD

The yield spread between U.S. Treasury bills and Eurodollars.

TEMPORAL METHOD

The temporal method translates assets valued in a foreign currency into the home currency using the exchange rate that exists when the assets are purchased. It is essentially the same as the *monetary-nonmonetary method* except in the treatment of physical assets that have been revalued. It applies the current exchange rate to all financial assets and liabilities, both current and long term. Physical, or nonmonetary, assets valued at historical cost are translated at historical rates. Because the various assets of a foreign subsidiary will in all probability be acquired at different times and exchange rates seldom remain stable for long, different exchange rates will probably have to be used to translate those foreign assets into the multinational's home currency. Consequently, the MNC's balance sheet may not balance.

EXAMPLE 116

Consider the case of a U.S. firm that on January 1, 20X1, invests $100,000 in a new Japanese subsidiary. The exchange rate at that time is $1 = ¥100. The initial investment is therefore ¥10 million, and the Japanese subsidiary's balance sheet looks like this on January 1, 20X1.

	Yen	Exchange Rate	U.S. Dollars
Cash	10,000,000	($1 = ¥100)	100,000
Owners' equity	10,000.000	($1 = ¥100)	100,000

Assume that on January 31, when the exchange rate is $1 = ¥95, the Japanese subsidiary invests ¥5 million in a factory (i.e., fixed assets). Then on February 15, when the exchange rate in $1 = ¥90, the subsidiary purchases ¥5 million of inventory. The balance sheet of the subsidiary will look like this on March 1, 20X1.

	Yen	Exchange Rate	U.S. Dollars
Fixed assets	5,000,000	($1 = ¥95)	52,632
Inventory	5,000,000	($1 = ¥90)	55,556
Total	10,000,000		108,187
Owners' equity	10,000,000	($1 = ¥100)	100,000

As can be seen, although the balance sheet balances in yen, it does not balance when the temporal method is used to translate the yen-denominated balance sheet tables back into dollars. In translation, the balance sheet debits exceed the credits by $8,187. How to cope with the gap between debits and credits is an issue of some debate within the accounting profession. It is probably safe to say that no satisfactory solution has yet been adopted.

A. Current U.S. Practice

U.S.-based MNCs must follow the requirements of *Statement 52*, "Foreign Currency Translation," issued by the U.S. Financial Accounting Standards Board (FASB) in 1981. Under *Statement 52*, a foreign subsidiary is classified either as a self-sustaining, autonomous subsidiary or as *integral* to the activities of the parent company. According to *Statement 52*, the local currency of a self-sustaining foreign subsidiary is to be its functional currency. The balance sheet for such subsidiaries is translated into the home currency using the exchange rate in effect at the end of the firm's financial year, whereas the income statement is translated using the average exchange rate for the firm's financial year. On the other hand, the functional currency of an integral subsidiary is to be U.S. dollars. The financial statements of such subsidiaries are translated at various historic rates using the temporal method (as we did in the example), and the dangling debit or credit increases or decreases consolidated earnings for the period.

See also CURRENT RATE METHOD; FASB No. 52.

TENOR

Time period of *drafts*.

See also DRAFT.

TERM STRUCTURE OF INTEREST RATES

The term structure of interest rates, also known as a *yield curve*, shows the relationship between length of time to maturity and yields of debt instruments. Other factors such as default risk and tax treatment are held constant. An understanding of this relationship is important to corporate financial officers who must decide whether to borrow by issuing long- or short-term debt. An understanding of yield-to-maturity for each currency is especially critical to an MNC's CFO. It is also important to investors who must decide whether to buy long- or short-term bonds. Fixed income security analysts should investigate the yield curve carefully in order to make judgments about the direction of interest rates. A yield curve is simply a graphical presentation of the term structure of interest rates. A yield curve may take any number of shapes. Exhibit 105 shows alternative yield curves: a flat (vertical) yield curve (Exhibit 105A), a positive (ascending) yield curve (Exhibit 105B), an inverted (descending) yield curve (Exhibit 105C), and a humped (ascending and then descending) yield curve (Exhibit 105D). For the yield curve whose shape changes over time, there are three major explanations, or theories of yield curve patterns: (1) the expectation theory, (2) the liquidity preference theory, and (3) the market segmentation, or "preferred habitat," theory.

EXHIBIT 105
Alternative Term-Structure Patterns

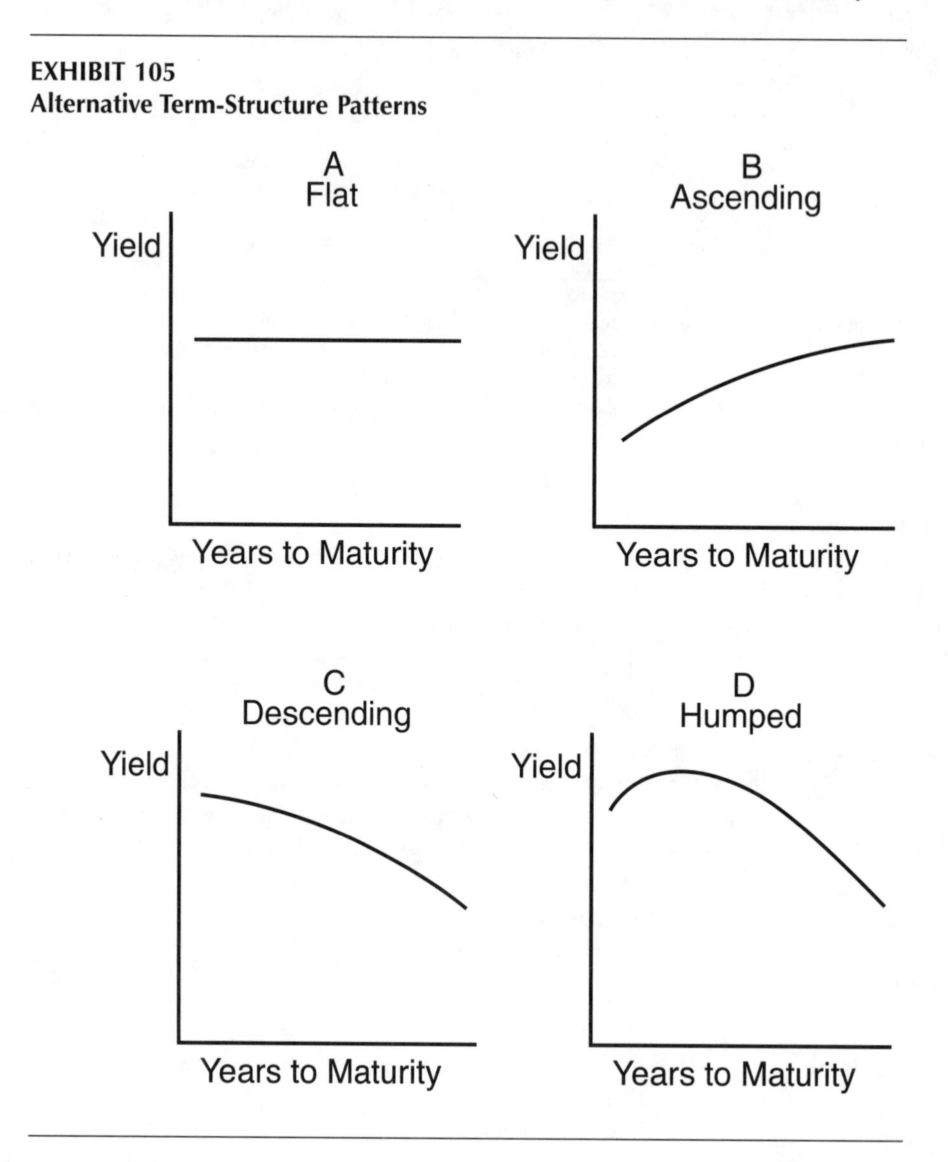

A. Expectation Theory

The expectation theory postulates that the shape of the yield curve reflects investors' expectations of future short-term rates. Given the estimated set of future short-term interest rates, the long-term rate is then established as the geometric average of future interest rates.

EXAMPLE 117

At the beginning of the first quarter of the year, suppose a 91-day T-bill yields a 6% annualized yield, and the expected yield for a 91-day T-bill at the beginning of the second quarter is 6.4%. Under the expectation theory, a 182-day T-bill is equivalent to having successive 91-day T-bills and thus should offer investors the same annualized yield. Therefore, a 182-day T-bill issued at the beginning of the first quarter of the year should yield 6.2%, which is an arithmetic mean (average) of successive 91-day T-bills.

$$1/2\ (6.00 + 6.40) = 1/2\ (12.40) = 6.20\%$$

Mathematically, a current long-term yield is a geometric average of current and successive short-term yields, or

$$(1 + {}_tR_n)^n = (1 + {}_tR_1)(1 + {}_{t+1}r_1)\ldots(1 + {}_{t+n-1}r_1)$$

where the subscripts to the left of the variable, t, $t + 1$, …, signify the period and the subscripts to the right, 1, 2, …, n signify the maturity of the debt instrument. R is the current yield, and r is a future (expected) yield. A positive (ascending) yield curve implies that investors expect short-term rates to rise, while a descending (inverted) yield curve implies that they expect short-term rates to fall.

EXAMPLE 118

Suppose a current 2-year yield is 9%, or ${}_tR_2 = .09$, and a current 1-year yield is 7%, or ${}_1R_t = .07$. Then the expected 1-year future yield ${}_{t+1}r_1$ is 0.11037, or 11.04%:

$$\begin{aligned}(1 + {}_tR_2)^2 &= (1 + {}_tR_1)(1 + {}_{t+1}r_1)\\(1.09)^2 &= (1.07)(1 + {}_{t+1}r_1)\\1.1881 &= (1.07)(1 + {}_{t+1}r_1)\\(1 + {}_{t+1}r_1) &= 1.1881/1.07\\{}_{t+1}r_1 &= 1.11037 - 1 = 0.11037 = 11.04\%\end{aligned}$$

B. Liquidity Preference Theory

The liquidity preference theory contends that risk-averse investors prefer short-term bonds to long-term bonds, because long-term bonds have a greater chance of price variation, i.e., carry greater interest rate risk. Accordingly, the theory states that rates on long-term bonds will generally be above the level called for by the expectation theory. Current long-term bonds should include a liquidity premium as additional compensation for assuming interest rate risk. This theory is nothing but a modification of the expectation theory. Mathematically, a current 2-year rate is a geometric average of a current and a future 1-year rate plus a liquidity risk premium L:

$$(1 + {}_tR_2)^2 = (1 + {}_tR_1)(1 + {}_{t+1}r_1) + L$$

Because of a liquidity premium, a yield curve would be upward-sloping rather than vertical when future short-term rates are expected to be the same as the current short-term rate.

C. Market Segmentation (Preferred Habitat) Theory

The market segmentation theory does not recognize expectations and emphasizes the rigidity in loan allocation patterns by lenders. Some lenders (such as banks) are required by law to lend primarily on a short-term basis. Other lenders (such as life insurance companies and pension funds) prefer to operate in the long-term market. Similarly, some borrowers need short-term money (e.g., to build up inventories), while others need long-term money (e.g., to purchase homes). Thus, under this theory, interest rates are determined by supply and

demand for loanable funds in each maturity market spectrum. The yield curve for U.S. dollar-denominated debt issues is available at the Federal Reserve Bank of New York website (www.ny.frb.org).
See also INTERNATIONAL YIELD CURVES.

THETA

See CURRENCY OPTION PRICING SENSITIVITY.

3-Ds

3-Ds stand for "dollar-denominated delivery." Virtual currency options are also called *3-Ds* (dollar-denominated delivery).
See VIRTUAL CURRENCY OPTIONS.

THREE-WAY ARBITRAGE

See TRIANGULAR ARBITRAGE.

TIME VALUE

1. Time value of money; present values (discounting) of a future sum of money or an annuity and future values (compounding) of a present sum of money or an annuity.
   See also DISCOUNTING.
2. The amount by which the option value exceeds the intrinsic value. The theoretical value of an option consists of an intrinsic value and a time value.
   See CURRENCY OPTION; OPTION.

TOKYO STOCK EXCHANGE

Tokyo Stock Exchange (TSE) is the largest stock exchange in Japan, with more than 80% of all transactions. Osaka is the second largest exchange, with about 15% of all transactions. By tradition, the TSE is an auction, order-driven market without market makers. Order clerks conclude trades by matching buyers and sellers without taking positions for their own accounts.

TOTAL RETURN

Total return (TR) is the most complete measure of an investment's profitability. Total return on an investment equals: (1) periodic cash payments (current income) and (2) appreciation (or depreciation) in value (capital gains or losses). Current income (C) may be bond interest, cash dividends, rent, etc. Capital gains or losses are changes in market value. A capital gain is the excess of selling price (P_1) over purchase price (P_0). A capital loss is the opposite. Return is measured considering the relevant time period (holding period), called a *holding period return.*

$$\text{Holding Period Return (HPR)} = \frac{\text{Current income} + \text{Capital gain(or loss)}}{\text{Purchase price}} = \frac{C + (P_1 - P_0)}{P_0}$$

EXAMPLE 119

Consider the investment in stocks A and B over a one period of ownership:

	Stock	
	A	B
Purchase price (beginning of year)	$100	$100
Cash dividend received (during the year)	$13	$18
Sales price (end of year)	$107	$97

The current incomes from the investment in stocks A and B over the one-year period are $13 and $18, respectively. For stock A, a capital gain of $7 ($107 sales price – $100 purchase price) is realized over the period. In the case of stock B, a $3 capital loss ($97 sales price – $100 purchase price) results.

Combining the capital gain return (or loss) with the current income, the total return on each investment is summarized below:

	Stock	
Return	A	B
Cash dividend	$13	$18
Capital gain (loss)	7	(3)
Total return	$20	$15

Thus, the return on investments A and B are:

$$\text{HPR(stock A)} = \frac{\$13 + (\$107 - \$100)}{\$100} = \frac{\$13 + \$7}{\$100} = \frac{\$20}{\$100} = 20\%$$

$$\text{HPR(stock B)} = \frac{\$18 + (\$97 - \$100)}{\$100} = \frac{\$18 - \$3}{\$100} = \frac{\$15}{\$100} = 15\%$$

See also ARITHMETIC AVERAGE RETURN VS. COMPOUND (GEOMETRIC) AVERAGE RETURN; RETURN RELATIVE.

TOTAL RETURN FROM FOREIGN INVESTMENTS

In general, the total dollar return on an investment can be broken down into three separate elements: dividend/interest income, capital gains (losses), and currency gains (losses).

A. Bonds

The one-period total dollar return on a foreign bond investment R can be calculated as follows:

$$\text{Total dollar return} = \text{Foreign currency bond return} \times \text{Currency gain(loss)}$$

$$1 + R = \left[1 + \frac{B_1 - B_0 - I}{B_0}\right](1 + \%C)$$

where

B_1 = foreign currency bond price at year-end
B_0 = foreign currency bond price at the beginning of the period
I = foreign currency bond coupon income
$\%C$ = percent change in dollar value of the foreign currency

---

EXAMPLE 120

Suppose the initial British bond price is £102, the coupon income is £9, the end-of-period bond price is £106, and the local currency appreciates by 8.64% against the dollar during the period. According to the formula, the total dollar return is 22.49%:

$$1 + R = [1 + (£106 - £102 + £9)/£102] \times (1 + 0.0864)$$
$$R = 0.2249 = 22.49\%$$

Note: The currency gain applies to both the local currency principal and to the local currency return.

B. Stocks

Using the same terminology the one-period total dollar return on a foreign stock investment can be calculated as follows:

$$\text{Total dollar return} = \text{Foreign currency stock return} \times \text{Currency gain(loss)}$$

$$1 + \text{R} = \left[1 + \frac{P_1 - P_0 - D}{P_0}\right](1 + \%\text{C})$$

where

P_1 = foreign currency stock price at year-end
P_0 = foreign currency stock price at the beginning of the period
D = foreign currency dividend income
$\%C$ = percent change in dollar value of the foreign currency

---

EXAMPLE 121

Suppose that, during the year, Honda Motor Company moved from ¥11,000 to ¥9,000, while paying a dividend of ¥60. At the same time, the exchange rate moved from ¥105 to ¥110. The total dollar return from this stock investment is a loss, which is computed as follows:

$$1 + R = [1 + (¥9{,}000 - ¥11{,}000 + ¥60)/¥11{,}000] \times (1 - 0.0455)$$
$$R = -0.2123 = -21.23\%$$

Note: The percent change in the yen rate is 0.00455 = (¥105 – ¥110)/ ¥110. In this example, the investor suffered both a capital loss on the foreign currency principal and a currency loss on the dollar value of the investment.

See also INTERNATIONAL RETURNS; TOTAL RETURN.

TRACKING STOCK

Issuing tracking stock is an increasingly popular corporate-financing technique. Tracking stock is a stock created by a company to follow, or track, the performance of one of its divisions—typically one that is in a line of business that is fast-growing and commands a higher industry price-to-earnings ratio than the parent's main business. Some companies distribute tracking stock to their existing shareholders. Others sell tracking stock to the public, raising additional cash for themselves. Some companies do both. Tracking stock, however, does not typically provide voting rights.

TRADE ACCEPTANCE

Trade acceptance is a time or date draft which has been accepted by the drawee (or the buyer) for payment at maturity. Trade acceptances differ from *bankers' acceptances* in that they are drawn on the buyer, carry only the buyer's obligation to pay, and cannot become bankers' acceptances or be guaranteed by a bank.

TRADE BALANCE

See BALANCE OF TRADE.

TRADE CREDIT INSTRUMENTS

Arrangements made to finance international trade credit are very much like the intracountry arrangements, but they also involve the extra complications of the international environment. The major trade credit instruments are:

- *Letter of credit*—a written statement made by a bank that it will pay a specified amount of money when certain trade conditions have been satisfied
- *Draft*—an order to pay someone (similar to a check)
- *Banker's acceptance*—a draft that has been accepted by a bank

An example will help illustrate these ideas.

EXAMPLE 122

Consider a New York firm that wants to import $200,000 worth of Japanese CD player components. The firm first gets the Japanese company to grant it 60 days' credit from the shipment date. Then the New York firm arranges a *letter of credit* through its New York bank, which is sent to the Japanese company. The Japanese company ships the equipment and presents a 60-day *draft* on the New York bank to its Japanese bank. Then the Japanese bank pays the Japanese company. The draft is then forwarded to the New York bank and, if all paperwork is in order, becomes a *banker's acceptance*, which is a $200,000 debt that the New York bank owes the Japanese bank. At the end of 60 days the New York importer pays the New York bank, which in turn pays the acceptance. In the interim, the Japanese bank could sell the acceptance on the open market. The final owner of the banker's acceptance would then present it to the New York bank for payment.

Note: There are at least four parties involved: an importer, an exporter, and their respective banks. Often there are other banks involved, too. The whole process has several detailed features and options associated with it. Finance companies and factors are also involved in financing trade credit.

See also BANKER'S ACCEPTANCE; DRAFT; LETTERS OF CREDIT.

TRADING AT A DISCOUNT

See FORWARD PREMIUM OR DISCOUNT.

TRADING AT A PREMIUM

See FORWARD DIFFERENTIAL.

TRANSACTION EXPOSURE

Transaction exposure is the extent to which the income from transactions is affected by fluctuations in foreign exchange values. This exposure arises whenever an MNC is committed to a foreign-currency-denominated transaction. Such exposure represents the potential gains or losses on the future settlement of outstanding obligations for the purchase or sale of goods and services at previously agreed prices and the borrowing or lending of funds in foreign currencies. An example would be a U.S. dollar loss, after the franc devalues, on payments received for an export invoiced in francs before that devaluation. Transaction exposure can be managed by contractual and operating *hedges*. The major contractual hedges use the forward, money, futures, and option markets, while operating strategies include the use of *currency swaps, back-to-back (parallel) loans,* and *leads and lags* in payment terms. Three contractual hedges are briefly explained below.

- *Forward-market hedge*. A forward hedge involves a forward contract and a source of funds to carry out that contract. The forward contract is entered into at the time the transaction exposure is created. Transaction exposure associated with a foreign currency can also be covered in the currency futures market.
- *Money-market hedge*. Like a forward-market hedge, a money-market hedge also employs a contract and a source of funds to fulfill that contract. In this case, however, the contract is a loan agreement. The MNC involved in the hedge borrows in one currency and exchanges the proceeds for another currency.
- *Option-market hedge*. An option-market hedge involves the purchase of a *call* (the right to buy) or *put* (the right to sell) option. This will allow the MNC to speculate the upside potential for depreciation or appreciation of the currency while limiting downside risk to a known (certain) amount.

EXAMPLE 123

Asiana Airlines has just signed a contract with Boeing to buy two new jet aircrafts for a total of $120,000,000, with payment in two equal installments. The first installment has just been paid. The next $60,000,000 is due three months from today. Asiana currently has excess cash of 50,000,000 won in a Seoul bank, from which it plans to make its next payment. It wishes to determine the method by which it could make its dollar payment and be assured of the largest remaining bank balance. The relevant data are given below.

	Value	Units
Beginning Seoul bank cash balance	90,000,000,000	won
Account payable in 90 days	$60,000,000	U.S. $
Spot rate	1100	won/$
Three-month forward rate	1095	won/$
Spot rate in 3 months (forecast)	1092	won/$
Korean 3-month interest rate	5.00%	per annum
U.S. 3-month interest rate	8.00%	per annum
OTC bank call option (90 days)		
Strike price	1090	won/$
Option premium (cost)	0.50%	per annum

Exhibits 106 and 107 provide evaluation of four alternatives at various spot rates.

EXHIBIT 106
Transaction Hedge/Payment Evaluation (Korean won)

Alternative		Expected Cost	Remaining Bank Balance
Unhedged		65,520,000,000	25,605,000,000
Forward hedge		65,700,000,000	25,425,000,000
Money-market hedge		65,514,705,882	25,610,294,118
OTC bank option	Premium	334,125,000	
	Exercise	65,400,000,000	25,390,875,000

Note: All costs are stated at end of 90-day period.

EXHIBIT 107
Graphic Generation of Hedging Alternatives (ending bank balance in won)

Spot rate	1084	1086	1088	1090	1092
Unhedged	26,085,000,000	25,965,000,000	25,845,000,000	25,725,000,000	25,605,000,000
Forward	25,425,000,000	25,425,000,000	25,425,000,000	25,425,000,000	25,425,000,000
Money market	25,610,294,118	25,610,294,118	25,610,294,118	25,610,294,118	25,610,294,118
Option	25,750,875,000	25,630,875,000	25,510,875,000	25,390,875,000	25,390,875,000

Exhibit 108 graphs expected bank balances of alternative strategies.

EXHIBIT 108
Hedge Valuation for Asiana Airlines (at various ending spot exchange rates)

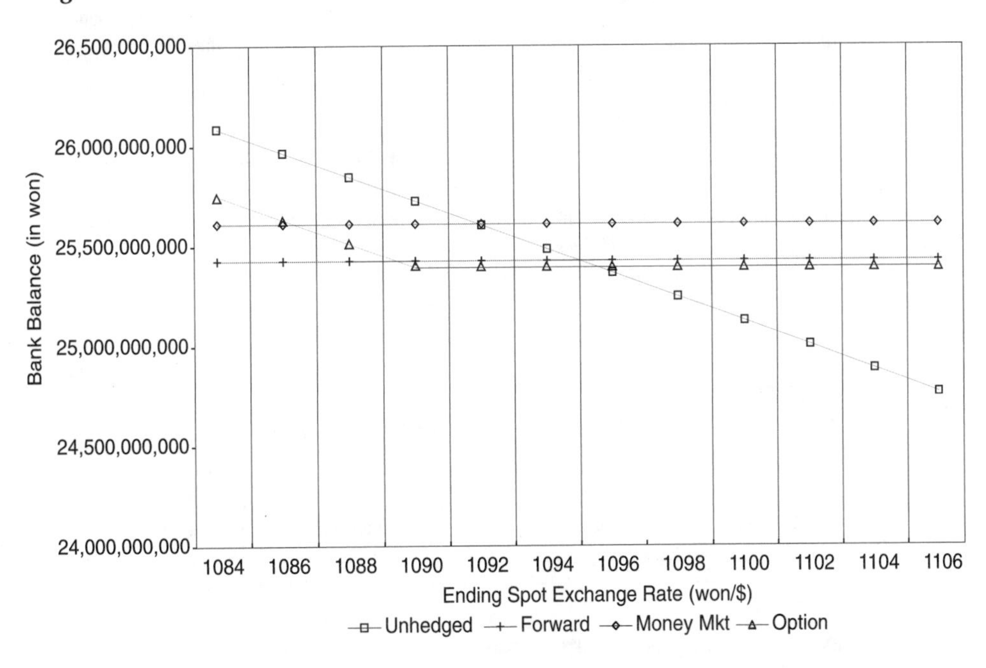

See also MONEY-MARKET HEDGE.

TRANSACTION RISK

Transaction risk is the risk resulting from *transaction exposure* and losses from changing foreign currency rates. It involves a receivable or a payable denoted in a foreign currency. See also TRANSACTION EXPOSURE.

TRANSFERABLE LETTER OF CREDIT

A *letter of credit* (L/C) under which the beneficiary (exporter) has the right to instruct the paying bank to make the credit available to one or more secondary beneficiaries. No L/C is transferable unless specifically authorized in the letter of credit. Further, it can be transferred only once. The stipulated documents are transferred alone with the L/C.

TRANSLATION EXPOSURE

Also called *accounting exposure*, the impact of an exchange rate change on the reported consolidated financial statements of an MNC. An example would be the impact of a French franc devaluation on a U.S. firm's reported income statement and balance sheet. The resulting translation (accounting) gain or losses are said to be *unrealized*—they are "paper" gains and losses. Exhibit 109 contrasts translation, transaction, and economic exposure. Exhibit 110 summarizes basic strategy for managing (*hedging*) translation exposure.

EXHIBIT 109
Comparison of Translation, Transaction, and Economic Exposure

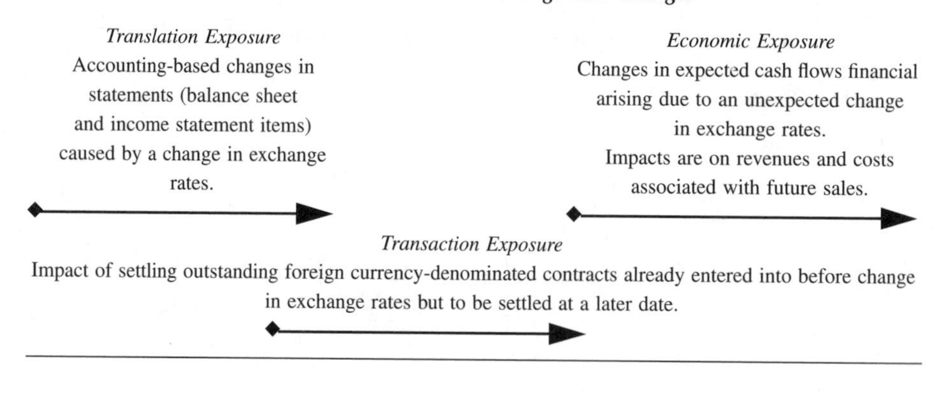

EXHIBIT 110
Basic Strategy For Managing (Hedging) Translation Exposure

	Assets	Liabilities
Hard currencies (Likely to appreciate)	Increase	Decrease
Soft currencies (Likely to depreciate)	Decrease	Increase

The strategy involves increasing hard-currency assets and decreasing soft-currency assets, while simultaneously decreasing hard-currency liabilities and increasing soft-currency liabilities. For example, if a devaluation appears likely, the basic strategy would be to reduce the level of cash, tighten credit terms (to reduce accounts receivable), increase local currency borrowing, delay accounts payable, and sell the weak currency forward.

See also ECONOMIC EXPOSURE; TRANSACTION EXPOSURE.

TRANSLATION GAIN OR LOSS

An *accounting* gain or loss resulting from changes caused by fluctuations in foreign currency-based receivables, payables, or other assets or liabilities.

TRANSLATION METHODS

See CURRENCY TRANSLATION METHODS.

TRANSLATION RISK

See TRANSLATION EXPOSURE.

TREYNOR'S PERFORMANCE MEASURE

Treynor's performance measure can be used to measure portfolio performance. It is concerned with systematic (beta) risk.

$$T_p = \frac{\text{Risk premium}}{\text{Portfolio's beta coefficient}}$$

EXAMPLE 124

An investor wants to rank two stock mutual funds he owns. The risk-free interest rate is 6%. Information for each fund follows:

Growth Fund	Return	Fund's Beta
A	14%	1.10
B	12	1.30

$$T_A = \frac{14\% - 6\%}{1.10} = 7.27 \text{ (First)}$$

$$T_B = \frac{12\% - 6\%}{1.30} = 4.62 \text{ (Second)}$$

Fund *A* is ranked first because it has a higher return relative to Fund *B*.

The index can be computed based on information obtained from financial newspapers such as *Barron's* and the *Wall Street Journal.*

See also SHARPE'S RISK-ADJUSTED RETURN.

TRIANGULAR ARBITRAGE

Also called *a three-way arbitrage*, triangular arbitrage eliminates exchange rate differentials across the markets for all currencies. This type of arbitrage involves more than two currencies. If the *cross rate* is not set properly, arbitrage may be used to capitalize on the discrepancy. When we consider that the bulk of foreign exchange trading involves the U.S. dollar, we note the role of comparing dollar exchange rates for different currencies to determine if the implied third exchange rates are in line. Since banks quote foreign exchange rates with respect to the dollar (the dollar is said to be the "numeraire" of the system), such comparisons are readily made. For instance, if we know the dollar price of pounds ($/£) and the dollar price of marks ($/DM), we can infer the corresponding pound price of marks (£/DM). Triangular arbitrage is a form of arbitrage seeking a profit as a result of price differences in foreign exchange among three currencies. This form of arbitrage occurs when the arbitrageur does not desire to operate directly in a two-way transaction, due to restrictions on the market or for any other reason. In this case, the arbitrageur moves through three currencies, starting and ending with the same one. *Note:* Like *simple, two-way arbitrage*, triangular arbitrage does not tie up funds. Also, the strategy is risk-free, because there is no uncertainty about the rates at which one buys and sells the currencies.

EXAMPLE 125

To simplify the analysis of arbitrage involving three currencies, let us temporarily ignore the bid–ask spread and assume that we can either buy or sell at one price. Suppose that in London $/£ = $2.00, while in New York $/DM = $0.40. The corresponding *cross rate* is the £/DM rate. Simple algebra shows that if $/£ = $2.00 and $/DM = 0.40, then £/DM = ($/DM)/($/£) = 0.40/2.00 = 0.2. If we observe a market where one of the three exchange rates—$/£, $/DM, £/DM—is out of line with the other two, there is an arbitrage opportunity.

Suppose that in Frankfurt the exchange rate is £/DM = 0.2, while in New York $/DM = 0.40, but in London $/£ = $1.90. Astute traders in the foreign exchange market would observe the discrepancy, and quick action would be rewarded. The trader could start with dollars and

1. Buy £1 million in London for $1.9 million as $/£ = $1.90.
2. The pounds could be used to buy marks at £/DM = 0.2, so that £1,000,000 = DM5,000,000.
3. The DM5 million could then be used in New York to buy dollars at $/DM = $0.40, so that DM5,000,000 = $2,000,000.
4. Thus, the initial $1.9 million could be turned into $2 million with the *triangular arbitrage* action earning the trader $100,000 (costs associated with the transaction should be deducted to arrive at the true arbitrage profit).

As in the case of the two-currency arbitrage covered earlier, a valuable product of this arbitrage activity is the return of the exchange rates to internationally consistent levels. If the initial discrepancy was that the dollar price of pounds was too low in London, the selling of dollars for pounds in London by the arbitrageurs will make pounds more expensive, raising the price from $/£ = $1.90 back to $2.00. (Actually, the rate would not return to $2.00, because the activity in the other markets would tend to raise the pound price of marks and lower the dollar price of marks, so that a dollar price of pounds somewhere between $1.90 and $2.00 would be the new equilibrium among the three currencies.)

EXAMPLE 126

Suppose the pound sterling is bid at $1.9809 in New York and the Deutsche mark at $0.6251 in Frankfurt. At the same time, London banks are offering pounds sterling at DM 3.1650. An astute trader would sell dollars for Deutsche marks in Frankfurt, use the Deutsche marks to acquire pounds sterling in London, and sell the pounds in New York. Specifically, the trader would

1. Acquire DM1,599,744.04 ($1,000,000/$0.6251) for $1,000,000 in Frankfurt,
2. Sell these Deutsche marks for £505,448.35 (1,599,744.04/DM3.1650) in London, and
3. Resell the pounds in New York for $1,001,242.64 (£505,448.35 × $1.9809).

Thus, a few minutes' work would yield a profit of \$1,242.64 (\$1,001,242.64 – \$1,000,000). In effect, the trader would, by arbitraging through the DM, be able to acquire sterling at \$1.9784 in London (\$0.6251 × 3.1650) and sell it at \$1.9809 in New York. Again, as can be seen in this example, the arbitrage transactions would tend to cause the Deutsche mark to appreciate vis-à-vis the dollar in Frankfurt and to depreciate against the pound sterling in London; at the same time, sterling would tend to fall in New York.

Opportunities for such profitable currency arbitrage have been greatly reduced in recent years, given the extensive network of people—aided by high-speed, computerized information systems—who are continually collecting, comparing, and acting on currency quotes in all financial markets. The practice of quoting rates against the dollar makes currency arbitrage even simpler. The result of this activity is that rates for a specific currency tend to be the same everywhere, with only minimal deviations due to transaction costs.

See also ARBITRAGE; COVERED INTEREST ARBITRAGE; FOREIGN EXCHANGE ARBITRAGE; SIMPLE ARBITRAGE.

TRIANGULATION

Triangulation is the method of conversion used under the new *euro* system. The conversion has to be made through the euro—for example, Dutch guilders to euros to francs, using the fixed conversion rates.

See also BILATERAL EXCHANGES; EURO.

TRUST RECEIPT

A trust receipt is an instrument that acknowledges that the borrower holds specified property in trust for the lender. The lender retains title. The goods are subject to repossession by the bank. The trust receipts are always used when merchandise is financed via acceptances under *letters of credit*. When the lender receives the sale proceeds, title is given up.

TWO-TIER FOREIGN EXCHANGE MARKET

An arrangement of two exchange markets—a formal market (at the official rate) for certain transactions and a free market for remaining transactions.

TWO-WAY ARBITRAGE

See SIMPLE ARBITRAGE.

TYPES OF OVERSEAS BANKING SERVICES

There are a number of organizational forms that banks may use to deliver international banking services to their customers. The primary forms are (1) *correspondent banks*, (2) *representative offices*, (3) *branch banks*, (4) *foreign subsidiaries and affiliates*, (5) *Edge Act corporations*, and (6) *international banking facilities (IBFs)*. Exhibit 111 shows a possible organizational structure for the foreign operations of U.S. banks. Though possible, all these forms need not exist for any individual bank. Exhibit 112 summarizes advantages and disadvantages of each type of form.

EXHIBIT 111
Organizational Structure for a U.S. Bank's International Operations

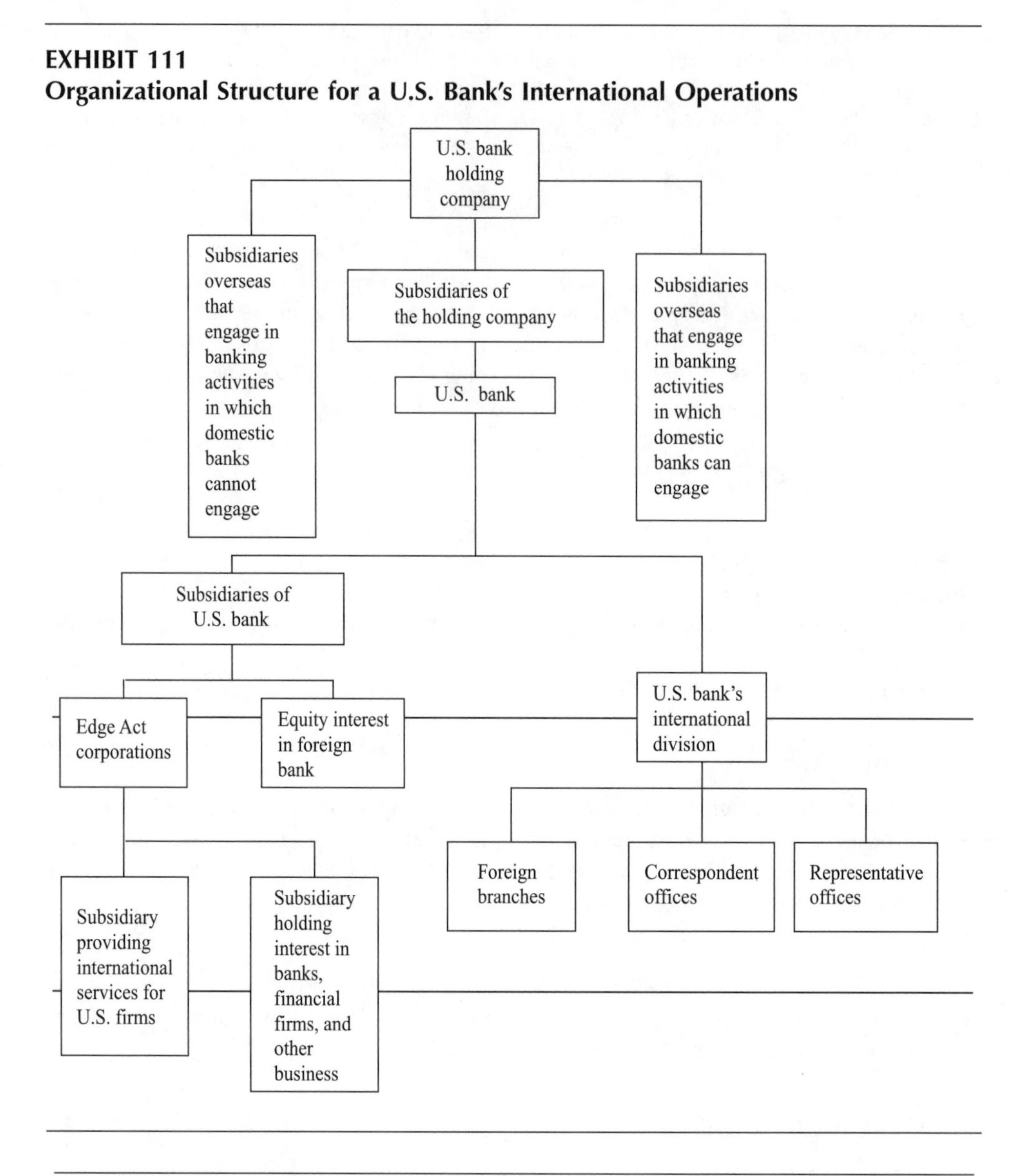

EXHIBIT 112
Advantages and Disadvantages of Types of Overseas Banking Services

Types	Advantages	Disadvantages
Correspondent banks	Minimal cost form of market entry No investment in staff or facilities Having multiple sources of business given and received Referrals to local banking opportunities Ability to cash in on local knowledge and contacts	Low priority given to the needs of U.S. customers Difficulty of obtaining due to capital restrictions Difficult to arrange certain types of credits Credit not provided regularly and extensively by the correspondent

Representative offices	Low-cost entry to foreign markets Efficient delivery of services Attracting additional business Preventing losses of current business	Inability to penetrate the foreign market more effectively Expensive because capital is not generated Difficult to attract qualified personnel Inability to conduct general banking activities
Foreign branches	Better control over foreign operations Enhanced ability to offer direct and integrated services to customers Improved ability to manage customer relationships Ability to conduct a full range of services	High-cost form of entry into a foreign market Difficult and expensive to train branch managers
Foreign subsidiaries and affiliates	Immediate access to local deposit markets Ability to use an established network of local contacts and clients	Expensive Highly risky Difficult to make work effectively

See also CONSORTIUM BANK; CORRESPONDENT BANK; EDGE ACT AND AGREEMENT CORPORATION; FOREIGN BRANCHES; FOREIGN SUBSIDIARIES AND AFFILIATES; INTERNATIONAL BANKING FACILITY; REPRESENTATIVE OFFICES.

U

UMBRELLA POLICY

See EXPORT-IMPORT BANK.

UNBUNDLING

1. A strategy that adopts more than one financial tool to transfer funds across countries.
2. Separating cash flows from a subsidiary to a parent company into their many separate components, such as royalties, lease payments, dividend, so as to increase the likelihood that some fund flows will be allowed during economically hard times.
3. A strategy of governments to try to force MNCs into sharing more of their benefits with the local country; for example, through shared ownership, required technology transfer, or local content requirements.

UNCONFIRMED LETTER OF CREDIT

A *letter of credit* (*L/C*) issued by one bank and not confirmed by another. Hence, an unconfirmed L/C is the obligation of only the issuing bank.

UNDERVALUED CURRENCY

1. A currency whose value a country seeks to keep below market to make its exports less expensive and more competitive.
2. A currency that has been oversold because of emotional or panic selling.

UNSYSTEMATIC RISK

Also called *diversifiable risk, company-specific risk*, or *controllable risk*, unsystematic risk in a portfolio is the amount of risk that can be removed by diversification.

See also BETA; SYSTEMATIC RISK.

V

VALUATION

1. The process of determining the intrinsic value of an asset, such as a security, business, or a piece of real estate. The process of determining security valuation involves finding the present value of an asset's expected future cash flows using the investor's required rate of return. Thus, the basic security valuation model can be defined mathematically as follows:

$$V = \sum_{t=1}^{n} \frac{CF_t}{(1+r)^t}$$

where

V = intrinsic value (or present value) of an asset
CF_t = expected future cash flows in period $t = 1, \ldots, n$
r = investor's required rate of return

2. Assessing the value of imported goods by customs to assess the appropriate duty charge.

VALUE DATE

Also called the *settlement date*.

1. The value date for spot exchange transactions is the date when value is given (i.e., funds are deposited) for those transactions between banks. It is set as the second working day after the transaction is concluded.
2. The point in time when a bank remittance actually becomes available to the payee for use.

VARIATION MARGIN

The amount to be paid to satisfy *maintenance margin*.

VEHICLE CURRENCY

A currency used in international trade to make quotes and payments, vehicle currency plays a central role in the foreign exchange market (e.g., the U.S. dollar and Japanese yen).

VIRTUAL CURRENCY OPTIONS

Virtual currency options, also *called 3-Ds* (dollar-denominated delivery), are options that do not require the payment or delivery of the underlying currency. Currently, 3-D options are available on the Deutsche mark and the Japanese yen. They are European-style options that mature anytime from one week to nine months, and they settle weekly.

See also CURRENCY OPTION.

VISIBLE TRADE

Also called the *balance of trade*, foreign trade in merchandise.

W

WACC

Weighted-Average Cost of Capital.
See COST OF CAPITAL.

WEAK CURRENCY

See SOFT CURRENCY.

WEAK DOLLAR

See DEPRECIATION OF THE DOLLAR.

WEIGHTED-AVERAGE COST OF CAPITAL (WACC)

See COST OF CAPITAL.

WEIGHTED-AVERAGE EXCHANGE RATE

The mean or average exchange rate used in translating income and expense accounts at the end of an accounting period, this rate takes into account the relative change of exchange rates during the period and adjusts the consolidated statement with this weighted-average rate.
See also TRANSLATION METHODS.

WHOLESALE BANKING

Banking services provided between merchant banks and other financial institutions.

WON

South Korea's currency.

WORKING CAPITAL GUARANTEE PROGRAM

See EXPORT-IMPORT BANK.

WORLD BANK

The World Bank (www.worldbank.org) is an integrated group of international institutions which provides financial and technical assistance to developing countries. The World Bank includes the *International Bank for Reconstruction and Development* and the International Development Association. World Bank affiliates, legally and financially separate, include the International Center for Settlement of Investment Disputes, the International Finance Corporation, and the Multilateral Investment Guarantee Agency. World Bank headquarters are in Washington, D.C.

WRITER

Also called a *grantor*, an individual who sells an option.

X

XENOCURRENCY

A currency that trades outside of its own borders.

Y

YANKEE BONDS

Dollar-denominated bonds issued within the United States by foreign banks and companies. See also FOREIGN BOND.

YANKEE CD

A certificate of deposit (CD) issued in the U.S. market by a branch of a foreign bank.

YANKEE STOCK OFFERINGS

Offerings of stock by non-U.S. MNCs in the U.S. markets.

YEN

Japanese currency. Its symbol is ¥.

YIELD

Also called *real return* or *real rate of return.*

1. Effective rate of return, or real return.
   See INTERNAL RATE OF RETURN.
2. The income earned on an investment, usually expressed as a percentage of the market price.
3. The percentage return earned on a common stock or preferred stock in dividends. It is figured by dividing the total of dividends paid in the preceding 12 months by the current market price. For example, a stock with a current market value of $40 a share which has paid $2 in dividends in the preceding 12 months is said to return 5% ($2/$40). If an investor paid $20 for the stock five years earlier, the stock would be returning him/her 10% on his/her original investment.
4. In the case of bonds, the *current yield* or *yield to maturity* (*YTM*).
5. The money earned on a loan, which is determined by multiplying the *annual percentage rate* (*APR*) by the amount of the loan over a stated time period.

YIELD CURVE

See TERM STRUCTURE OF INTEREST RATES.

YIELD TO CALL

The yield of a bond, if it is held until the call date. This yield is valid only if the security is called prior to maturity. The calculation of yield to call is based on the coupon rate, length of time to the call date, and the market price. In general, bonds are callable over several years and normally are called at a small premium.

YIELD TO MATURITY

Also called *effective yield*, yield to maturity (YTM) is the real return to be received from interest income plus capital gain (or loss) assuming the bond is held to maturity. The YTM incorporates the stated rate of interest on the bond as well as any discount or premium that may have been generated when bought.

YTM

See YIELD TO MATURITY.

YUAN

China's currency.

Z

ZAIRE

The currency of Congo (kinshasa) and Zaire.

ZERO-COUPON BOND

A bond sold at a deep discount, the zero-coupon bond pays no periodic interests (hence the name), so the total yield is obtained entirely as capital gain on the final maturity date. The interest is added to the principal semiannually and both the principal and the accumulated interest are paid at maturity. Although a fixed rate is implicit in the discount and the specific maturity, they are not fixed income securities in the traditional sense because they provide for no periodic income. Although the interest on the bond is paid at maturity, accrued interest, though not received, is taxable yearly as ordinary income. Zero coupon bonds have two basic advantages over regular coupon-bearing bonds: (1) a relatively small investment is required to buy these bonds and (2) a specific yield is assured throughout the term of the investment.

ZERO SUM GAME

1. A conflict situation in which a financial gain by one country results in an economic loss by another.
2. A situation in which an option holder and an option writer are facing each other. The profit from selling a call is the mirror image of the profit from buying the call.

APPENDIX

- INFORMATION SOURCES—ARTICLES
- INFORMATION SOURCES—STATISTICS
- USEFUL INTERNATIONAL BUSINESS AND FINANCE WEBSITES
- USEFUL WEBSITES FOR OPTIONS AND FUTURES
- WORLD CENTRAL BANKS
- MONETARY UNITS
- CHICAGO MERCANTILE EXCHANGE (CME) CONTRACT SPECIFICATIONS FOR CURRENCY FUTURES AND OPTIONS
- PHILADELPHIA STOCK EXCHANGE CURRENCY OPTIONS SPECIFICATIONS
- TABLES

INFORMATION SOURCES—ARTICLES

The following periodicals and services contain articles of current interest regarding in international finance, money, and banking.

1. Business International Corp. publishes
   - *Business International Money Report*, a weekly survey of important events, techniques, ideas, and data in international finance;
   - *Business Europe, Business Asia*, and *Business Latin America*, weekly surveys of business conditions in their respective areas;
   - *Financing Foreign Operations*, a loose-leaf reference volume, updated periodically, that provides an excellent survey of domestic (for most nations) and international financing sources; and
   - *Investment, Licensing and Trade*, a survey of these topics for most countries contained in the same type of periodically updated reference volume. Business International also publishes special studies on topics ranging from trade potential with China to the effects of the Foreign Corrupt Practices Act on the U.S. ability to compete overseas to the exchange rates that MNCs use in budgeting and performance evaluation.

2. *Euromoney* (www.euromoney.com) is a monthly magazine that contains articles, written primarily by professionals in the field, dealing with current issues and events in international finance. It concentrates on the Euromarkets, also providing data on them.
3. *World Financial Markets* and *The Morgan Guaranty Survey*, published monthly by Morgan Guaranty Trust, are excellent sources of discussions and statistics on current happenings in world financial markets.
4. *The Economist*, published weekly, is probably the best overall source of information on important political and economic events around the world.
5. *International Finance*, published by Chase Manhattan Bank, deals with a variety of international finance topics.

The following academic journals, published quarterly, focus on international finance, money, trade, banking, economics, and business.

1. *Columbia Journal of World Business (CJWB)* contains interesting articles, written by both academicians and practitioners, on all topics related to international business.
2. *Journal of International Business Studies* is a more technical and academically oriented publication than the *CJWB*.
3. *Journal of International Money and Finance* is an academically rigorous journal that specializes in international finance.
4. *Journal of International Financial Management and Accounting* deals with international finance and accounting issues.
5. *Global Finance Journal* focuses on issues of international financial management.

Other sources of articles related to international finance, money, and business include:

1. *The Financial Times* (London) (www.ft.com)
2. *The Wall Street Journal*
3. *The New York Times*
4. *Journal of Commerce*
5. *Japan Economic Journal*
6. *The Financial Post* (Toronto)
7. *Handelsblatt* (in German)
8. *Banque* (in French)
9. *Federal Reserve Bulletin*
10. *First Chicago World Report*
11. *Far Eastern Economic Review*
12. *Business Week*
13. *Fortune*

When you are seeking articles or information about a particular topic, a company, or country, the following indexes and abstracts might prove useful.

1. *Business Periodicals Index*
2. *F & S International*
3. *Journal of Economic Literature*
4. *Public Affairs Information Service*
5. *The Wall Street Journal Index*
6. *The New York Times Index*

Further, all the Federal Reserve Banks publish periodicals that frequently contain stories dealing with international finance and economics. These periodicals are available free for the asking. The Fed is also a good source of data on exchange rates, interest rates, capital and trade flows, international bank lending, and other international statistics. The following is a partial list.

1. *Business Conditions*, Federal Reserve Bank of Chicago
2. *International Letter*, Federal Reserve Bank of Chicago
3. *Monthly Review*, Federal Reserve Bank of New York
4. *New England Economic Review*, Federal Reserve Bank of Boston

INFORMATION SOURCES—STATISTICS

The following are useful statistical sources on international finance, money, and banking.

GENERAL

1. *International Financial Statistics* (International Monetary Fund)
2. *OECD Financial Statistics* (Organization for Economic Cooperation and Development)
3. *OECD Industrial Production*
4. *OECD Main Economic Indicators*
5. *General Statistics* (European Economic Community)
6. *Social Statistics* (ECC)
7. *Agricultural Statistics* (ECC)
8. *Monthly Bulletin of Statistics* (The United Nations)
9. *Survey of Current Business* (U.S. Department of Commerce)
10. *International Economic Indicators and Competitive Trends* (U.S. Department of Commerce)
11. *Balance of Payments Yearbook* (IMF)
12. *Business International Money Report*
13. *Key Figures of European Securities*
14. *U.S. Bureau of the Census*
15. *Federal Trade Commission (FTC)*
16. *International Economic Policy (IEP) Country Officers*
17. *International Trade Administration (ITA)*
18. *U.S. Department of Labor—Foreign Labor Trends*
19. *U.S. Export Assistance Centers (USEACs)*
20. *U.S. International Trade Commission (USITC)*
21. *U.S. Small Business Administration (SBA)*
22. Central Bank bulletins for most countries

FOREIGN EXCHANGE RATES

1. *The Wall Street Journal*
2. Federal Reserve Bank of New York *Monthly Review*
3. *Selected Interest and Exchange Rates* (Federal Reserve System)
4. *World Currency Charts* (American International Investment Corporation)
5. *Federal Reserve Bulletin*
6. *International Financial Statistics* (IMF)
7. *Pick's Currency Yearbook*
8. *Euromoney* (www.euromoney.com)
9. *World Financial Markets*
10. *International Monetary Market Yearbook* (Chicago Mercantile Exchange)
11. *Business International Money Reports*

EURODOLLARS, EUROBONDS, EUROBANKS

1. *Euromoney* (www.euromoney.com)
2. *International Bond Guide* (White Weld)
3. *Annual Report* (Bank for International Settlements—www.ibs.com)
4. *World Financial Markets*
5. *Borrowing in International Capital Markets* (World Bank)
6. *Quarterly Bulletin* (Bank of England)

7. *The Week in Eurobonds* (Kidder, Peabody)
8. *The Money Manager*
9. *The Financial Times* (www.ft.com)
10. *Capital International Perspectives* (Capital International)
11. *OECD Financial Statistics*

CAPITAL FLOWS AND RESTRICTIONS

1. *World Financial Markets*
2. *Annual Report on Exchange Restrictions* (IMF)

U.S. BANKS INTERNATIONAL OPERATIONS

1. *Federal Reserve Bulletin*
2. *Foreign Government Treatment of U.S. Commercial Banks* (Report to Congress, U.S. Treasury)
3. *The Largest 500 Banks in the World*, Annual (The Banker)

The Central Banks of most countries also publish bulletins that contain useful statistics. Many of the major domestic and foreign banks publish weekly or monthly economic letters.

USEFUL INTERNATIONAL BUSINESS AND FINANCE WEBSITES

Web Address	Primary Focus
Stock and Futures Exchanges	
www.cme.com	Chicago Mercantile Exchange
www.phlx.com	Philadelphia Stock Exchange
www.stockex.co.uk/aim	London Stock Exchange
www.tse.or	Tokyo Stock Exchange
www.simex.com/sg	Singapore Exchange
Statistical Data and Analysis	
www.census.gov	For the latest economic, financial, socioeconomic, and political surveys and statistics
www.yardeni.com	Dr. Yadeni's Economics Network containing economic analysis of major economics and statistical charts
www.uncc.edu/lis/library/reference/intbus/vibehome.htm	Internet resources of international business and economic information
www.pei-intl.com/quotes/watch/htm	Princeton's Global MarketWatch providing current financial market trends and indicators
www.bloomberg.com	Latest information from financial markets around the world.
Currency and Interest Rates	
www.xe.net/currency/iso_4217.html	Currency ISO Symbols
www.knewmoney.com	Foreign currency investing
www.xe.net/currency/table.html	Cross currency table calculator
www.oanda.com	Currency converters
www.x-rates.com	Exchange Rates
www.ny.frb.org/pihome/maktrates www.bog.frb.fed.us/releases/h15/current	Federal Reserve Bank of New York (Foreign exchange rates and Eurocurrency interest rates)
www.ms.com/gef.html www.bmo/economic/fxrates.html www.bofa.com/corporate/corporate/html	Currency forecasting services (Morgan Stanley, Bank of Montreal, Bank of America)
www.stls.frb.org/fred	The Fed's data bank for numerous economic and financial time series, e.g., on balance of payment statistics, interest rates, and foreign exchange rates
www.pacific.commerce.ubc.ca/xr/euro	Information about Euro
www.europa.eu.int/euro	Information on Euro
www.emgmkts.com	Emerging market currencies
www.singstat.gov.sg	Comparison between the actual values of foreign currencies and the value that should exist under conditions of purchasing power parity (PPP)
www.jpmorgan.com/MarketDataInd/Forex/CurrIndex.html	Exchange rate indexes (JP Morgan Currency Indexes)

(Continued)

Web Address	Primary Focus
Euro Sources	
www.euro.fee.be	European Federation of Accountants
www.europarl.eu.int/	European Parliament
www.europa.euint/euro/	European Commission
ue.eu.int	Council of European Union
www.ecb.int/	European Central Bank
www.sec.gov/interps/legal/slbcim6.htm	SEC Staff Legal Bulletin no. 6
Accounting Practices and Rules	
www.seg.gov	U.S. Securities & Exchange Commission (SEC)
www.sec.gov/cgi-bin/srch-edgar	SEC's Edgar Files (MNC's financial data)
raw.rutgers.edu/raw/fasb	FASB accounting practices
Ratings	
www.ratings.com/criteria/sovereigns.index	Bond ratings
www.moodys.com/repldata/ratings/ratsov.html	
www.transparency.de	Transparency International's Corruption Perception Index
www.imd.ch/wcy/wcy_online	The Institute for Management's World Competitiveness Analysis rankings of about 50 countries
www.adr-dmg.com/adr-dmg/welcome.html	ADR activities (Deutsche Morgan Grenfell)
Banks and Financial Management	
www.bis.org	Bank for International Settlements
www.bis.org/cbanks.html	Monetary policies and economics performance of central banks
www.ino.com	Currency movements for option strategies
www.us.kpmg.com/ecs/tp.html	Transfer pricing (KPMG-U.S.)
www.barclays.co.uk/psmd/prodcuts.htm#business	Factoring
www.bankamerica.com.corporate	
www.citibank.com/corpbank	
www.bmo.com/cebsite/cashmenu.htm	Working capital management
Clearinghouse Associations	
www.theclearinghouse.org www.chips.org	Clearinghouse associations (New York Clearinghouse Association, Clearing House Interbank Payments System)
www.tma-net.org	Treasury management
www.weforum.org/	World Economic Forum
www.opic.gov	Long-term political risk insurance and project financing (Overseas Private Investment Corporation)
World Organizations	
www.oecd.org	OECD
www.exim.gov	U.S. Export-Import Bank
www.ebrd.org	European Bank for Reconstruction and Development

Web Address	Primary Focus
www.nafta.net/	NAFTAnet home page
www.worldbank.org/	The World Bank
www.imf.org/	International Monetary Fund (IMF)
Foreign Trade	
www.exporthotline.com/	Export Hotline
tradeport.org/	TradePort—International Trade
www.commerce.ca.gov/international/	California Trade and Commerce Agency
www.intl-trade.com/library.html	International Trade Law Library
www.sbaonline.sba.gov/oit/	Small Business Administration—Office International Trade
www.stat-usa.gov/	U.S. Department of Commerce
www.exportinstitute.com/	Export Institute
www.earthone.com/internat.html	Association for International Business
www.aernet.com/	American Export Register
www.polb.com/	Port of Long Beach
www.isbc.com/	International Small Business Consortium
www.tradeport.org/	International Trade and Export Assistance
www.ita.doc.gov/	International Trade Administration
Letter of Credit Services	
www.bankamerica.com/corp www.citibank.com www.barclays.com www.deutschebank.com www.unionbank.com www.swissbank.com	Letter of credit and other trade financing services
Tax Practices	
www.arthurandersen.com www.taxnews.com/tnn_public www.ey.com/tax www1.ey.com:81/offshore.htm www.dttus.com/dtti www.kpmg.com www.pw.com www/pw.com/uk	Tax Practices for International Business (Arthur Andersen, Coopers & Lybrand, Ernst & Young, Deloite Touche, Tohmastsu International, KPMG, Price Waterhouse, Price Waterhouse United Kingdom)
Project Financing	
www.milbank.com/projfin.html www.fccm.com www.npfi.com.hk www.ge.com/gec/index.html worldbank.org/rmcup/capintro.htm	Project Finance and Lending (Milbank, First Chicago Capital Markets, Nomura Project Finance International, General Electric Finance International, World Bank)

(*Continued*)

Web Address	Primary Focus
Central Banks and International Offshore Financial Centers	
www.investorlinks.com/directory/global.html	Links to central banks around the world
www.internationalbanking.barclays.com/	Offshore Financial Centers (Barclays Offshore Banking, Bermuda, British Virgin Islands, Cayman Islands, Guernsey, Labuan, Malta, Vanuatu)
www.bergiz.com	
www.elanbvi.com/bviaffil2.html	
www.cayman.com	
www.alexpicot.demon.co.uk/alexpicot.htm	
www.maybank.com.my/maybank.labuan.html	
www.u-net.com/metcowww	
www.vanuatu.net.vu/bdo/htm#offshore	

USEFUL WEBSITES FOR OPTIONS AND FUTURES

The following is a list designed to educate the investor on the rewards and risks associated with investing options, futures, and financial derivatives.

Web Address	Primary Focus
www.tfc-charts.w2d.com	Futures quotations
www.cboe.com	The home page for the Chicago Board Options Exchange (CBOE)
www.cbot.com	The home page for the Chicago Board of Trade
www.kcbt.com	The home page for the Kansas City Board of Trade
www.options-iri-com/options/basic/basic.htm	An excellent comprehensive site for learning about options
www.futuresmag.com/library/contents.htm	Covers various aspects of derivatives trading such as strategies and market analyses
www.ino.com	For the latest information and prices of options and financial futures as well as the corresponding historic price charts
www.adtrading.com	Applied Derivatives Trading Magazine has articles on options and other derivatives; its beginners corner is for new investors
www.pacificex.com/options	Good information on specific options, such as LEAPS and index options
www.optionscentral.com	Both education and trading material, as well as links to other option sites
www.worldlinkfutures.com/trad.htm	Provides an electronic course on futures and options for beginners
www.eftc.gov/cftc_information.htm	This commodity trading futures corporation site has information on the regulation and trading of futures
www.ahandyguide.com/cat1/f/f263/htm	Provides multiple links to a variety of web sites on futures trading
www.margil.com/mrgl101.htm	Educational resources and links on futures trading

WORLD CENTRAL BANKS

Argentina	Banco Central de la Republica Argentina
Armenia	Central Bank of Armenia
Aruba	Centrale Bank van Aruba
Australia	Reserve Bank of Australia
Austria	Oesterreichische Nationalbank
Bahrain	Bahrain Monetary Agency
Belgium	Nationale Bank van Belgie
Benin	Banque Centrale des Etats de l'Afrique de l'Ouest
Bolivia	Banco Central de Bolivia
Bosnia	Central Bank of Bosnia and Herzegovina
Brazil	Banco Central do Brasil
Bulgaria	Bulgarian National Bank
Burkina Faso	Banque Centrale des Etats de l'Afrique de l'Ouest
Canada	Bank of Canada
Chile	Banco Central de Chile
Colombia	Banco de la Republica
Costa Rica	Banco Central de Costa Rica
Côte d'Ivoire	Banque Centrale des Etats de l'Afrique de l'Ouest
Croatia	Croatian National Bank
Cyprus	Central Bank of Cyprus
Czech Republic	Ceska Narodni Banka
Denmark	Danmarks Nationalbank
East Caribbean	The East Caribbean Central Bank
Ecuador	Banco Central del Ecuador
El Salvador	The Central Reserve Bank of El Salvador
Estonia	Eesti Bank
European Union	European Central Bank
Finland	Suomen Pankki
France	Banque de France
Germany	Deutsche Bundesbank
Greece	Bank of Greece
Guatemala	Banco de Guatemala
Guinea Bissau	Banque Centrale des Etats de l'Afrique de l'Ouest
Hong Kong	Hong Kong Monetary Authority
Hungary	National Bank of Hungary
Iceland	Central Bank of Iceland
India	Reserve Bank of India
Indonesia	Bank of Indonesia
Ireland	Central Bank of Ireland
Israel	Bank of Israel
Italy	Banca d'Italia
Jamaica	Bank of Jamaica
Japan	Bank of Japan
Jordan	Central Bank of Jordan
Kenya	Central Bank of Kenya
Korea	Bank of Korea
Kuwait	Central Bank of Kuwait
Latvia	Bank of Latvia
Lebanon	Banque du Liban
Lithuania	Lietuvos Bankas
Luxembourg	Banque Centrale du Luxembourg
Macedonia	National Bank of the Republic of Macedonia

Malaysia	Bank Negara Malaysia
Mali	Banque Centrale des Etats de l'Afrique de l'Ouest
Malta	Central Bank of Malta
Mauritius	Bank of Mauritius
Mexico	Banco de Mexico
Moldova	The National Bank of Moldova
Mozambique	Bank of Mozambique
Netherlands	De Nederlandsche Bank
Netherlands	Bank van de Nederlandse Antillen
New Zealand	Reserve Bank of New Zealand
Niger	Banque Centrale des Etats de l'Afrique de l'Ouest
Norway	Norges Bank
Peru	Banco Central de Reserva del Peru
Poland	National Bank of Poland
Portugal	Banco de Portugal
Qatar	Qatar Central Bank
Russia	Central Bank of Russia
Senegal	Banque Centrale des Etats de l'Afrique de l'Ouest
Singapore	Monetary Authority of Singapore
Slovakia	National Bank of Slovakia
Slovenia	Bank of Slovenia
South Africa	The South African Reserve Bank
Spain	Banco de España
Sri Lanka	Central Bank of Sri Lanka
Sweden	Sveriges Riksbank
Switzerland	Schweizerische Nationalbank
Tanzania	Bank of Tanzania
Thailand	Bank of Thailand
Togo	Banque Centrale des Etats de l'Afrique de l'Ouest
Tunisia	Banque Centrale de Tunisie
Turkey	Türkiye Cumhuriyet Merkez Bankasi
Ukraine	National Bank of Ukraine
United Kingdom	Bank of England
United States	The Federal Reserve
Zambia	Bank of Zambia
Zimbabwe	Bank of Zimbabwe

MONETARY UNITS

Country, Island, or Territory	Currency	Symbol
Afghanistan	afghani	Af
Albania	lek	L
Algeria	dinar	DA
American Samoa	dollar	$
Angola	kwanza	Kz
Anguilla	dollar	EC$
Antarctica	krone	NKr
Argentina	peso	$
Australia	dollar	A$
Austria	schilling	S
Bahamas	dollar	B$
Bahrain	dinar	BD
Bangladesh	taka	Tk
Barbados	dollar	Bds$
Belgium	franc	BF
Belize	dollar	BZ$
Benin	franc	CFAF
Bhutan	ngultrum	Nu
Bolivia	boliviano	BS
Botswana	pula	P
Bouvet Island	krone	NKr
Brazil	real	R$
British Indian Ocean Territory	rupee	Mau Rs
British Virgin Islands	dollar or pound	$ or £
Brunei	ringitt	B$
Bulgaria	leva	Lv
Burkina Faso	franc	CFAF
Burundi	franc	FBu
Cameroon	franc	CFAF
Canada	dollar	Can$
Canton and Enderbury Islands	dollar	A$
Cape Verde Island	escudo	C.V.Esc.
Central African Republic	franc	CFAF
Chad	franc	CFAF
Chile	peso	Ch$
China	yuan	Y
Christmas Island	dollar	A$
Cocos (Keeling) Islands	dollar	A$
Colombia	peso	Col$
Comoros	franc	CF
Congo	franc	CFAF
Cook Islands	dollar	NZ$
Costa Rica	colon	slashed C
Cyprus	pound	£C
Czech Republic	koruna	Kĉ
Denmark	krone	Dkr
Djibouti	franc	DF
Dominica	dollar	EC$
Dominican Rep.	peso	RD$
Dronning Maud Land	krone	NKr

Country, Island, or Territory	Currency	Symbol
Ecuador	sucre	S/
Egypt	pound	£E
El Salvador	colon	¢
Equatorial Guinea	franc	CFAF
Ethiopia	birr	Br
European Union	euro	€
Färö Islands	krone	Dkr
Falkland Islands	pound	£F
Fiji	dollar	F$
Finland	markka	mk
France	franc	F
French Guiana	franc	F
French Polynesia	franc	CFPF
Gabon	franc	CFAF
Gambia	dalasi	D
Germany	deutsche mark	DM
Ghana	cedi	¢
Gibraltar	pound	£G
Greece	drachma	Dr
Greenland	krone	Dkr
Grenada	dollar	EC$
Guadeloupe	franc	F
Guam	dollar	$
Guatemala	quetzal	Q
Guinea-Bissau	franc	CFAF
Guinea	syli	FG
Guyana	dollar	G$
Haiti	gourde	G
Heard and McDonald Islands	dollar	A$
Honduras	lempira	L
Hong Kong	dollar	HK$
Hungary	forint	Ft
Iceland	króna	IKr
India	rupee	Rs
Indonesia	rupiah	Rp
Iran	rial	Rls
Iraq	dinar	ID
Ireland	pound or punt	IR£
Israel	new shekel	NIS
Italy	lira	Lit
Ivory Coast	franc	CFAF
Jamaica	dollar	J$
Japan	yen	¥
Johnston Island	dollar	$
Jordan	dinar	JD
Kampuchea	new riel	CR
Kenya	shilling	K Sh
Kiribati	dollar	A$
Korea, North	won	Wn
Korea, South	won	W

(*Continued*)

Country, Island, or Territory	Currency	Symbol
Kuwait	dinar	KD
Laos	new kip	KN
Latvia	lat	Ls
Lesotho	loti (pl., maloti)	L (pl., M)
Liberia	dollar	$
Libya	dinar	LD
Liechtenstein	franc	SwF
Luxembourg	franc	LuxF
Macao	pataca	P
Madagascar	franc	FMG
Malawi	kwacha	MK
Malaysia	ringgit	RM
Maldives	rufiyaa	Rf
Mali	franc	CFAF
Malta	lira	Lm
Martinique	franc	F
Mauritania	ouguiya	UM
Mauritius	rupee	Mau Rs
Midway Islands	dollar	$
Mexico	peso	Mex$
Monaco	franc	F
Mongolia	tugrik	Tug
Montserrat	dollar	EC$
Morocco	dirham	DH
Mozambique	metical	Mt
Myanmar	kyat	K
Nauru	dollar	A$
Namibia	dollar	$
Nepal	rupee	NRs
Netherlands	guilder	f.
Netherlands Antilles	guilder	Ant.f.
New Caledonia	franc	CFPF
New Zealand	dollar	NZ$
Nicaragua	gold cordoba	C$
Niger	franc	CFAF
Nigeria	naira	double-dashed N
Niue	dollar	NZ$
Norfolk Island	dollar	A$
Norway	krone	NKr
Oman	rial	RO
Pakistan	rupee	Rs
Panama	balboa	B
Panama Canal Zone	dollar	$
Papua New Guinea	kina	K
Paraguay	guarani	slashed G
Peru	new sol	S/.
Philippines	peso	dashed P
Pitcairns Island	dollar	NZ$
Poland	zloty	z dashed l
Portugal	escudo	Esc
Puerto Rico	dollar	$
Qatar	riyal	QR
Reunion	franc	F

Country, Island, or Territory	Currency	Symbol
Romania	leu	L
San Marino	lira	Lit
Saudi Arabia	riyal	SRIS
Seychelles	rupee	SR
Sierra Leone	leone	Le
Singapore	dollar	S$
Slovakia	koruna	Sk
Solomon Island	dollar	SI$
Somalia	shilling	So. Sh.
South Africa	rand	R
Spain	peseta	Ptas
Sri Lanka	rupee	SLRs
St. Kitts and Nevis	dollar	EC$
St. Lucia	dollar	EC$
St. Vincent and Grenada	dollar	EC$
Sudan	pound	LSd
Suriname	guilder	Sur.f.
Swaziland	ilangeni (pl., emalangeni)	L (pl., E)
Sweden	krona	Sk
Switzerland	franc	SwF
Syria	pound	£S
Taiwan	new dollar	T$
Tanzania	shilling	TSh
Thailand	baht	Bht or Bt
Togo	franc	CFAF
Tokelau	dollar	NZ$
Tonga	pa'anga	PT
Trinidad and Tobago	dollar	TT$
Tunisia	dinar	D
Turkey	lira	TL
Turks and Caicos Islands	dollar	$
Tuvalu	dollar	A$
Uganda	shilling	USh
United Arab Emeriates	dirham	Dh
United Kingdom	pound	£
United States of America	dollar	$
Uruguay	peso uruguayo	$U
Vanuatu	vatu	VT
Vatican	lira	Lit
Venezuela	bolivar	Bs
Vietnam	new dong	D
Virgin Islands	dollar	$
Wake Island	dollar	$
Wallis and Futuna Islands	franc	CFPF
Western Sahara	peseta	Ptas
Western Samoa	tala	WS$
Yemen	rial	YRls
Yugoslavia	dinar	Din
Zambia	kwacha	K
Zimbabwe	dollar	Z$

CHICAGO MERCANTILE EXCHANGE (CME) CONTRACT SPECIFICATIONS FOR CURRENCY FUTURES AND OPTIONS

(www.cme.com/clearing/spex/cscurrency.htm)

Commodity Size	Hours*	Months	Codes Clr/Tick	Minimum Fluctuation in Price	Limit	Strike Price Interval/Notes
Australian Dollar 100,000 Australian Dollars	RTH: 7:20 a.m.–2:00 p.m. (9:16 a.m.)^ GLOBEX2: 2:30 p.m.–7:05 a.m. Mon–Thurs 5:30 p.m.–7:05 a.m. Sun	Mar, Jun, Sep, Dec GLOBEX2: First six quarterly months	AD/AD	0.0001 (1 pt) ($10.00/pt) ($10.00) Certain transactions: 0.00005 (1/2 pt = $5.00)	NO LIMIT BETWEEN 7:20 a.m.–7:35 a.m. [RTH] Expanding limits: See rule 3004 GLOBEX2: 400 points	See note +++,**
Australian Dollar Options	RTH: 7:20 a.m.–2:00 p.m. (2:00 p.m.)^	Quarterly, serial months & weekly expiration options +	AD/KA Calls AD/JA Puts Weekly Exp. Options: 1AC/5AC Calls 1AP/5AP Puts	0.0001 (1 pt) ($10.00/pt) ($10.00) cab = $5.00 0.00005 (1/2 pt) when: • Premium <5 ticks • Spreads w/net premium of <5 tick value • Nongeneric currency option combinations with total premuim of <10 tick value • Weekly options	Option ceases trading when corresponding futures locks limit	$/AD $0.01 intervals e.g., $0.76, $0.77, plus .005 intervals for the first 7 listed expirations, e.g., $0.775, $0.780

Commodity Size	Hours*	Months	Codes Clr/Tick	Minimum Fluctuation in Price	Limit	Strike Price Interval/Notes
Brazilian Real 100,000 Real	RTH: 7:20 a.m.–2:00 p.m. (2:00 p.m.)^ GLOBEX2: 2:30 p.m.–7:05 a.m. Mon–Thurs 5:30 p.m.–7:05 a.m. Sun	All twelve calendar months GLOBEX2: First four consecutive calendar months and the next two months in the March quarterly cycle	BR/BR	.0005 (1/2 pt) ($10.00/pt) ($5.00) Certain transactions: 0.00005 (1/2 pt = $5.00)	Expanding limits do not apply to first and second contract months GLOBEX2: Trading halted when primary futures contract is locked at 0.2000 above or below the Reference RTH Price. GLOBEX2 price limit applies to all contract months	N/A
Brazilian Real Options	RTH: 7:20 a.m.–2:00 p.m. (2:00)^	All twelve calendar months & weekly expiration options	BRC Calls BRP Puts Weekly Exp. Options: 1RC/5RC Calls 1RP/5RP Puts	0.0001 (1 pt) ($10.00/pt) ($10.00) cab = $5.00 0.00005 (1/2 pt) when: • Premium <5 ticks • Spreads w/net premium of <5 tick value • Nongeneric currency option combinations with total premuim of <10 tick value • Weekly options	Option ceases trading when corresponding futures locks limit except for options where the underlying futures contract of the option is not subject to currency price limits	$/BR 0.01 intervals, e.g., 1.0600, 1.0700, 1.0800 and for the first 7 listed expirations only additional strikes at $0.005 intervals, e.g., $1.08500, $1.09000 See notes **,****, ++++
British Pound 62,500 British Pounds	RTH: 7:20 a.m.–2:00 p.m. (9:16)^ GLOBEX2: 2:30 p.m.–7:05 a.m. Mon– Thurs 5:30 p.m.–7:05 a.m. Sun	Mar, Jun, Sep, Dec. GLOBEX2: First six March quarterly months	BP/BP	0.0002 (2 pt) ($6.25/pt) ($12.50) Certain transactions: 0.0001 (1 pt =$6.25)	NO LIMIT BETWEEN 7:20 a.m.–7:35 a.m. [RTH] Expanding limits: See rule 3004 GLOBEX2: 800 points	See notes +++,**

(*Continued*)

Commodity Size	Hours*	Months	Codes Clr/Tick	Minimum Fluctuation in Price	Limit	Strike Price Interval/Notes
British Pound Options	RTH: 7:20 a.m.–2:00 p.m. (2:00 p.m.)^ GLOBEX2: 2:30 p.m.–7:05 a.m. Mon–Thurs 5:30 p.m.–7:05 a.m. Sun	Quarterly serial months & weekly expiration options + GLOBEX2: One quarterly & two serial months	BP/CP Calls BP/PP Puts Weekly Exp. Options: 1BC/5BC Calls 1BP/5BP Puts	.0002 (2 pt) ($6.25/pt) ($12.50) cab = $6.25 0.0001 (1 pt) when: • Premium <5 ticks • Spreads w/net premium of <5 tick value • Nongeneric currency option combinations with total premuim of <10 tick value • Weekly options	Option ceases trading when corresponding futures lock limit GLOBEX2: Same	$/BP $0.020 intervals e.g., $1.440, $1.460, $1.480, etc., plus strikes at $0.01 intervals for the first 7 listed expirations e.g., 1.43, 1.44, 1.45, etc. See note**
Canadian Dollar 100,000 Canadian Dollars	RTH: 7:20 a.m.–2:00 p.m. (9:16)^ GLOBEX2: 2:30 p.m.–7:05 a.m. Mon–Thurs 5:30 p.m.– :05 a.m. Sun	Mar, Jun, Sep, Dec GLOBEX2: First six March quarterly months	C1/CD	0.0001 (1 pt) ($10.00/pt) ($10.00) Certain transactions: 0.00005 (1/2 pt = $5.00)	NO LIMIT BETWEEN 7:20 a.m.– 7:35 a.m. [RTH] Expanding limits: See rule 3004 GLOBEX2: 400 points	See notes +++,**

Commodity Size	Hours*	Months	Codes Clr/Tick	Minimum Fluctuation in Price	Limit	Strike Price Interval/Notes
Canadian Dollar Options	RTH: 7:20 a.m.–2:00 p.m. (2:00 p.m.)^ GLOBEX2: 2:30 p.m.–7:05 a.m. Mon–Thurs 5:30 p.m.–7:05 a.m. Sun	Quarterly Serial months & weekly expiration options + GLOBEX2: One quarterly & two serial months	C1/CV Calls C1/PV Puts	0.0001 (1 pt) ($10.00/pt) ($10.00) cab = $5.00 0.00005 (1/2 pt) when: • Premium <5 ticks • Spreads w/net premium of <5 tick value • Nongeneric currency option combinations with total premuim of <10 tick value • Weekly options	Option ceases trading when corresponding futures locks limit GLOBEX2: Same	$/CD $0.005 intervals e.g., $0.800, $0.805 See note **
Deutsche Mark 125,000 Deutsche Marks	RTH: 7:20 a.m.–2:00p.m. (9:16 a.m.)^ GLOBEX2: 2:30p.m.–2:00 p.m. next day Mon–Thurs 5:30 p.m.–2:00 p.m. next day Sun & Holidays	Mar, Jun, Sep, Dec GLOBEX2: Frist six March quarterly months	D1/DM	0.0001 (1 pt) ($12.50/pt) ($12.50) Certain transactions: 0.00005 (1/2 pt = $6.25)	NO LIMIT BETWEEN 7:20 a.m.–7:35 a.m. [RTH] Expanding limits: See rule 3004 GLOBEX2: 400 points	See notes +++,**
Deutsche Mark Options	RTH: 7:20 a.m.–2:00 p.m. (2:00 p.m.)^ GLOBEX2: 2:30 p.m.–7:05 a.m. next day Mon–Thurs	Quarterly serial months & weekly expiration options + GLOBEX2: One quarterly & two serial months	D1/CM Calls D1/PM Puts Weekly Exp. Options: 1DC/5DC Calls 1DP/5DP Puts	0.0001 (1 pt) ($12.50/pt) ($12.50) cab = $6.25 0.00005 when: • Premium <5 ticks • Spreads w/net premium of <5 tick value • Nongeneric currency option combinations with total premuim of <10 tick value • Weekly options	Option ceases trading when corresponding futures locks limit GLOBEX2: Same	$/DM $0.01 intervals, e.g., $0.63, $0.64 plus $0.005 intervals for first 7 listed expirations e.g., $0.635, $0.640. See note **

(Continued)

Commodity Size	Hours*	Months	Codes Clr/Tick	Minimum Fluctuation in Price	Limit	Strike Price Interval/Notes
EuroFX 125,000 Euros	RTH: 7:20 a.m.–2:00 p.m. (9:16 a.m.)^ GLOBEX2: 2:30 p.m.– 7:05 a.m. Mon–Thurs 5:30 p.m.– 7:05 a.m. Sun	Mar, Jun, Sep, Dec GLOBEX2: First six March quarterly months	EC/EC	0.0001 (1 pt) ($12.50/pt) ($12.50) Certain transactions: 0.00005 (1/2 pt = $6.25)	NO LIMIT BETWEEN 7:20 a.m.– 7:35 a.m. [RTH] and 1:45 p.m.– 2:00 p.m. [RTH] Expanding limits: See rule +++.**	
Euro FX Options	RTH: 7:20 a.m.–2:00 p.m. (2:00 p.m.)^ GLOBEX2: 2:30 p.m.– 7:05 a.m. Mon–Thurs 5:30 p.m.– 7:05 a.m. Sun	Quarterly serial months & weekly expiration options + GLOBEX2: March quarterly cycle	EC/EC Calls EC/EP Puts Weekly Exp. Options: 1X/2X/3X/ 4X/5X	0.0001 (1 pt) ($12.50/pt) ($12.50) cab = $6.25 For transactions at <5 ticks 0.00005 ($6.25)	Option ceases trading when corresponding futures locks limit GLOBEX 2: Same	$0.01 per Euro, e.g., $1.0500, $1.0600, $1.0700, etc., and for the first 7 listed option expirations only, additional strike prices at intervals of $0.005, e.g., $1.055, $1.065, $1.075, etc.
E-Mini Euro 62,500 Euros	GLOBEX 2: (9:16 a.m.)^ 2:30 p.m.– 2:00 p.m. Next day Mon–Thurs 5:30 p.m.– 2:00 p.m. Sun & Most Holidays	Mar, Jun, Sept, Dec.	E7/E7	0.0001 (1 pt) ($6.25/pt) $6.25 0.00005 (1/2 tick) ($3.125) for intracurrency spread transactions	$.1600/Euro	N/A
E-Mini Euro Option Launch:TBD	TBD	TBD	E7 Puts E7 Calls	0.0001 ($6.25/pt) ($6.25) Cab = $3.125	N/A	$0.01 per Euro, e.g., $1.0500, $1.0600, $1.0700 and for the first 7 listed option expirations only, additional strike prices at intervals of $0.005, e.g., $1.055, $1.065, $1.075

Commodity Size	Hours*	Months	Codes Clr/Tick	Minimum Fluctuation in Price	Limit	Strike Price Interval/Notes
French Francs 500,000 French Francs	RTH: 7:20 a.m.–2:00 p.m. (9:16 a.m.)^ GLOBEX2: 2:30 p.m.– 7:05 a.m. Next day Mon–Thurs 5:30p.m.– 7:05 a.m. Sun	Mar, Jun, Sep, Dec GLOBEX 2: First six March quarterly months	FR/FR	.00002 (2 pt) ($5.00/pt) ($10.00) Certain transactions: 0.00001 (1 pt = $5.00)	NO LIMIT BETWEEN 7:20 a.m.–7:35 a.m. [RTH] Expanding limits: See rule 3004 GLOBEX2: 1000 points	See notes +++,**
French Francs Options	RTH: 7:20 a.m. –2:00 p.m. (2:00 p.m.)^ GLOBEX2: 2:30 p.m. –7:05 a.m. Next day Mon–Thurs 5:30 p.m.– 7:05 a.m. Sun	Quarterly serial months & weekly expiration options +	FR/1F Calls FR/1F Puts Weekly Exp. Options: 1L/5L Calls 1L/5L2 Puts	.00002 (2 pt) ($5.00/pt) ($10.00) cab = $5.00 when: • Premium <5 ticks • Spreads w/net premium of <5 tick value • Nongeneric currency option combinations with total premuim of <10 tick value • Weekly options	Option ceases trading when corresponding futures are limit bid/offered	US$ per French Franc 0.00250 intervals, e.g., 0.18000, 0.18250, 0.18500 See note **
Japanese Yen 12,500,000 Japanese Yen	RTH: 7:20 a.m.–2:00 p.m. (9:16 a.m.)^ GLOBEX2: 2:30 p.m. –7:05 a.m. Mon–Thurs 5:30 p.m. –7:05 a.m. Sun	Mar, Jun, Sep, Dec GLOBEX2: First six March quarterly months	J1/JY	0.000001 (1 pt) ($12.50/pt) ($12.50) Certain transactions: 0.000005 (1/2 pt = $6.25)	NO LIMIT BETWEEN 7:20 a.m.– 7:35 a.m. [RTH] Expanding limits: See rule 3004 GLOBEX2: 400 points	See notes +++,**

(*Continued*)

Commodity Size	Hours*	Months	Codes Clr/Tick	Minimum Fluctuation in Price	Limit	Strike Price Interval/Notes
Japanese Yen Options	RTH: 7:20 a.m.–2:00 p.m. (2:00 p.m.)^ GLOBEX2: 2:30 p.m. –7:05 a.m. Mon–Thurs 5:30 p.m.– 7:05 a.m. Sun	Quarterly serial months & weekly expiration options +	J1/CJ Calls J1/PJ Puts Weekly Exp. Options: 1JC/5JC Calls 1JC/5JP Puts	0.000001 (1 pt) ($12.50/pt) ($12.50) cab = $6.25 0.0000005 (1/2 pt) when: • Premium <5 ticks • Spreads w/net premium of <5 tick value • Nongeneric currency option combinations with total premuim of <10 tick value • Weekly options	Option ceases trading when corresponding futures locks limit GLOBEX2: Same	$/JY $0.0001 intervals, e.g., $0.0072, $0.0071 plus $0.00005 intervals for the first 7 listed expirations, e.g., $0.00725, $0.00730 See note**
E-Mini Japanese Yen 6,250,000 Yen	2:30 p.m.– 2:00 p.m. Next day Mon–Thurs (9:16 a.m.)^ 5:30 p.m.– 2:00 p.m. next day Sun & Most Holidays	Mar, Jun, Sep, Dec	J7/J7	.000001 (1pt) ($6.25/pt) ($6.25) .0000005 (1/2 tick) for intracurrency spread transactions	$.0008/Yen	N/A
E-Mini Japanese Yen Options Launch: TBD	TBD	TBD	J7/J7 Calls J7/J7 Puts	.000001 (1 pt) ($6.25/pt) ($6.25) cab = $3.125	N/A	$.0001/¥ e.g., $.0093, $.0094 Plus $.00005/¥ intervals for the first 7 listed expirations, e.g., $.00935, $.0094
Mexican Peso 500,000 New Mexican Pesos	RTH: 7:20 a.m.–2:00 p.m. (9:16 a.m.)^ GLOBEX2: 2:30 p.m.– 7:05 a.m. Mon–Thurs 5:30 p.m.– 7:05 a.m. Sun	Mar, Jun, Sep, Dec GLOBEX2: Same	MP/MP	0.000025 (2 1/2 pts) ($5.00/pt) ($12.50)	Expanding limits: See rule 3004 GLOBEX2: Trading halted when primary futures contract is locked at 0.02000 above or below the Reference RTH	N/A

Commodity Size	Hours*	Months	Codes Clr/Tick	Minimum Fluctuation in Price	Limit	Strike Price Interval/Notes
Mexican Peso Options	RTH: 7:20 a.m.–2:00 p.m. (2:00 p.m.)^	Quarterly serial months & weekly expiration options +	MPC/Calls MPP/Puts Weekly Exp. Options: 1MC/5MC Calls 1MP/5MP Puts	0.000025 (2 1/2 pts) ($5.00/pt) ($12.50) cab = $6.25	Options ceases trading when corresponding future locks limit	$/MP $0.01 intervals, e.g., $0.12, $0.13, $0.14 See notes **, #
New Zealand Dollar 100,000 New Zealand Dollars	RTH: 7:20 a.m.–2:00 p.m. (9:16 a.m.)^ GLOBEX2: 2:30 p.m.–7:05 a.m. Mon–Thurs 5:30 p.m.–7:05 a.m. Sun	Mar, Jun, Sep, Dec GLOBEX2: First six March quarterly months	NE/NE	0.0001 (1 pt) ($10.00/pt) ($10.00) Certain transactions: 0.00005 (1/2 pt = $5.00)	NO LIMIT BETWEEN 7:20 a.m.–7:35 a.m. [RTH] Expanding limits: See rule 3004 GLOBEX2: 500 points	See notes +++,**
New Zealand Dollar Options	RTH: 7:20 a.m.–2:00 p.m. (2:00 p.m.)^	Quarterly serial months & weekly expiration options +	NE/NE Calls NE/NE Puts Weekly Exp. Options: 1ZC/5ZC Calls 1ZC/ 5ZP Puts	0.0001 (1 pt) ($10.00/pt) ($10.00) cab = $5.00 0.00005 (1/2 pt) when: • Premium <5 ticks • Spreads w/net premium of <5 tick value • Nongeneric currency option combinations with total premuim of <10 tick value • Weekly options	Options ceases trading when corresponding futures locks limit	$0.01 per New Zealand Dollar, e.g., $0.70, $0.71, $0.72 and for the first 7 listed options expirations only, additional strikes at intervals of $0.005, e.g., $0.705, $0.715, $0.725
Russian Ruble 500,000 Russian Ruble	RTH: 7:20 a.m.–2:00 p.m. (11:00 a.m. Moscow Time)^ GLOBEX2: 2:30 p.m.–7:05 a.m. Mon–Thurs 5:30 p.m.–7:05 a.m. Sun	Mar, Jun, Sep, Dec GLOBEX2: First six March quarterly months	RU/RU	0.000025 (2 1/2 pts) ($5.00/pt) ($12.50)	NO LIMIT BETWEEN 7:20 a.m.–7:35 a.m. [RTH] Expanding limits: See rule 3004 GLOBEX2: $.02000 per Russian Ruble	See notes +++,**

(Continued)

Commodity Size	Hours*	Months	Codes Clr/Tick	Minimum Fluctuation in Price	Limit	Strike Price Interval/Notes
Russian Ruble Options	RTH: 7:20 a.m.–2:00 p.m. (2:00 p.m.)^ (2 Serial Mos. & Weeklies) (See GLOBEX2)^^^	Quarterly, serial months & weekly expiration options + Mar, Jun, Sep, Oct, Nov, Dec	RU/RU Calls RU/RU Puts Weekly Exp. Options: 1U,2U,3U, 4U,5U	0.000025 (2 1/2 pts) ($5.00/pt) ($12.50) cab = $6.25	Options ceases trading at the same time and date as underlying futures	$0.005 per Russian Ruble, e.g., $.165, $.175, $.185
South African Rand 500,000 South African Rand	RTH: 7:20 a.m.–2:00 p.m. (9:16 a.m.)^ GLOBEX2: 2:30 p.m.–7:05 a.m. Mon–Thurs 5:30 p.m.–7:05 a.m. Sun	All 12 calendar months GLOBEX2: First 13 consecutive calendar months plus the next two March quarterly cycle months	RA/RA	0.000025 (2 1/2 pts) ($5.00/pt) ($12.50)	NO LIMIT BETWEEN 7:20 a.m.–7:35 a.m. or 1:45 p.m. to 2:00 p.m. [RTH] Expanding limits: See rule 3004 GLOBEX2: Trading halted when primary futures contract is locked at 0.02500 above or below the Reference RTH Price	N/A
South African Rand Options	RTH: 7:20 a.m.–2:00 p.m. (2:00 p.m.)^	All 12 calendar months & weekly expiration options +++++	RA/RA Calls RA/RA Puts Weekly Exp. Options: 1NC/5NC Calls 1NP/5NP Puts	0.000025 (2 1/2 pts) ($5.00/pt) ($12.50) cab = $6.25	Option ceases trading when corresponding future locks limit	$/RA at intervals of $0.00500, e.g., $0.21500, $0.22000 and for the first 7 listed options expirations only Additional strikes at intervals of $0.00250, e.g., $0.21750, $0.22250, $0.22750 See notes **,+++++
Swiss Franc 125,000 Swiss Francs	RTH: 7:20 a.m.–2:00 p.m. (9:16 a.m.)^ GLOBEX2: 2:30 p.m.–7:05 a.m. Mon–Thurs 5:30 p.m.–7:05 a.m. Sun	Mar, Jun, Sep, Dec GLOBEX2: First six March quarterly months	E1/SF	0.0001 (1 pt) ($12.50/pt) ($12.50) Certain transactions: 0.00005 (1/2 pt = $6.25)	NO LIMIT BETWEEN 7:20 a.m.–7:35 a.m. [RTH] Expanding limits: See rule 3004	See notes +++,**

Commodity Size	Hours*	Months	Codes Clr/Tick	Minimum Fluctuation in Price	Limit	Strike Price Interval/Notes
Swiss Franc Options	RTH: 7:20 a.m.–2:00 p.m. (2:00 p.m.)^ GLOBEX2: 2:32 p.m.–7:03 a.m. Mon–Thurs 5:36 p.m.–7:03 a.m. Sun	Quarterly, serial months & weekly expiration options + GLOBEX 2: One quarterly & two serial months	E1/CF Calls E1/PF Puts Weekly Exp. Options: 1SC/5SC Calls 1SP/5SP Puts	0.0001 (1 pt) ($12.50/pt) ($12.50) 0.00005 (1/2 pt) when: • Premium <5 ticks • Spreads w/net premium of <5 tick value • Nongeneric currency option combinations with total premuim of <10 tick value • Weekly options	Option ceases trading when corresponding future locks limit	$/SF $0.01, e.g., $0.72, $0.73 plus $0.005 intervals for the first 4 listed expirations, e.g., $0.725, $0.730 See note **

Notes to Contract Specifications

Options also eligible for strike listings at 1/2 the normal strike price interval under certain circumstances. Please see Rulebook.

Termination of trading for certain interest rate contracts shall be 11:00 a.m. London time on the second London business day immediately preceding the third Wednesday of the contract month. This is 5:00 a.m. Chicago time, except when Daylight Savings Time is in effect in either, but not both, London or Chicago.

The underlying futures contract for serial month options is the Eurodollar (Euroyen) futures contract in the March quarterly cycle with an expiration date most closely following the option's expiration date. For example, the December 1997 Eurodollar (Euroyen) futures contract is the underlying contract for the November 1997 options.

Spot Butter trading occurs on the last business day of the week (usually Friday), except if a holiday falls on Thursday, then trading is conducted on the business day before the holiday.

* Times in parentheses indicate close on last day of trading (Central Time). Please note that pursuant to Rules 3902.G and 3302.H, the Eurodollar and Libor contracts will close on the last day of trading at 11:00 a.m. London Time on the second London bank business day immediately preceding the third Wednesday of the contract month. This is 5:00 a.m. Chicago time except when Daylight Savings time is in effect in either, but not both London or Chicago.

** See special provisions of contract in Rulebook.

*** In the third and fourth nearest contract month of the March quarterly cycle for the following products: quarterly S&P Midcap 400, quarterly Russell 2000, quarterly Mexican Par Brady Bond, quarterly Argentine FRB Bond, quarterly Argentine Par Brady Bond, quarterly Brazilian C Bond, quarterly Brazilian EI Bond, quarterly S&P 500/BARRA Growth and quarterly S&P 500/BARRA Value, the exercise prices shall be an integer divisible by 5. In all other months, the exercise prices will be a number divisible by 2.5.

**** The underlying futures contract for Brazilian Real monthly options is the futures contract with the same contract month as the monthly option. For Brazilian Real weekly expiration options, the underlying futures contract is the nearest futures contract in the consecutive contract month cycle that has not yet terminated.

(*Continued*)

+ The underlying futures contract for both serial and quarterly options as well as for weekly expiration options is the futures contract with greater than two Exchange business days before its LDT when the options expire. Please note that a weekly option will expire during each week in which no serial or quarterly option is scheduled to expire, however, weekly currency options may be listed up to five weeks prior to their expiration.

++ In the third and fourth nearest contract month in the S&P 500 March quarterly cycle, the exercise prices, with the exception of quarter point strike prices, shall be an integer divisible by 10. In all other months, the exercise prices will be an integer divisible by 5.

+++ Currency futures trades may also occur during RTH in 1/2 tick increments for intracommodity spreads executed as simultaneous transactions pursuant to Rule 542.A.

++++ The One-Month LIBOR, Feeder Cattle, quarterly S&P 500, quarterly E-Mini S&P 500, quarterly S&P 500/BARRA Growth Index, quarterly S&P 500/BARRA Value Index, quarterly Midcap 400, quarterly E-mini S&P 500, quarterly Nikkei 225, quarterly NASDAQ 100 Index, quarterly IPC Index, quarterly Eurodollar, quarterly Euroyen, quarterly FT-SE 100, monthly Brazilian Real, quarterly Mexican Par Brady Bond, quarterly Argentine FRB Bond, quarterly Argentine Par Brady Bond, quarterly Brazilian C Bond, quarterly Brazilian EI Bond, and the monthly Goldman Sachs contracts are cash-settled on the settlement day.

+++++For monthly options, the underlying futures contract is the same as the option contract month. For weekly expiration options that expire before the monthly option, the underlying futures contract is the same as the option contract month. For weekly options which expire after the monthly options, the underlying futures contract is the futures contract of the next consecutive calendar month.

^ Times in parentheses indicate close on last day of trading (Central Time). RTH = REGULAR TRADING HOURS.

^^ The Reopening Process: Once trading in the primary securities market resumes after a trading halt, trading in CME domestic Stock Index futures contracts shall resume only after 50 percent of the stocks underlying the S&P 500 Stock Price Index (selected according to capitalization weights) have reopened.

^^^ Times in parentheses indicate close on last day of trading (Central Time). Please note that pursuant to Rule 3025.E, the contract will close on the last day of trading at 2:00 A.M. (Chicago Time), but may be either 1:00 a.m. or 3:00 a.m. (Chicago Time) when Daylight Savings Time is in effect in either, but not both, Moscow or Chicago.

^^^^ Flex® Contracts are not traded within 10 minutes of the daily opening and within 30 minutes of the daily close.

PHILADELPHIA STOCK EXCHANGE CURRENCY OPTIONS SPECIFICATIONS

(www.phlx.com/products/standard.html)

	Australian Dollar	British Pound	Canadian Dollar	Deutsche Mark	Euro	Japanese Yen	Swiss Franc
TICKER SYMBOLS (1) *(American/ European)*							
Mid-month Options	XAD/CAD	XBP/CBP	XCD/CCD	XDM/CDM	XEU/ECU	XJY/CJY	XSF/CSF
Half-point Strike (Three near-term months only)	XAZ/CAZ	n.a./n.a.	n.a./n.a.	XDZ/CDZ	n.a./n.a.	XJZ/CJZ	XSZ/CSZ
Month-end Options (2)	ADW/EDA	BPW/EPO	CDW/ECD	DMW/EDM	XEW/ECW	JYW/EJY	SFW/ESW
Half-point Strike	AZW/EAW	n.a./n.a.	n.a./n.a.	DMZ/EDZ	n.a./n.a.	JYZ/EJZ	SFZ/ESZ
Long-term Options							
13 to 24 Months	n.a.	n.a./YPX	n.a.	n.a./YDM	n.a.	n.a./YJY	n.a.
CONTRACT SIZE	50,000	31,250	50,000	62,500	62,500	6,250,000	62,500
POSITION & EXERCISE LIMITS (3)	200,000	200,000	200,000	200,000	200,000	200,000	200,000
BASE CURRENCY	USD	USD	USD	USD	USD	USD	USD
UNDERLYING CURRENCY	AUD	GBP	CAD	DEM	EURO	JPY	CHF
EXERCISE PRICE INTERVALS							
Three Nearest Months	1¢	1¢	.5¢	.5¢	2¢	.005¢	.5¢
6, 9, and 12 Months	1¢	2¢	.5¢	1¢	2¢	.01¢	1¢
Over 12 Months	n.a.	4¢	n.a.	2¢	n.a.	.02¢	n.a.
PREMIUM QUOTATIONS	Cents per unit	Cents per unit	Cents per unit	Cents per unit	Cents per unit	Hundredths of a cent per unit	Cents per unit
MINIMUM PREMIUM CHANGE	\$.(00)01 per unit = \$5.00	\$.(00)01 per unit = \$3.125	\$.(00)01 per unit = \$5.00	\$.(00)01 per unit = \$6.25	\$.(00)01 per unit = \$6.25	\$.(0000)01 per unit = \$6.25	\$.(00)01 per unit = \$6.25
MARGIN	USD	USD	USD	USD	USD	USD	USD

* The Philadelphia Stock Exchange (PHLX) also trades currency options with customizable contract terms.

Expiration Months

Mid-month Options: March, June, September, and December + two near-term months

Month-end Options: Three nearest months

Long-term Options (2): 18 and 24 months (June and December)

Expiration Date/(4) Last Trading Day

Providing it is a business day, otherwise the day immediately prior

Mid-month Options: Friday before the third Wednesday of expiring month

Month-end Options: Last Friday of the month

Long-term Options: Friday before the third Wednesday of expiring month

Expiration Settlement Date (4)

Mid-month Options: Third Thursday of expiring month, except for March, June, September, and December expirations which are the third Wednesday.

Month-end Options: Thursday following the last Friday of the month.

Long-term Options: Third Thursday of expiration month, except for March, June, September and December expirations which are the third Wednesday.

Exercise Style

Mid-month Options:

Dollar Based: American and European

Cross Rates: European

Month-end Options:

Dollar Based: American and European

Cross Rates: European

Long-term Options:

European

Trading Hours

2:30 a.m. to 2:30 p.m. Philadelphia time, Monday through Friday.

Trading hours for Canadian dollar are 7:00 a.m. to 2:30 p.m. Philadelphia time, Monday through Friday.

Issuer and Guarantor

The Options Clearing Corporation (OCC)

(1) Fluctuations in the underlying may cause a "wrap around" situation whereby alternate symbols may be added. Contact the PHLX for information on alternate symbols that may be in use.

(2) When a Long-term Option has only 12 months remaining until expiration, it converts to the one year European-style Mid-month Option. Also the PHLX may not have listed all Long-term Options for all currencies or all expiration months. Contact the PHLX for information on which Long-term Options are available.

(3) For purposes of Position Limits, Standardized Contracts are aggregated with Customized Contracts.

(4) Changes may occur due to holidays. Please contact the PHLX for more information.

TABLE 1
Future Value of $1

Periods	4%	6%	8%	10%	12%	14%	20%
1	1.040	1.060	1.080	1.100	1.120	1.140	1.200
2	1.082	1.124	1.166	1.210	1.254	1.300	1.440
3	1.125	1.191	1.260	1.331	1.405	1.482	1.728
4	1.170	1.263	1.361	1.464	1.574	1.689	2.074
5	1.217	1.338	1.469	1.611	1.762	1.925	2.488
6	1.265	1.419	1.587	1.772	1.974	2.195	2.986
7	1.316	1.504	1.714	1.949	2.211	2.502	3.583
8	1.369	1.594	1.851	2.144	2.476	2.853	4.300
9	1.423	1.690	1.999	2.359	2.773	3.252	5.160
10	1.480	1.791	2.159	2.594	3.106	3.707	6.192
11	1.540	1.898	2.332	2.853	3.479	4.226	7.430
12	1.601	2.012	2.518	3.139	3.896	4.818	8.916
13	1.665	2.133	2.720	3.452	4.364	5.492	10.699
14	1.732	2.261	2.937	3.798	4.887	6.261	12.839
15	1.801	2.397	3.172	4.177	5.474	7.138	15.407
20	2.191	3.207	4.661	6.728	9.646	13.743	38.338
30	3.243	5.744	10.063	17.450	29.960	50.950	237.380
40	4.801	10.286	21.725	45.260	93.051	188.880	1469.800

TABLE 2
Future Value of an Annuity of $1

Periods	4%	6%	8%	10%	12%	14%	20%
1	1.000	1.000	1.000	1.000	1.000	1.000	1.000
2	2.040	2.060	2.080	2.100	2.120	2.140	2.220
3	3.122	3.184	3.246	3.310	3.374	3.440	3.640
4	4.247	4.375	4.506	4.641	4.779	4.921	5.368
5	5.416	5.637	5.867	6.105	6.353	6.610	7.442
6	6.633	6.975	7.336	7.716	8.115	8.536	9.930
7	7.898	8.394	8.923	9.487	10.089	10.730	12.916
8	9.214	9.898	10.637	11.436	12.300	13.233	16.499
9	10.583	11.491	12.488	13.580	14.776	16.085	20.799
10	12.006	13.181	14.487	15.938	17.549	19.337	25.959
11	13.486	14.972	16.646	18.531	20.655	23.045	32.150
12	15.026	16.870	18.977	21.395	24.133	27.271	39.580
13	16.627	18.882	21.495	24.523	28.029	32.089	48.497
14	18.292	21.015	24.215	27.976	32.393	37.581	59.196
15	20.024	23.276	27.152	31.773	37.280	43.842	72.035
20	29.778	36.778	45.762	57.276	75.052	91.025	186.690
30	56.085	79.058	113.283	164.496	241.330	356.790	1181.900
40	95.026	154.762	259.057	442.597	767.090	1342.000	7343.900

TABLE 3
Present Value of $1

Periods	4%	5%	6%	8%	10%	12%	14%	16%	18%	20%	22%	24%	26%	28%	30%
1	0.962	0.952	0.943	0.926	0.909	0.893	0.877	0.862	0.847	0.833	0.820	0.806	0.794	0.781	0.769
2	0.925	0.907	0.890	0.857	0.826	0.797	0.769	0.743	0.718	0.694	0.672	0.650	0.630	0.610	0.592
3	0.889	0.864	0.840	0.794	0.751	0.712	0.675	0.641	0.609	0.579	0.551	0.524	0.500	0.477	0.455
4	0.855	0.823	0.792	0.735	0.683	0.636	0.592	0.552	0.516	0.482	0.451	0.423	0.397	0.373	0.350
5	0.822	0.784	0.747	0.681	0.621	0.567	0.519	0.476	0.437	0.402	0.370	0.341	0.315	0.291	0.269
6	0.790	0.746	0.705	0.630	0.564	0.507	0.456	0.410	0.370	0.335	0.303	0.275	0.250	0.227	0.207
7	0.760	0.711	0.665	0.583	0.513	0.452	0.400	0.354	0.314	0.279	0.249	0.222	0.198	0.178	0.159
8	0.731	0.677	0.627	0.540	0.467	0.404	0.351	0.305	0.266	0.233	0.204	0.179	0.157	0.139	0.123
9	0.703	0.645	0.592	0.500	0.424	0.361	0.308	0.263	0.225	0.194	0.167	0.144	0.125	0.108	0.094
10	0.676	0.614	0.558	0.463	0.386	0.322	0.270	0.227	0.191	0.162	0.137	0.116	0.099	0.085	0.073
11	0.650	0.585	0.527	0.429	0.350	0.287	0.237	0.195	0.162	0.135	0.112	0.094	0.079	0.066	0.056
12	0.625	0.557	0.497	0.397	0.319	0.257	0.208	0.168	0.137	0.112	0.092	0.076	0.062	0.052	0.043
13	0.601	0.530	0.469	0.368	0.290	0.229	0.182	0.145	0.116	0.093	0.075	0.061	0.050	0.040	0.033
14	0.577	0.505	0.442	0.340	0.263	0.205	0.160	0.125	0.099	0.078	0.062	0.049	0.039	0.032	0.025
15	0.555	0.481	0.417	0.315	0.239	0.183	0.140	0.108	0.084	0.065	0.051	0.040	0.031	0.025	0.020
16	0.534	0.458	0.394	0.292	0.218	0.163	0.123	0.093	0.071	0.054	0.042	0.032	0.025	0.019	0.015

17	0.513	0.436	0.371	0.270	0.198	0.146	0.108	0.080	0.060	0.045	0.034	0.026	0.020	0.015	0.012
18	0.494	0.416	0.350	0.250	0.180	0.130	0.095	0.069	0.051	0.038	0.028	0.021	0.016	0.012	0.009
19	0.475	0.396	0.331	0.232	0.164	0.116	0.083	0.060	0.043	0.031	0.023	0.017	0.012	0.009	0.007
20	0.456	0.377	0.312	0.215	0.149	0.104	0.073	0.051	0.037	0.026	0.019	0.014	0.010	0.007	0.005
21	0.439	0.359	0.294	0.199	0.135	0.093	0.064	0.044	0.031	0.022	0.015	0.011	0.008	0.006	0.004
22	0.422	0.342	0.278	0.184	0.123	0.083	0.056	0.038	0.026	0.018	0.013	0.009	0.006	0.004	0.003
23	0.406	0.326	0.262	0.170	0.112	0.074	0.049	0.033	0.022	0.015	0.010	0.007	0.005	0.003	0.002
24	0.390	0.310	0.247	0.158	0.102	0.066	0.043	0.028	0.019	0.013	0.008	0.006	0.004	0.003	0.002
25	0.375	0.295	0.233	0.146	0.092	0.059	0.038	0.024	0.016	0.010	0.007	0.005	0.003	0.002	0.001
26	0.361	0.281	0.220	0.135	0.084	0.053	0.033	0.021	0.014	0.009	0.006	0.004	0.002	0.002	0.001
27	0.347	0.268	0.207	0.125	0.076	0.047	0.029	0.018	0.011	0.007	0.005	0.003	0.002	0.001	0.001
28	0.333	0.255	0.196	0.116	0.069	0.042	0.026	0.016	0.010	0.006	0.004	0.002	0.002	0.001	0.001
29	0.321	0.243	0.185	0.107	0.063	0.037	0.022	0.014	0.008	0.005	0.003	0.002	0.001	0.001	0.001
30	0.308	0.231	0.174	0.099	0.057	0.033	0.020	0.012	0.007	0.004	0.003	0.002	0.001	0.001	
40	0.208	0.142	0.097	0.046	0.022	0.011	0.005	0.003	0.001	0.001					

TABLE 4
Present Value of an Annuity of $1

Periods	4%	5%	6%	8%	10%	12%	14%	16%	18%	20%	22%	24%	26%	28%	30%
1	0.962	0.952	0.943	0.926	0.909	0.893	0.877	0.862	0.847	0.833	0.820	0.806	0.794	0.781	0.769
2	1.886	1.859	1.833	1.783	1.736	1.690	1.647	1.605	1.566	1.528	1.492	1.457	1.424	1.392	1.361
3	2.775	2.723	2.673	2.577	2.487	2.402	2.322	2.246	2.174	2.106	2.042	1.981	1.868	1.816	1.816
4	3.630	3.546	3.465	3.312	3.170	3.037	2.914	2.798	2.690	2.589	2.494	2.404	2.320	2.241	2.166
5	4.452	4.330	4.212	3.993	3.791	3.605	3.433	3.274	3.127	2.991	2.864	2.745	2.635	2.532	2.436
6	5.242	5.076	4.917	4.623	4.355	4.111	3.889	3.685	3.498	3.326	3.167	3.020	2.885	2.759	2.643
7	6.002	5.786	5.582	5.206	4.868	4.564	4.288	4.039	3.812	3.605	3.416	3.242	3.083	2.937	2.802
8	6.733	6.463	6.210	5.747	5.335	4.968	4.639	4.344	4.078	3.837	3.619	3.421	3.241	3.076	2.925
9	7.435	7.108	6.802	6.247	5.759	5.328	4.946	4.607	4.303	4.031	3.786	3.566	3.366	3.184	3.019
10	8.111	7.722	7.360	6.710	6.145	5.650	5.216	4.833	4.494	4.192	3.923	3.682	3.465	3.269	3.092
11	8.760	8.306	7.887	7.139	6.495	5.988	5.453	5.029	4.656	4.327	4.035	3.776	3.544	3.335	3.147
12	9.385	8.863	8.384	7.536	6.814	6.194	5.660	5.197	4.793	4.439	4.127	3.851	3.606	3.387	3.190
13	9.986	9.394	8.853	7.904	7.103	6.424	5.842	5.342	4.910	4.533	4.203	3.912	3.656	3.427	3.223
14	10.563	9.899	9.295	8.244	7.367	6.628	6.002	5.468	5.008	4.611	4.265	3.962	3.695	3.459	3.249
15	11.118	10.380	9.712	8.559	7.606	6.811	6.142	5.575	5.092	4.675	4.315	4.001	3.726	3.483	3.268
16	11.652	10.838	10.106	8.851	7.824	6.974	6.265	5.669	5.162	4.730	4.357	4.033	3.751	3.503	3.283

17	12.166	11.274	10.477	9.122	8.022	7.120	6.373	5.749	5.222	4.775	4.391	4.059	3.771	3.518	3.295
18	12.659	11.690	10.828	9.372	8.201	7.250	6.467	5.818	5.273	4.812	4.419	4.080	3.786	3.529	3.304
19	13.134	12.085	11.158	9.604	8.365	7.366	6.550	5.877	5.316	4.844	4.442	4.097	3.799	3.539	3.311
20	13.590	12.462	11.470	9.818	8.514	7.469	6.623	5.929	5.353	4.870	4.460	4.110	3.808	3.546	3.316
21	14.029	12.821	11.764	10.017	8.649	7.562	6.687	5.973	5.384	4.891	4.476	4.121	3.816	3.551	3.320
22	14.451	13.163	12.042	10.201	8.772	7.645	6.743	6.011	5.410	4.909	4.488	4.130	3.822	3.556	3.323
23	14.857	13.489	12.303	10.371	8.883	7.718	6.792	6.044	5.432	4.925	4.499	4.137	3.827	3.559	3.325
24	15.247	13.799	12.550	10.529	8.985	7.784	6.835	6.073	5.451	4.937	4.507	4.143	3.831	3.562	3.327
25	15.622	14.094	12.783	10.675	9.077	7.843	6.873	6.097	5.467	4.948	4.514	4.147	3.834	3.564	3.329
26	15.983	14.375	13.003	10.810	9.161	7.896	6.906	6.118	5.480	4.956	4.520	4.151	3.837	3.566	3.330
27	16.330	14.643	13.211	10.935	9.237	7.943	6.935	6.936	5.492	4.964	4.525	4.154	3.839	3.567	3.331
28	16.663	14.898	13.406	11.051	9.307	7.984	6.961	6.152	5.502	4.970	4.528	4.157	3.840	3.568	3.331
29	16.984	15.141	13.591	11.158	9.370	8.022	6.983	6.166	5.510	4.975	4.531	4.159	3.841	3.569	3.332
30	17.292	15.373	13.765	11.258	9.427	8.055	7.003	6.177	5.517	4.979	4.534	4.160	3.842	3.569	3.332
40	19.793	17.159	15.046	11.925	9.779	8.244	7.105	6.234	5.548	4.997	4.544	4.166	3.846	3.571	3.333

Cumulative Normal Probability Tables

The probability that a drawing from a unit normal distribution will produce a value less than the constant d is

$$\text{Prob}\ (\tilde{z} < d) = \int_{-\infty}^{d} \frac{1}{\sqrt{2\pi}} e^{-z^2/2} dz = N(d)$$

Range of d: −2.49:5 ≤ d ≤ 0.00

d	−0.00	−0.01	−0.02	−0.03	−0.04	−0.05	−0.06	−0.07	−0.08	−0.09
−2.40	0.00820	0.00798	0.00776	0.00755	0.00734	0.00714	0.00695	0.00676	0.00657	0.00639
−2.30	0.01072	0.01044	0.01017	0.00990	0.00964	0.00939	0.00914	0.00889	0.00866	0.00842
−2.20	0.01390	0.01355	0.01321	0.01287	0.01255	0.01222	0.01191	0.01160	0.01130	0.01101
−2.10	0.01786	0.01743	0.01700	0.01659	0.01618	0.01578	0.01539	0.01500	0.01463	0.01426
−2.00	0.02275	0.02222	0.02169	0.02118	0.02068	0.02018	0.01970	0.01923	0.01876	0.01831
−1.90	0.02872	0.02807	0.02743	0.02680	0.02619	0.02559	0.02500	0.02442	0.02385	0.02330
−1.80	0.03593	0.03515	0.03438	0.03362	0.03288	0.03216	0.03144	0.03074	0.03005	0.02938
−1.70	0.04457	0.04363	0.04272	0.04182	0.04093	0.04006	0.03920	0.03836	0.03754	0.03673
−1.60	0.05480	0.05370	0.05262	0.05155	0.05050	0.04947	0.04846	0.04746	0.04648	0.04551
−1.50	0.06681	0.06552	0.06426	0.06301	0.06178	0.06057	0.05938	0.05821	0.05705	0.05592
−1.40	0.08076	0.07927	0.07780	0.07636	0.07493	0.07353	0.07215	0.07078	0.06944	0.06811
−1.30	0.09680	0.09510	0.09342	0.09176	0.09012	0.08851	0.08691	0.08534	0.08379	0.08226
−1.20	0.11507	0.11314	0.11123	0.10935	0.10749	0.10565	0.10383	0.10204	0.10027	0.09853
−1.10	0.13567	0.13350	0.13136	0.12924	0.12714	0.12507	0.12302	0.12100	0.11900	0.11702
−1.00	0.15866	0.15625	0.15386	0.15150	0.14917	0.14686	0.14457	0.14231	0.14007	0.13786
−0.90	0.18406	0.18141	0.17879	0.17619	0.17361	0.17106	0.16853	0.16602	0.16354	0.16109
−0.80	0.21186	0.20897	0.20611	0.20327	0.20045	0.19766	0.19489	0.19215	0.18943	0.18673
−0.70	0.24196	0.23885	0.23576	0.23270	0.22965	0.22663	0.22363	0.22065	0.21770	0.21476
−0.60	0.27425	0.27093	0.26763	0.26435	0.26109	0.25785	0.25463	0.25143	0.24825	0.24510
−0.50	0.30854	0.30503	0.30153	0.29806	0.29460	0.29116	0.28774	0.28434	0.28096	0.27760
−0.40	0.34458	0.34090	0.33724	0.33360	0.32997	0.32636	0.32276	0.31918	0.31561	0.31207
−0.30	0.38209	0.37828	0.37448	0.37070	0.36693	0.36317	0.35942	0.35569	0.35197	0.34827
−0.20	0.42074	0.41683	0.41294	0.40905	0.40517	0.40129	0.39743	0.39358	0.38974	0.38591
−0.10	0.46017	0.45620	0.45224	0.44828	0.44433	0.44038	0.43644	0.43251	0.42858	0.42465
0.00	0.50000	0.49601	0.49202	0.48803	0.48405	0.48006	0.47608	0.47210	0.46812	0.46414
0.00	0.50000	0.50399	0.50798	0.51197	0.51595	0.51994	0.52392	0.52790	0.53188	0.53586
0.01	0.53983	0.54380	0.54776	0.55172	0.55567	0.55962	0.56356	0.56749	0.57142	0.57535
0.20	0.57926	0.58317	0.58706	0.59095	0.59483	0.59871	0.60257	0.60642	0.61026	0.61409
0.30	0.61791	0.62172	0.62552	0.62930	0.63307	0.63683	0.64058	0.64431	0.64803	0.65173
0.40	0.65542	0.65910	0.66276	0.66640	0.67003	0.67364	0.67724	0.68082	0.68439	0.68793
0.50	0.69146	0.69497	0.69847	0.70194	0.70540	0.70884	0.71226	0.71566	0.71904	0.72240
0.60	0.72575	0.72907	0.73237	0.73565	0.73891	0.74215	0.74537	0.74857	0.75175	0.75490
0.70	0.75804	0.76115	0.76424	0.76730	0.77035	0.77337	0.77637	0.77935	0.78230	0.78524
0.80	0.78814	0.79103	0.79389	0.79673	0.79955	0.80234	0.80511	0.80785	0.81057	0.81327
0.90	0.81594	0.81859	0.82121	0.82381	0.82639	0.82894	0.83147	0.83398	0.83646	0.83891
1.00	0.84134	0.84375	0.84614	0.84850	0.85083	0.85314	0.85543	0.85769	0.85993	0.86214
1.10	0.86433	0.86650	0.86864	0.87076	0.87286	0.87493	0.87698	0.87900	0.88100	0.88298
1.20	0.88493	0.88686	0.88877	0.89065	0.89251	0.89435	0.89617	0.89796	0.89973	0.90147
1.30	0.90320	0.90490	0.90658	0.90824	0.90988	0.91149	0.91309	0.91466	0.91621	0.91774
1.40	0.91924	0.92073	0.92220	0.92364	0.92507	0.92647	0.92785	0.92922	0.93056	0.93189
1.50	0.93319	0.93448	0.93574	0.93699	0.93822	0.93943	0.94062	0.94179	0.94295	0.94408
1.60	0.94520	0.94630	0.94738	0.94845	0.94950	0.95053	0.95154	0.95254	0.95352	0.95449
1.70	0.95543	0.95637	0.95728	0.95818	0.95907	0.95994	0.96080	0.96164	0.96246	0.96327

1.80	0.96407	0.96485	0.96562	0.96637	0.96712	0.96784	0.96856	0.96926	0.96995	0.97062
1.90	0.97128	0.97193	0.97257	0.97320	0.97381	0.97441	0.97500	0.97558	0.97615	0.97670
2.00	0.97725	0.97778	0.97831	0.97882	0.97932	0.97982	0.98030	0.98077	0.98124	0.98169
2.10	0.98214	0.98257	0.98300	0.98341	0.98382	0.98422	0.98461	0.98500	0.98537	0.98574
2.20	0.98610	0.98645	0.98679	0.98713	0.98745	0.98778	0.98809	0.98840	0.98870	0.98899
2.30	0.98928	0.98956	0.98983	0.99010	0.99036	0.99061	0.99086	0.99111	0.99134	0.99158
2.40	0.99180	0.99202	0.99224	0.99245	0.99266	0.99286	0.99305	0.99324	0.99343	0.99361